Fire Plan Review and Inspection Guidelines

Mike Thrapp

Fire Plan Review and Inspection Guidelines

ISBN: 978-1-58001-663-6

Cover Design: Dianna Hallmark
Cover Art Director: Dianna Hallmark
Publications Manager: Mary Lou Luif
Project Editor: Jodi Henkiel
Manager of Development: Scott Stookey
Illustrator/Interior Design: Lisa Jachymiak

COPYRIGHT© 2008 by M.C. Thrapp

ALL RIGHTS RESERVED. This publication is a copyrighted work owned by M.C. Thrapp. Portions of this publication are reprinted from the *International Fire Code* (IFC*)*, *International Building Code* (IBC) and other I-Codes. These Codes are owned and copyrighted by the International Code Council. Without advance written permission from the copyright owner, no part of this book may be reproduced, distributed or transmitted in any form or by any means, including, without limitation, electronic, optical or mechanical means (by way of example and not limitation, photocopying, or recording by or in an information storage and retrieval system). For information on permission to copy material exceeding fair use, please contact: ICC Publications, 4051 W. Flossmoor Rd, Country Club Hills, IL 60478, Phone 888-ICC-SAFE (422-7233).

The information contained in this document is believed to be accurate; however, it is being provided for informational purposes only and is intended for use only as a guide. Publication of this document by the ICC should not be construed as the ICC engaging in or rendering engineering, legal or other professional services. Use of the information contained in this workbook should not be considered by the user to be a substitute for the advice of a registered professional engineer, attorney or other professional. If such advice is required, it should be sought through the services of a registered professional engineer, licensed attorney or other professional.

Trademarks: "International Code Council" and the "ICC" logo are trademarks of International Code Council, Inc.

First Printing: March 2008

Printed in the United States of America

About this Textbook

This manual is in its 23rd year of assisting fire protection designers, contractors, service, and fire personnel with applying the requirements for codes and national standards.

The performance of any fire protection system carries a great expectation of reliability. The expectation is to minimize or eliminate the loss of life and/or property damage. This manual has been developed to assist the user in the design and installation of quality fire protection systems.

I appreciate the companies providing systems and services that assist in the protection of life and property. I commend those that maintain the high standards of the fire and life safety industry.

Sincerely,

Mike C. Thrapp
Fire Marshal

About the Author

Mike Thrapp's fire service career spans 27 years with 20 years serving at different capacities in fire prevention. He has served as Fire Marshal in Anchorage, Alaska, and Eugene, Oregon, for a total of 15 years.

While in Anchorage, he held an adjunct faculty member position for the Fire Science Program at the University of Alaska Anchorage. He obtained his ICBO Plans Examiner and Fire Code certifications in 1986 and currently possesses the Building Plans Examiner, and Fire Inspector and Plans Examiner certifications from the International Code Council. He is also certified as a Plans Examiner–A Level and a Plans Examiner–Fire and Life Safety in the State of Oregon. He received a Bachelor of Arts in Management and completed the National Fire Academy's Executive Fire Officer Program in 1994.

Disclaimer

Important!

This manual is designed as a resource, guideline and policy document. The user is responsible to research and comply with all applicable codes and standards associated with the design, installation and maintenance of the systems mentioned in this manual.

The user of this document shall indemnify, defend, save and hold the author harmless from any and all claims, lawsuits, or liability including attorneys' fees and costs allegedly arising out of, in connection with, or incident to any loss, damage or injury to persons or property occurring during the course of or as a result of the user's performance pursuant to using this manual.

Acknowledgments

Since 1985, this guide has evolved from a 28-page document to its present volume. Its origin began in Anchorage, Alaska for local use and it migrated to Eugene, Oregon, where I am the City's Fire Marshal and its use and publication has now expanded nationally by ICC, beyond what I imagined.

Over the years, I am appreciative of those who have provided ideas, constructive criticism, and recommendations.

Many fire protection companies, office associates, peers, and designers have offered ideas and comments that have been incorporated into this book.

Despite my lack of patience, Faye Forhan recommended I wait and permit her to proofread the document. Thank goodness I waited because she found the need for numerous corrections, and I have appreciated her work.

Several years passed and the book increased in content. Liz, who wishes remain anonymous, proofread this document and many more corrections had to be made. Her time commitment, tenacity and attention to detail are evident and lauded.

This edition of the book received contribution and detailed scrutiny and review by ICC staff member Scott Stookey. His input, greatly respected and extolled, has taken the book to a new level. ICC staff Page Dougherty did the initial review of the original content and recommended taking it to the national level. His efforts are also appreciated.

Each code change and subject additions that caused major revisions initiated a daunting task of editing and without everyone's assistance this book would not exist.

Thank you!
Mike Thrapp

Fire Plan Review and Inspection Guidelines

Preface

The basic requirements for plan review, inspection and testing of the various fire protection systems are addressed in this manual. This document discusses major elements of the 2006 *International Fire Code* and the 2006 *International Building Code* relevant to fire protection systems and their applicable standards.

Section 105.7 of the 2006 *International Fire Code*® (IFC®) establishes the minimum requirements for construction permits. The IFC requires construction permits for the installation or modification of private fire hydrants, automatic-fire extinguishing systems, fire alarm and detection systems, standpipe systems, fire pumps and related equipment. Construction permits are also required for operations involving the storage, use, handling and dispensing of hazardous materials, including flammable and combustible liquids, compressed gases and those materials regulated in Chapter 27 of the IFC.

Chapter 45 of the IFC adopts by reference over 100 standards written by the membership of the National Fire Protection Association (NFPA). NFPA publishes a variety of standards, but many of the standards adopted in IFC and IBC Chapter 9 address the design, installation, testing and maintenance of water-based and alternative fire-extinguishing agents, including dry chemical, wet chemical, carbon dioxide and clean agents. In addition, many of the adopted NFPA standards have extensive requirements for storage tanks used for the underground or above-ground storage of flammable and combustible liquids, spray finishing using flammable finishes and motor vehicle fuel-dispensing. In most cases, Section 105.7 of the IFC also requires construction permits for these various systems.

Fire Plan Review and Inspection Guidelines was developed to assist code officials responsible for the plan review and approval of shop drawings and calculations submitted for IFC construction permits. This text contains plan review and inspection guidelines for all of the fire protection systems required by Chapter 9 of the 2006 IFC. This text focuses on plan review of the various types of fire protection systems prescribed by the IFC and IBC.

The manuscript is organized to be as consistent as practical with the number convention used in the Chapter 9 of the IFC and IBC. Plan review worksheets are organized based on the type of fire protection system specified in the IFC and IBC. For automatic sprinkler systems, worksheets are divided into specific subject matter such as the requirements for NFPA 13R or 13D, and separate worksheets for the plan review of automatic sprinkler systems designed for the protection of storage occupancies are also included.

To assist code officials facilitate plan reviews, each plan review worksheet follows a format that is consistent with the requirements of the relevant IFC requirements or the applicable NFPA standard. Additionally, this text includes a compact disk of each of the plan review forms that can be electronically completed and printed. This provides a format that is consistent for each plan review which benefits the design professional, installing contractor and code official. It also provides a means for jurisdictions to electronically store and retrieve plan review comments, which can save jurisdictions the burden of archiving paper files of plan review and construction inspection records generated during the installation or modification of these systems.

Users of the IFC and IBC understand that the International Code Council code development cycle is on a different time schedule when compared to the NFPA process. However, design professionals and installing contractors generally wish to use the latest edition of a particular NFPA standard because the new requirements can benefit them with a more cost effective installation that provides improved safety for the building owner. For automatic sprinkler systems, fire detection and alarm systems and fire pumps, this text contains plan review worksheets for the 2002 editions of NFPA 13, 13R, 13D, 20 and 72 as well as worksheets for the 2007 editions of these NFPA standards.

To promote consistency in construction inspections, each system plan review worksheet also has a construction inspection checklist to assist inspectors ensure that the fire protection system or hazardous material process is installed and tested in accordance with the 2006 IFC and the applicable NFPA standard.

The content of the plan review worksheets and the construction inspection checklists are not all inclusive. It is the user's responsibility to research and apply the pertinent codes and standards applicable for plans and calculations to confirm compliance.

The following fire protection systems are addressed in this text:
  Automatic Sprinkler Systems including
    NFPA 13
    NFPA 13 sprinkler requirements for the protection of high piled storage
    NFPA 13R

NFPA 13D

Alternative Automatic Fire-Extinguishing Systems including

    Clean agent systems

    Carbon dioxide systems

    Halon systems

    Water mist systems

Thermal Fire Protection and Life Safety Systems including

    Kitchen hood suppression systems

    Emergency and standby power systems

    Fire detection and alarm systems

    Fire pumps

    Private fire service mains

    Private water tank systems

    Standpipe systems

For each type of system the following forms are provided to facilitate the plan review and construction inspections:

    Plan submittal requirements; including a plan review worksheet

    New installation acceptance inspection and testing instructions, and a worksheet

    Installer certification form

    References for the pertinent codes and/or standards

    Miscellaneous information and sample jurisdiction policies

    Pre and post installation requirements

TABLE OF CONTENTS

About this Textbook iii
About the Author iv
Disclaimer v
Acknowledgments vii
Preface ix

General Information

1. Service Agency Approval Requirements 1
2. Codes and Standards 2
3. Recommended Policies for Fire Protection Systems 5
4. When are Permits or Plan Reviews Required? 7
5. Sample Plan Review and the Acceptance Test Process 9
6. Building Code Study Content Recommendations 11
7. The Most Common Plan Review Deficiencies 12
8. Fire Code Plan Review Worksheet 13
9. Final Inspection Worksheet 17
10. Sample Service Interval and Report Policy 20

Building, Site, and Water Supply Subjects

11. Fire Watch Recommended Policy 22
12. Emergency Plans, Overview and Pamphlet 23
13. Site Review Checklist 39
14. Turnarounds and Secondary Emergency Access Roads 41
15. Fire Access Road Turnarounds 43
16. Emergency Access Lock Box 49
17. Private Fire Service Mains 51
18. Private Water Tank Systems 73
19. Emergency and Standby Power 83

Fire Protection and Detection System Specifics and Checklists

20. Sprinkler Systems (13, 13R, and 13D) 93

21. Water Mist Systems 161
22. Carbon Dioxide Systems.. 173
23. Halon Systems 183
24. Clean Agent Systems... 195
25. Kitchen Hood Suppression Systems.. 205
26. Standpipe and Hose Cabinets 215
27. Fire Alarm Systems.. 229
28. Fire Pump... 257

Miscellaneous Plan Review Subjects 279
29. Spray Finish, Powder Coating, and Electrostatic Worksheet............ 281
30. Motor Vehicle Fuel Dispensing Station Worksheet 284
31. Aboveground Fuel Tank, Plan Review Worksheet.......................... 287
32. Underground Fuel Tanks, Plan Review/Inspection Worksheet........ 289
33. Medical Gases, Plan Review/Inspection Worksheet....................... 291

Index 293
Acronym Table 296

Service Agency Approval Requirements

The authority having jurisdiction should ensure each agency and their representatives (where applicable) involved in the installation and servicing of fire protection equipment have the appropriate contractor license(s) and/or certification(s) as required by local or state requirements.

Codes and Standards

The following references are shortened in content to provide the general intent when plan review and maintenance are required. Associated codes, standards, and references are listed, which will assist in these activities.

General

2006 International Fire Code, Sections 901.2 and 901.5 Approval and Testing
Plans and specifications shall be submitted to the fire department for review and approval prior to construction.

2006 International Fire Code, Section 901.4 Maintenance
All fire protection systems shall be maintained in accordance with recognized standards at all times. A copy of all service reports shall be sent to the Fire Prevention Division.

2006 International Fire Code, Section 901.4.3 Special Hazards
When occupancies of an especially hazardous nature or where special hazards exist, the chief is authorized to require additional protection for these situations.

Codes and Standards Containing Specific Requirements

Some of the standards listed are not found in the IFC and IBC referenced standards. The standards may be applicable under the authority of IFC Section 102.7, Subjects not regulated by this code.

A copy of the following books is recommended when establishing a reference library.

    2006 International Building Code (IBC)
    2006 IBC Code and Commentary

    2006 International Fire Code (IFC)
    2006 IFC Code and Commentary

    2006 International Mechanical Code (IMC)
    2006 IMC Code and Commentary

A current edition of the *National Fire Codes* published by the National Fire Protection Association

State, county or city amendments

Private Fire Service Mains
    IFC
    NFPA 24 and 25

Private Water Tank Systems
IFC
NFPA 22

Automatic Sprinkler Systems
IBC
IFC
NFPA 13, 13D, 13R, and 25
Manufacturer's listing requirements

Standpipes
IBC
IFC
NFPA 14 and 25

Fire Pumps
IBC
IFC
NFPA 20 and 25
Manufacturer's listing requirements

Fire Alarm Systems
IFC
IBC
IMC
NFPA 72
Manufacturer's listing requirements

Kitchen Fire Suppression Systems
IFC
NFPA 13, 17A, and 25 MC
Manufacturer's listing requirements and the manufacturer's design manual

Clean Agent Systems
NFPA 2001 and 72
Manufacturer's listing requirements and the manufacturer's design manual

Carbon Dioxide Systems
NFPA 12 and 72
Manufacturer's listing requirements and the manufacturer's design manual

Halon Systems
NFPA 12A
Manufacturer's listing requirements

Water Mist Systems
NFPA 13, 25, and 750
Manufacturer's listing requirements and the manufacturer's design manual

Portable Fire Extinguishers
IFC
NFPA 10

Emergency and Standby Power Systems
IFC 604
IFC Chapter 34 if a generator fueled using fuel oil or diesel is installed
Manufacturer's listing requirements

Recommended Books
Brock, Pat D., Fire Protection Hydraulics and Water Supply Analysis, Fire Protection Publications: Oklahoma State University, 2000.

Bukowski, Richard W, P.E., and Moore, Wayne D., P.E., Fire Alarm Signaling Systems Handbook, NFPA: Massachusetts, 2003.

Dubay, Christian, P.E., Automatic Sprinkler Systems Handbook, NFPA: Massachusetts, 2002 and 2007.

Fire Protection Handbook, NFPA: Massachusetts, 2008.

Gagnon, Robert M., Designer's Guide to Automatic Sprinkler Systems, SFPE: Bethesda, Maryland; NFPA: Massachusetts, 2005.

Gagnon, Robert M., and Kirby, Ronald H., A Designer's Guide to Fire Alarm Systems, NFPA: Massachusetts, 2003.

Gagnon, Robert M., Design of Water-Based Fire Protection Systems, Delmar Publishers: International Thomson Publishing Co., 1997.

Hickey, Harry E., Hydraulics for Fire Protection, NFPA: Massachusetts, 1980.

International Code Council, 2006 IFC Fire Protection Systems: International Code Council Inc., 2006.

Loss Prevention Data: For the Fire Service, Vol 1 and 2, Factory Mutual Global.

Richardson, Lee F., P.E., and Moore, Wayne D., P.E., National Fire Alarm Code Handbook, NFPA: Massachusetts, 2002 and 2007.

The SFPE Handbook of Fire Protection Engineering, NFPA: Massachusetts, 2002.

Fire Prevention Division
Recommended Policies for Fire Protection Systems

Some design, installation, and maintenance related issues are not clearly defined in the codes. The following policy statements have been developed over the years and are provided to serve as examples that state expectations and clarify some of the more ambiguous areas found in the codes. Some of these statements are not based on code but on problems experienced in the field and they are intended to stimulate ideas for local policies. Each jurisdiction is encouraged through policy to document the intended approach in order to provide clear expectations and achieve a higher level of consistency.

General

1. Even though the IBC or IFC may not require a system to be installed, systems or components being installed must be designed and installed in accordance with national standards and the manufacture specifications, and must go through the permit, plan review and inspection process (IFC 901.4.2).

2. When a fire protection system is taken out of service as a result of repairs, alterations or other reasons, a fire watch (policy in back of manual) shall be provided.

   Include at least the following information on service reports that are sent to the Fire Prevention office:
   - business and/or building name
   - correct address that indicates where the system is located
   - date of the inspection
   - type of system and manufacturer

3. Work done without a permit is subject to a "Stop Work Order" in accordance with Section 111 of the IFC, permit appropriate fees and investigation fees charged, and civil penalties being levied (depends on legal remedies available to a local jurisdiction).

4. To assist fire department personnel in determining the status of a system, the inspection tag must indicate if the system is a noncompliant system. Please indicate if the system is noncompliant on the regular inspection tag.

5. All service reports for fire protection/detection systems shall be submitted to the local Fire Marshal's Office.

Fire Alarm

1. Design the circuits/systems to handle at least a 10 percent above device/power demand for future expansions (policy may not be enough, it may require a local amendment).

2. New systems tie into existing systems, e.g., clean agent, halon, kitchen hood, etc.

3. Part of the servicing of a system should include assessing whether the sound pressure level (dBA) is adequate (compliant with NFPA standards) throughout the building.

4. Smoke detectors shall be tested for sensitivity in accordance with Section 907.20.3 of the IFC and NFPA standards.
5. Device wiring shall not be T-tapped, unless allowed by the manufacturer listing criteria.
6. Secondary power is required on all systems, except for antiquated 120 volt systems as outlined:
    1. 120 volt system serving a one-story building;
    2. 120 volt system with a maximum three horns and four pull stations; where occupants do not sleep in the building; and
    3. Group A-occupancies with an occupant load less than 300 (IFC 907.2.1).
7. Noncompatible smoke detectors will be replaced with compatible devices (NFPA 72 4.4.2).
8. Zones on a Fire Alarm Control Units (FACU) and remote annunciators shall be clearly identified.
9. New systems: FACUs will be located at the entrance of the building or provide an annunciator panel at the entrance and also indicate the location of the main panel.
10. Security systems require fire department review when a fire detection or alarm system is extended from the security system panel. The security panel shall listed for the combination use (pertains to commercial buildings).

Automatic Sprinkler Systems

1. Residential sprinkler systems shall be hydraulically calculated and be designed in accordance with NFPA 13D or 13R.
2. Sprinkler systems for speculation warehouses shall be a minimum ordinary hazard Group 2. The design area density shall be a minimum of 0.17/3000 sq. ft.
3. All cooler/freezers shall be sprinklered when located in a sprinklered building.
4. In sprinklered buildings, sprinkler coverage shall be provided in rooms protected by halon, clean agent or other alternative fire extinguishing systems. Exception: if it is an existing condition and no system or room alterations are made.
5. A backflow prevention device shall be installed as required by local or state plumbing codes.
6. Fire Department Connections (FDC) shall be located on the main entrance side of the building or the street side, whichever is the fire department response location as determined by the plan reviewer.
7. The exterior water flow alarm and signage shall be located directly above the FDC approximately 10 ft. above grade.

Fire Pump

A test header shall be provided for new installations and systems that are modified or have no means for testing.

Portable Extinguishers

Apartments or condos (Group R-2): if extinguishers are not desirable in the public areas (hallways, exit balconies, lobbies or stairways) because of theft or vandalism, then extinguishers, 2-A:10-B:C minimum, shall be provided in each dwelling unit.

When are Permits or Plan Reviews Required?

Generally, permits must be obtained for work being performed on fire protection systems. At the local level a jurisdiction may want to establish specific criteria to determine when a permit or plan review is necessary and the following conditions are sample recommendations.

General

All new system installations whether, the system is required by code or not.

Automatic Sprinkler System

1. When there is a relocation of 15 or more or an addition of 10 or more sprinkler heads to a system.
2. Anytime conventional sprinklers and piping are replaced with flexible piping and sprinklers.
3. Changes to piping that require seismic bracing.
4. Anytime there are changes to areas that are considered the most demanding design density flow area.
5. Increase of building area and/or an increase of the system design density.
6. At the discretion of the fire official that may consider that there is sufficient change to the system and that the minimum design density requirements and/or seismic bracing requirements must be verified.

Note: Sprinkler head addition limitation to a system is cumulative. Meaning: if 10 sprinklers require a permit or plan review and 9 sprinklers are added one month and 1 sprinkler a month later, it does not negate obtaining a permit or plan review. As soon as 10 sprinklers, cumulative, are modified or added, then a plan review and construction permit should be required. A tracking system is necessary for this.

Fire Alarm

1. Replacement of the Fire Alarm Control Unit.

    Provide: battery calculations, verification that sound pressure level measured in dBA are adequate, system components are compatible, and use pertinent sections of the plan review worksheet included in this manual.

2. Addition of 11 or more fire alarm initiating devices.

    Provide: battery calculations, verification that sound pressure level measured in dBA are adequate, system components are compatible, that circuit capacities are not exceeded with new devices, wiring diagram showing the connection between new and existing systems, and use pertinent sections of the plan review worksheet.

3. Addition of 6 or more fire alarm indicating devices, including horns and strobes.
   Provide: battery calculations, verification that sound pressure level measured in dBA are adequate, system components are compatible, line voltage calculations for lines with new devices, wiring diagram showing connection between new and existing systems, and use pertinent sections of plan review worksheet
4. If a permit is not required, provide a system certification form and a basic as-built plan showing the location of the new devices and which circuit was augmented.

Note: The device addition limitation to a system is cumulative for an annual period. For example, if 7 devices require a permit and 6 devices are added one month and 6 devices a month later; it does not negate obtaining a permit. As soon as 7 devices, cumulative, are reached then a permit should be required. A tracking system is necessary for this.

Alternative Fire-Extinguishing System
1. Any addition to the system.
2. Relocation of a system.

Kitchen Hood Suppression
1. Addition of one or more nozzles.
2. Relocation of a system.

Sample Plan Review and the Acceptance Inspection Process

Before a system installation is performed, plans will be submitted to the jurisdiction for review and approval. The following is suggested criteria for the permitting process, review, and inspection process.

1. Provide at least three (3) sets of plans:
   - one for the fire department.
   - one for the building department.
   - one for the job site.
2. When plans are disapproved:
   - the applicant will be notified to pick up the plans for correction.
   - a plan review comment sheet noting the deficiencies will be attached.
   - the applicant will resubmit corrected plans with the original comment sheet.
3. When the plans are approved:
   - the applicant will be notified to pick up the plans.
4. After the system installation is completed:
   - the applicant is to call in an inspection request preferably one business day in advance.
   - the contractor is to use the acceptance test check list as a guide for pre-testing the system.
   - the contractor is to have the Installation Certification form completed and on-site prior to inspection (refer to each system's Contractor's prefinal inspection guide).
5. System preinspection and testing: Use the inspection worksheet found in the pertinent system section before requesting a fire inspection.
   - **Private Fire Service Water Main and Hydrant Systems**: Perform a pressure test to verify system integrity and flush the system before requesting a fire inspection.
   - **Automatic Sprinkler Systems**: Perform a pressure test and inspection of the system to verify system integrity and proper installation before requesting a fire inspection.
   - **Standpipe Systems**: Perform a pressure test and inspection of the system including a flow test to verify system integrity and proper installation before requesting a fire inspection.
   - **Fire Pump Systems**: Perform a test of the system to verify the fire pump, pressure maintenance pump, and controller functions properly. Record the water flow and water pressure results before requesting a fire inspection.

- **Fire Alarm Systems**: Perform a functional test, inspection, and sound pressure (dBA) level test. Provide the results of the sound pressure test for each room with doors closed. The results will be verified. A fire alarm sound pressure (dBA) level pretest form is provided in the fire alarm section.
- **Kitchen Hood Suppression Systems**: Perform a test of the system to verify the fire protection system activates and functions properly before requesting a fire inspection.
- **Alternative Fire Extinguishing Systems**: Perform a functional test of the system to verify the system functions properly before requesting for a fire inspection.
- **Emergency Power Systems**: Perform a load test of the system and the transfer switch to verify the system functions properly before requesting a fire inspection.

Building Code Study Content Recommendations For Fire Plan Review

A comprehensive building code study expedites the plan review process. It provides essential information about a project, which enables the plan reviewer to review and apply the correct sections of the relevant codes to that project.

Code study information establish the requirements for fire department water flow demands, the number of hydrants, fire detection and suppression systems, egress requirements, and occupant load restrictions.

The following information should be included in a basic code study for new building projects:

1. Group and division designation of the occupancy classifications and uses within the building.

2. Construction type of the building.

3. Number of stories, square footage and occupancy type for each floor, and the height of the building.

4. Fire area of specific occupancies. (See definition for "Fire area" in IFC Section 902.1)

5. Calculations for determining increases in basic allowable areas.

6. Occupant load matrix for occupant use areas and different occupancy classifications.

7. Egress width calculations.

8. Indicate if automatic sprinklers protection is provided.

9. Fire protection systems that will be incorporated into the building should be indicated on the design drawings or specifications.

The Most Common Plan Review Deficiencies

The following design items most frequently cause the rejection of plans.

1. Incomplete building code studies for new buildings and tenant improvements.

2. Site plans that lack complying emergency fire access roads.

3. Site plans that do not provide available water flow information, size and location of water mains, and the distance and locations to hydrants.

4. Fire department connections that are not in an accessible location near the front entry point of the building.

5. Smoke detector locations are not detailed nor provided in all the required locations.

6. Fire extinguisher classification type and mounting locations that are not detailed.

7. Fire extinguisher cabinets that are recessed into fire-rated assemblies that do not provide a listed or approved detail of how the wall assembly fire rating is maintained.

8. Emergency access lock box mounting location is not detailed.

9. Hazardous materials inventory statements that are not provided.

10. Complete information and design details for high-piled storage areas that are not provided as required by Section 2301.3 of the IFC.

11. Emergency lighting designs seldom provide the required average of 1 footcandle at the floor level along the exit path.

12. Using "As approved or determined by the Fire Marshal" or "To be determined by…" instead of determining and designing the requirement on the plans.

Fire Plan Review and Inspection Guidelines

Fire Code Plan Review Worksheet
2006 IFC

Date:_____

Permit: _____

Address: _____

Reviewer: _____

New Building _____
Tenant Improve. _____
Remodel _____
Addition _____
Site Work _____
FP System _____
Other _____

Reference numbers following worksheet statements represent an IFC code section unless otherwise specified.

NA |OK| N   **NA** – not applicable   **OK** – acceptable   **N** – need, not acceptable

1. __ | __ | __ Ensure a comprehensive code study is provided:
   __ | __ | __ occupancy type,
   __ | __ | __ square footage,
   __ | __ | __ number of stories
   __ | __ | __ construction type,
   __ | __ | __ allowable area formula,
   __ | __ | __ occupant load determination (based on net or gross)
   __ | __ | __ rooms are identified for use and/or process
   __ | __ | __ fire protection systems: sprinklers__, fire alarm__, kitchen hood__, main/hydrant__, other__

Site Access:
2. __ | __ | __ Fire access road complies with the plans, 503.1.1.
3. __ | __ | __ Fire access road signage is required and plan details are provided and stripping (cross-hatching) may be necessary when determined by the fire official, 503.3.
4. __ | __ | __ Security gate, when provided, is detailed and the opening method complies with fire department policy, IFC 503.6.
5. __ | __ | __ Emergency turn-around is in accordance with the plans, 503.2.5.
6. __ | __ | __ Fire access road is designed to support an apparatus weighing 75,000 lb. gross vehicle weight, Appendix D102.
7. __ | __ | __ Fire access road is an all weather driving surface such as asphalt, concrete, chip-seal (oil matting) or similar, 503.2.3. Grass pavers with fire lane signage is permitted if a low concrete curb is provided along the edges to outline and identify the driving area.
8. __ | __ | __ Any portion of the building exterior wall shall be within 150 ft. of the fire access road, 503.1.
9. __ | __ | __ Grades are not to exceed 10 percent, 503.2.7 and Appendix D.
10. __ | __ | __ As an alternate for grade exceeding 10 percent: Access grade shall not exceed 10 percent but if it does, the first portion of the grade shall be limited to 10 percent for a length of 200 ft. and then 15 percent to 20 percent for a maximum of 200 ft., repeat the cycle as necessary, 503.2.7 and Appendix D and IFC 102.7, .8.
11. __ | __ | __ The dead-end fire access road(s) in excess of 150 ft. and shall be provided with a turn-around, 503.2.5.
12. __ | __ | __ The turning radius for the emergency apparatus road(s) is in accordance with local policy, 503.2.4.
13. __ | __ | __ All fire access roads shall be constructed and maintained prior to and during construction, 1410.

Site Water:
14. __ | __ | __ This project requires_____ GPM fire flow, a minimum of (number of hydrants) _____ hydrants spaced an average of _____ feet, 508.1 and Appendices B and C.
15. __ | __ | __ When a hydrant water flow report is required, the test shall be performed by the water purveyor. Existing reports may be used if a tested has been performed within _____ of the date of the permit application or as requried by the fire official.

Fire Plan Review and Inspection Guidelines

16. __| _ | __The most remote exterior portion of a nonsprinklered building is within 400 ft. or for a sprinklered building within 600 ft. of a hydrant, provide hydrant(s) in accordance with 508, Appendix C.
17. __| _ | __Does the nearest hydrant, as shown on the site plan, exceed the allowable distance from the property? Appendix C. This is a different requirement than Item 16.
18. __| _ | __Prior to the installation of private fire service water main/hydrants systems, plans shall be submitted for a permit, review, and approval, 901.2.
19. __| _ | __As an alternate materials and method, the building may required to be protected throughout by an approved automatic sprinkler system if the required minimum water flow is not available, 104.7.

Emergency Plans:
20. __| _ | __Required for some occupancies such as A, B, E, H, I, R and some M's as based on specific criteria, 404.2. Obtain, review, and approve the plan prior to issuing a Certificate of Occupancy.

Exterior:
Address
21. __| _ | __Large apartment complex, 5 or more buildings, large address map is detailed at each street entry point and large address placard is detailed to be approved for size and content (should include complex site layout, foot print of each building, address or ID for each building, at least 24 in. x 36 in.), or as required by the fire official, 505.1.
Lock Box
22. __| _ | __Detailed on plans near the front entry and to left of entry (when facing the entry), the lock box should be mounted at about 6 ft. above ground level or as required by the fire official, 506.1.
Vehicle Impact
23. __| _ | __Guard posts are detailed on the plans for hydrants, tanks, generators, gas meters, etc. subject to vehicular damage, 312.1.
24. __| _ | __Guard post details comply with IFC 312.2; 4 in. diameter steel posts, filled with concrete, spaced up to 4 ft. from other posts, set 3 ft. deep, located 3 ft. or more from object, 312.2.
25. __| _ | __Other approved barriers comply based on details of location and specification data sheets, 312.3.
Hydrants, private
26. __| _ | __Flush and pressure test criteria are on the plans, proper port height and that the direction of the ports to enable fire engine hook-up are detailed, 508.4.
27. __| _ | __Specification data sheet is provided to verify model and brand of hydrant complies with city and fire department specifications.
Water Mains/Hydrants, private
28. __| _ | __Required in accordance with IFC 508, Appendices B and C, designed in accordance with NFPA 24, and consult the appropriate worksheet in this book.
Fire Department Connection for Sprinkler and/or Standpipes
29. __| _ | __Location is in accordance with fire department policy and IFC 912, design is in accordance with NFPA 13 and/or 14, and consult the appropriate worksheet in this book.

Exit System:
Exit Signs and Emergency Lights
30. __| _ | __Internal or external illuminated exit signs are provided, IFC 1011.2.
31. __| _ | __Exit signs are at directional changes in hallways, visible from any direction of egress travel, and where necessary, IFC 1011.1. (Exception: main exterior exit doors that are obviously an exit and rooms or areas only requiring one exit).
32. __| _ | __Exit signs are flush or perpendicular to mounting surface as required for occupant to read signage.
33. __| _ | __Battery backup or emergency power is provided; emergency lighting for occupancies and rooms required when there are two or more exits, IFC 1006.3.
34. __| _ | __Signage detail is provided for the landing areas of vertical stairways more than 3 stories. Signage states location-East, West, etc., terminus-roof access or not, floor number, and floors served by stairs, and is 5 ft. above the floor landing 1020.1.6. Signs shall be an approved legible permanent design.
Hardware
35. __| _ | __Manually operated flush or bolt locks are not permitted on egress doors, IFC 1008.1.8.4.
36. __| _ | __Panic hardware required for each door in means of egress for Group A and E occupancies with 50 or more occupants and any Group H 1008.1.9.

Fire Plan Review and Inspection Guidelines

Occupant load
37. _ | _ | __Occupant load calculations are provided for the rooms and assembly areas, and occupant load signs are processed for assembly areas, 1004.

Egress Width and Doorways
38. _ | _ | __Egress width calculations are provided for egress paths and assembly areas, 1005.
39. _ | _ | __Doorways are of adequate width for the occupant load.
40. _ | _ | __The number of doors for each room or area complies with 1015, Table 1015.1, and 1019.

Travel Distance
41. _ | _ | __Travel distance for the occupancy is in accordance with 1016, Table 1016.1, Table 1019.2.

Fire Extinguishers:
42. _ | _ | __Type, size, mounting height, and travel distance are appropriate for the hazard and are detailed on the plans, 906.
43. _ | _ | __Room use or type of hazard and/or process is detailed on the plans.
44. _ | _ | __For a hazardous type business, 50 ft. travel distance or 30 ft. depending on the size of the extinguisher, check the plans.

Fire Protection Systems:
Automatic Sprinkler Systems
45. _ | _ | __Required in accordance with IFC 903? Designed in accordance with NFPA 13 and consult the appropriate worksheet in this book.
46. _ | _ | __Room use or type of hazard/process is detailed on the plans.
47. _ | _ | __Valves, pipe, hanger, and sprinkler data sheets and hydraulic calculations are provided.
48. _ | _ | __The location of the fire department connection is in accordance with IFC 912 and the design is in compliance with NFPA 13.

Standpipes
49. _ | _ | __Required in accordance with IFC 905? Design is in accordance with NFPA 14 and consult the separate worksheet in this book.

Fire Pumps
50. _ | _ | __Follow the worksheet in this book.

Fire Alarms
51. _| _ | __Required in accordance with IFC 907? Design per NFPA 72, and consult the separate worksheet in this book.
52. _| _ | __Room use or type of hazard and/or process is detailed on the plans.

Alternate Clean Agent Gas or Halon
53. _| _ | __Required in accordance with IFC 904? Consult the separate worksheet in this book.
54. _| _ | __Room use or type of hazard and/or process is detailed on the plans.
55. _| _ | __Equipment and devices data sheets, voltage and battery calculations are provided.
56. _| _ | __Concentration and discharge calculations are provided.

Kitchen Hood System
57. _| _ | __Required in accordance with IFC 904.2.1? Design is in accordance with the manufacturer's design and installation manual, and consult the separate worksheet in this book.

Emergency Alarm
58. _| _ | __Required in accordance with IFC 908.

Use and Process Specifics:
Group R Occupancies
59. _| _ | __Smoke detectors are located in sleeping rooms, hallway and adjacent room if vaulted ceiling, also at the top of stairs and lower level, 907.3.2.

Assembly Use
60. _| _ | __Occupant loads certificates for assembly rooms are created, 1004.3.
61. _| _ | __Exit door hardware; panic hardware is on all exit doors except main entry, no other security bolt hardware allowed with panic hardware, 1008.1.9.
62. _| _ | __Data sheets for fire-resistive decorative materials are provided, 804, 806, 807.

Fire Plan Review and Inspection Guidelines

Battery Rooms
63. __| _ | __Spill control method and materials are detailed, IFC 608.5.
64. __| _ | __When ventilation is required, calculations are provided to verify ventilation rate maintains the room atmosphere to 1 percent by volume of hydrogen or a minimum ventilation rate of 1 cfm/ft^2 of floor area, 608.6.1.
65. __| _ | __Door signed stating room contains lead-acid batteries, energized electrical circuits, and corrosive liquids are detailed on the plans, 608.7.
66. __| _ | __Battery systems are seismically braced and room smoke detection is provided, 608.8 and 608.9.

Fuel-Dispensing Station
67. __| _ | __Details for dispensing unit or island, and emergency shut-off location are provided, and tank construction is in accordance with IFC Chapter 22 and 34. Consult the worksheet in this book.

Hazardous Materials
68. __| _ | __Hazardous Materials, IFC 27 always applies then the chapter specifics.
69. __| _ | __Spill control and secondary containment, IFC 2704.2, 2705.1.3, 2705.2.1.3, 2705.2.1.4, 3404.3.7.3, 3404.3.7.8.2, 3404.4.3, 3405.3.7.5.3 and 3405.3.7.5.6.3.
70. __| _ | __Spill control and secondary containment shall be liquid tight, i.e. compatible floor coating, 2704.2.1(1) and (2), 2704.2.2.1 (1) and (2).
71. __| _ | __Cryogenics, IFC 32.
72. __| _ | __Compressed Gases, IFC 30.
73. __| _ | __Flammable Combustible Liquids, IFC 34.
74. __| _ | __Highly Toxic and Toxic Materials, IFC 37.
75. __| _ | __Oxidizers, IFC 40.
76. __| _ | __Unstable and Water Reactive Materials, IFC 43 and 44.
77. __| _ | __Consult the various worksheets in this book.

High-piled Storage
78. __| _ | __A policy consideration for speculation buildings with potential to house high-pile storage but not designed for high-piled storage is requiring demarcation lines (12 ft. above floor level) and verbiage on walls ("No storage permitted above this line by the authority of the Fire Marshal's Office.") and letter from owner agreeing that building is not designed for nor will it be used for high-piled storage, 2304.3.
79. __| _ | __The 14 items in IFC 2301 are detailed or provided, 2301.1.
80. __| _ | __Aisle width and rack locations are on the plans, 2301.1.
81. __| _ | __Exterior fire access doors are on the plans, 2301.1.
82. __| _ | __Detection and suppression systems are provided, 2301.1.
83. __| _ | __Smoke removal system or heat vents or curtain boards are detailed on the plans, 2301.1.
84. __| _ | __Need product hazard classifications, quantities, location, etc, consult additional information in this book and use the Inventory Information and Rack Storage Information sheets, 2301.1.

LPG Tanks
85. __| _ | __Location on a site plan is provided, IFC 38 and NFPA 58. Consult the worksheet in this book.

Medical Gases
86. __| _ | __Quantity, location and room construction are provided and detailed, IFC 30 and consult the worksheet in this book.

Paint Booth, Powder Coating, Electrostatic Spray Operations
87. __| _ | __Manufacture data sheets for the booth and equipment are provided, IFC 15 and consult the worksheet in this book.
88. __| _ | __Construction and ventilation details for spray booths, areas or rooms are provided.

Welding Operations
89. __| _ | __Location is detailed, signage detailed, type of gases, storage location and quantity of gases, IFC 26.

Fire Plan Review and Inspection Guidelines

Fire Marshal's Office Final Inspection Worksheet

Date:_____ Permit:_____ Address:_____

This a basic inspection guide and is not all inclusive.

NA |OK| N   **NA** – not applicable   **OK** – acceptable   **N** – need, not acceptable

Inspection Preparation:
1. __ | _ | __Confirm inspection request from contractor.
2. __ | _ | __Verify preinspections were done and relevant forms are on-site, e.g., dbA log, certifications, flow test log, etc.
3. __ | _ | __Permit file and plans are reviewed for requirements noted on plans and other documents.
4. __ | _ | __Occupant load certificate is in file for delivery during final inspection.

Exterior:
Access
5. __ | _ | __Fire lane complies with the plans, IFC 503.1.1.
6. __ | _ | __Fire lane signage is provided and striped as necessary, IFC 503.3.
7. __ | _ | __Gates operate as required, IFC 503.5.
8. __ | _ | __Emergency turn-arounds are per plans, IFC 503.2.5.
Address
9. __ | _ | __Visible from the street, minimum 4 in. high and ½ in. wide lettering, and contrasting color, IFC 505.1.
10. _ | _ | __Large apartment complex, 5 or more buildings, has large address map at street entry points.
Lock Box
11. _ | _ | __Installed per plans near front entry on left of entry (when facing entry), usually mounted at about 6 ft. but may vary depending on department policy, 506.1.
12. _ | _ | __Building keys are provided and locked into the box or deferred later for crew installation.
Vehicle Impact
13. _ | _ | __Guard posts are provided per plans or for hydrants, tanks, generators, gas meters, etc. subject to vehicular damage, IFC 312.1.
14. _ | _ | __Guard posts are constructed per code, 4 in. diameter, steel, filled with concrete, 4 ft. from other posts, set 3 ft. deep, located 3 ft. or more from object, IFC 312.2.
15. _ | _ | __Other approved barriers comply with plans, IFC 312.3.
Hydrants, private
16. _ | _ | __Flush, pressure test, conduct a flow test to verify the required GPM for the project is available, ensure proper port height and that the direction of the ports enable fire engine hook-up, IFC 508.4.
17. _ | _ | __Obtain certification, flow test results for each hydrant, and documents indicate hydrant locations.
Water Mains, private
18. _ | _ | __Refer to the appropriate worksheet.

Exit System:
Landings
19. _ | _ | __Landings at exterior exits are provided and not greater than ½ in. below threshold, IFC 1008.1.5.
Exit Signs and Emergency Lights
20. _ | _ | __Internal or external illuminated exit signs are provided, IFC 1011.2.
21. _ | _ | __Exit signs are at directional changes in hallways, visible from any direction of egress travel, and where necessary, IFC 1011.1. (Exception: main exterior exit door that are obviously an exit and rooms or areas only requiring one exit.)
22. _ | _ | __Battery backup or emergency power is provided for emergency lighting for occupancies required to have two or more exits, IFC 1006.3.
23. _ | _ | __Signage (minimum 12 in.) is on each landing of vertical stairways more than 3 stories. Signage states location-East, West, etc (1 in. lettering), terminus-roof access or not (1 in), floor number (5 in. lettering), and floors served by stairs, 1-12 (1 in. lettering), IFC 1020.1.6.
Hardware
24. _ | _ | __Manually operated flush or bolt locks not permitted on egress doors, IFC 1008.1.8.4.

Fire Plan Review and Inspection Guidelines

25. _|_ | _ Panic hardware required for each door in means of egress for Group A and E occupancies with 50 or more occupants and any Group H occupancy, IFC 1008.1.9.

Fire Extinguishers:
26. _|_ | _ Tagged, sealed, mounted and visible and the location and installation complies with the plans.

Fire Protection Systems:
   Fire Doors
27. _|_ | _ Roll down fire doors: detectors or fusible links on both sides, do drop down test or verify it was done by building inspector. Note on inspection report where fire doors are located, info will be put in fire protection system database for tracking annual servicing.
   Sprinklers
28. _|_ | _ Use the appropriate worksheet.
   Standpipes
29. _|_ | _ Use the appropriate worksheet.
   Fire Alarms
30. _|_ | _ Use the appropriate worksheet.
   Kitchen Hood System
31. _|_ | _ Use the appropriate worksheet.
   Alternate Clean Agent Gas or Halon
32. _|_ | _ Use the appropriate worksheet.
   Fire Pumps
33. _|_ | _ Use the appropriate worksheet.
   Gas and Chemical Leak Detection
34. _|_ | _ Use the appropriate worksheet.
   Fire Pumps
35. _|_ | _ Use the appropriate worksheet.
   Special Hazard Detection or Suppression Systems
36. _|_ | _ Use the appropriate worksheet.

Use and Process Specifics:
   Group R Occupancies
37. _|_ | _ Access, address, fire extinguishers, private hydrants were verified in worksheet above.
38. _|_ | _ Exit lighting and signage are verified in worksheet items above.
39. _|_ | _ Smoke detectors; green light (house power) is on, located in sleeping rooms, hallway and adjacent room if vaulted ceiling, also at the top of stairs and lower level, test with the circuit breaker off.
   Assembly Use
40. _|_ | _ Occupant loads certificates for assembly rooms are delivered and mounting location instructions are provided, IFC 1004.3.
41. _|_ | _ Access, address, fire extinguishers, private hydrants were verified in worksheet above.
42. _|_ | _ Exit lighting and signage are verified in worksheet above.
43. _|_ | _ Exit door hardware: panic hardware is on all exit doors except main entry, no other security bolt hardware allowed with panic hardware, 1008.1.9.
44. _|_ | _ Decorative materials are fire-resistive in accordance with IFC 803, 804, 806, and 807.
   Above-ground Storage Tanks for Flammable and Combustible Liquids
45. _|_ | _ Use the appropriate worksheet.
   Battery Rooms
46. _|_ | _ Spill control method and materials in place, IFC 608.5.
47. _|_ | _ Documentation to verify ventilation rate of at least 1 cfm/ft^2 for floor area, IFC 608.6.
48. _|_ | _ Door signed stating room contains lead-acid batteries, energized electrical circuits, and corrosive liquids, IFC 608.7.
49. _|_ | _ Battery systems are seismically braced and room smoke detection is provided, IFC 608.7 and 608.8.
   Fuel-Dispensing Station
50. _|_ | _ Use the appropriate worksheet.
   HPM Workstations, Gas Cabinets, Silane Gas and Ozone System
51. _|_ | _ Use the appropriate worksheet.

Fire Plan Review and Inspection Guidelines

High-Piled Storage
52. _ | _ | __ Aisle width and rack locations in compliance with plans, IFC 2306.9.
53. _ | _ | __ Fire access doors are accessible, IFC 2306.6.1.
54. _ | _ | __ Detection and suppression systems are in compliance with the plans, IFC 2306.4, 2306.5.
55. _ | _ | __ Smoke removal system or heat vent or curtain board are in compliance with the plans, IFC 2306.7.

LPG Tanks
56. _ | _ | __ Followed and used the appropriate worksheet.

Medical Gases
57. _ | _ | __ Followed and used the appropriate worksheet.

Paint Booth
58. _ | _ | __ Followed and used the appropriate worksheet.

Battery Rooms
59. _ | _ | __ Spill control method and materials in place, IFC 608.5.
60. _ | _ | __ Documentation to verify ventilation rate of at least 1 cfm/ft^2 for floor area, IFC 608.6.

Warehouse
61. _ | _ | __ Exit lighting and signage are verified in worksheet above.
62. _ | _ | __ Miscellaneous requirements: vents, sprinklers, smoke detection, rack layout, complies with the plans.
63. _ | _ | __ Access doors are provided.

Sample Service Interval and Service Report Policy

The sample body of text below can be used as content for a policy or as the body of a letter to designers and contractors.

After the inspection or service of a fire protection system, a copy of the inspection service report shall be forwarded, according to an IFC local amendment or policy, to the:

Office of the Fire Marshal
xxxx Plan Review St.
Design City, US 00001

Service reports will be entered into a database, which initiates an inspection on systems identified as having code deficiencies. For accurate data input, service reports must have the following information:

- name of the business with the fire protection system
- address of the business location
- be specific when listing a deficiency, e.g., its location, the device or element that is deficient
- date of the inspection
- type, model, and manufacturer of system

If a system is repaired, please send in a completed service report so the department database can be updated. Also, before a sprinkler or fire alarm system is disabled, notify and coordinate the event with the building representative in accordance with Section 907.7 of the IFC. The building shall be provided with a fire watch from a security agency. Fire watch criteria will be provided upon request.

Included in this document are sample inspection and service report forms. Each company may use these forms or an approved equivalent form.

Every system inspected or serviced requires an inspection tag attached to the system.

Inspection and service shall be performed in accordance with NFPA Standards. The following matrix is a sample guide for the service frequency that integrates the requirements of the IFC, NFPA, and modifications to frequencies by the authority having jurisdiction. The service report shall be sent to the Fire Marshal's Office address noted above.

Inspection/Test Intervals and Corresponding Service Report

	3 mo.	6 mo.	1 yr.	5 yr.	6 yr.	12 yr.	Code
Clean agent			X				IFC 904.10.1
Carbon dioxide			X				NFPA 14 4.8.3 IFC 904.8.1
Emergency & standby power systems			X**				** recommended, IFC 604.3.1
Fire alarms			X				IFC 907.20.2
Detector sensitivity			X	X			IFC 907.20.3
Fire door assemblies			X				IFC 703.4
Fire pumps			X				NFPA 25 8.3.3
Halon		X					NFPA 12A 6.1.1 IFC 904.9.1
Dry chemical		X					IFC 904.6.1
Gas or liquid leak detection systems			X				IFC 2703.2.9.2
Kitchen systems		X					IFC 904.11.6
Portable extinguisher			X		X	X	NFPA 10 6.3.1, 6.3.3, 7.2
Private hydrant			X				IFC 508.5.3
Private water tank							Varies, NFPA 25 Table 9.1
Smoke control system (dedicated)		X					IFC 909.20.4
Smoke control system (nondedicated)			X				IFC 909.20.5
Automatic sprinklers			X annual report				Varies. See NFPA 25 Table 5.1
Standpipes	X		X annual report	X			Varies NFPA 25 Table 6.1
Water mist	X	X	X annual report				Varies NFPA 750 13.2.2

Fire Watch
Sample informational and policy bulletin.

Office of the Fire Marshal

Fire Watch

References: 2006 International Fire Code Sections 101.2, 102.7, 102.8, and 901.7

1. **Purpose:**
   To provide fire watch requirements for an owner or responsible party when adequate egress is not available, when demolition of a building with hazardous conditions exists, or when a fire alarm or sprinkler system are in disrepair or nonfunctional. Fire watch exception: If a fire alarm or sprinkler system is only non-functional for one day during regular business hours (8:00 AM to 5:00 PM).

2. **Scope:**
   This policy applies to all facilities in which the Fire Marshal's Office has authority.

3. **Background:**
   Due to a lack of specifics in the International Fire Code, this policy was developed to clarify and specify fire watch requirements.

4. **Requirements:**
   Fire Marshal's Office specifics:
   - A. The duration and times per day for a fire watch shall be assigned by the Fire Marshal's Office.
   - B. Conduct periodic patrols of the entire facility. Patrol the facility every 15 minutes if the facility has people sleeping, is an institutional facility or an occupied assembly facility. Facilities not meeting the previous conditions shall be patrolled every 30 minutes.
   - C. The Office of the Fire Marshal may designate additional safety duties for fire watch personnel.
   - D. Fire watch personnel shall be preapproved by the Office of the Fire Marshal.

   Duties:
   - A. Fire watch personnel shall have access to one approved means of communication; know the exact address of the property, and how to report a fire or other emergency condition by calling 9-1-1.
   - B. Fire watch personnel shall be familiar with the buildings and property and have an accepted written plan for patrolling the property.
   - C. Fire watch personnel shall be trained in the use of fire extinguishers shall have access to all facility fire extinguishers and know their location.
   - D. Fire watch personnel shall have knowledge of and be trained in the facility's evacuation plan in the event of a fire. They shall be able to communicate with non-English speaking residents well enough to give an evacuation order.
   - E. Fire watch personnel shall not be permitted, while on duty, to perform any other duties.
   - F. Fire watch personnel shall not be impaired and shall remain awake and alert at all times.
   - G. Fire watch personnel shall keep a log of fire watch related activities. The log shall include; address of the facility, time of each patrol, name of the fire watch person, notes for other related activities performed. The log shall be faxed to (xxx-xxxx) or delivered daily to the Office of the Fire Marshal.

PREPARED BY:		
EFFECTIVE DATE:	REVISED DATE:	BY:

Fire Plan Review and Inspection Guidelines

Emergency Plans
2006 IFC Chapter 4

Emergency plans apply to many different occupancies. These plans are to be submitted to the Fire Department Plan Reviewer during the plan review process for approval. Detailed requirements are found in Chapter 4. A sample Fire Evacuation Planning Guide follows this short list.

Which occupancies require emergency plans?
Groups A, E, H, I, R-1, R-2, R-4, High-rise, B and M with 500 or more occupants or more than 100 occupants above or below the level of exit discharge, covered malls, underground buildings, and buildings with atriums which contain a Group A, E, or M occupancies.

What are emergency plans?
Emergency plans consist of procedures for:
- reporting emergencies
- use of alarms
- notification of occupants and emergency responders in the event of alarm system malfunctions
- isolating the fire
- evacuating each fire area and the building
- relocating ambulatory and nonambulatory persons

Emergency plans specify:
- training frequency and subjects
    - staff duties during emergencies
        * familiarization with assigned duties, evacuation routes, areas of refuge, exterior assembly areas, leading groups or assisting individuals
    - how to recognize and respond to fires
    - how to initiate a fire alarm signal
    - how to deal with restrained and nonambulatory occupants
    - how to use fire protection appliances, fire extinguishers
- fire drill frequency

Emergency plans provide:
- location of egress diagrams within the building
    * diagrams show two evacuation routes for each room or area and floor
- location of areas of refuge
- location of exterior assembly areas for the accounting of the evacuees
- location of portable extinguishers
- location of manual fire alarm pull stations
- location of fire alarm control panels
- an emergency guide for apartment buildings which includes:
    - the function of single station smoke detectors
    - how to use accessible fire protection appliances i.e., fire extinguishers
    - emergency evacuation plan for each dwelling unit

Fire Plan Review and Inspection Guidelines

EMERGENCY PLANNING AND PREPAREDNESS

Office of the Fire Marshal
Reference: 2006 IFC

Introduction

This document was created to assist in providing the requirements for reporting emergencies, coordinating with emergency personnel, organizing emergency plans, and procedures to manage emergencies. It is important that every business or building have a plan for what to do in the event of an emergency. Formulating a plan will reduce the chances of confusion, panic or injury. The individuals in charge of building safety should take the time to create a plan, practice it and share it with the building occupants.

TABLE OF CONTENTS

1. General information
2. Where fire safety and emergency plans are required
3. Fire evacuation plans
4. Fire safety plans
5. Specific use/occupancy emergency plan requirements
6. Emergency drill frequency
7. Recordkeeping
8. Surveying building
9. Designating an exterior meeting spot
10. Employee duties, assignments and training
11. Pick your fire safety team
12. Reporting emergencies
13. Procedures for persons unable to use exit stairs
14. If you are unable to leave the floor
15. Sample floor plans

1. General Information

- In the event of a fire, the owner or occupant shall immediately notify the fire department via 9-1-1.
- Upon activation of a fire alarm signal, employees or staff shall immediately notify the fire department via 9-1-1.
- Nothing prohibits the sounding of the fire alarm signal for conducting evacuation drills but call the nonemergency number for 9-1-1 to inform them of the drill.
- It is illegal to interfere with fire operations.
- Fire safety and evacuation plans shall be available to occupants.

2. Where Fire Safety and Evacuation Plans are Required.

Section 404 of the International Fire Code, Section 404 requires a number of occupancies to have a fire safety and evacuation plan. Those occupancies that require such a plan are:

- **Group A:** Excluding assembly occupancies used exclusively for purposes of religious worship and have an occupant load less than 2,000, a fire safety and evacuation plan is required in buildings with assembly rooms with a capacity of 50 individuals or more.

- **Group B:** Buildings used for office, professional or service-type transactions with 500 or more occupants or more than 100 occupants above or below the level of exit discharge.

- **Group E:** Educational facilities used through the 12th grade and day care facilities.

- **Group H:** Buildings involving high-hazard uses involving manufacturing, processes, generation or storage of materials, which constitute a physical or health hazard exceeding maximum allowable quantities.

- **Group I:** Nurseries for children, hospitals, sanitariums, nursing homes, homes for children, half-way houses, group homes, and health-centers for patients unable to evacuate themselves as a result of medical treatments. Mental hospitals and sanitariums, jails, prisons, reformatories and buildings where personal liberties of occupants are similarly restrained.

- **Group R-1 and R-4:** Buildings used for hotels, motels, boarding houses, dormitories, fraternities and sororities, residential care for 5 to 16 persons.

- **Group R-2:** college and university buildings.

- **High-rise:** Buildings with an occupied floor level 75 ft. above the lowest level of fire department access.

- **Group M:** Retail sales and display buildings with an occupant load 500 or more or occupants above or below the level of exit discharge.

- **Covered malls:** Exceeding 50,000 sq. ft. in aggregate floor area.

Fire Plan Review and Inspection Guidelines

- ✓ **Underground buildings:** All.

- ✓ **Buildings with an atrium that contain a Group A, E or M occupancy:** An atrium is an opening in the building connecting two or more stories and closed at the top.

3. *Fire Evacuation Plans*

Section 404.3.1 of the IFC requires fire evacuation plans to contain the following information:

- ✓ Emergency exit routes and information whether the evacuation includes the entire building or just selected floors.

- ✓ Procedures for employees who must operate critical equipment before evacuating.

- ✓ Procedures for where employees and occupants should meet and how they should be accounted for after the evacuation.

- ✓ The assignment of personnel responsible for assisting others and rendering medical aid and how personnel will be identified during the emergency.

- ✓ The preferred and an alternate method of notifying the occupants of an emergency or fire.

- ✓ The preferred and an alternate method of reporting an emergency or fire to 9-1-1.

- ✓ The assignment of personnel who can be contacted for additional information or explanation of duties under the plan and how personnel will be identified during the emergency.

- ✓ A description of the emergency voice/alarm communication system alert tone and the voice messages, where provided.

4. *Fire Safety Plans*

Section 404.3.2 of the IFC requires fire safety plans to include the following minimum information:

- ✓ A procedure for reporting a fire or other emergency.

- ✓ The strategy and procedures for notifying, relocating or evacuating occupants.

- ✓ Site plans indicating: 1) where occupants assemble after evacuation, 2) location of nearest hydrants, and 3) normal fire vehicle access roads so people do not obstruct the use of Items 2 and 3.

- ✓ Floor plans showing: 1) exits and the routes to get to the exits, 2) accessible exit routes, 3) areas of refuge (special areas designed into the building for the location of people that have accessibility needs), 4) manual fire alarm boxes, 5) fire extinguishers, 6) occupant-use hose stations, and 7) the location of fire alarm annunciators and controls.

- ✓ A list of major fire hazards associated with the building, its use and processes.

- ✓ The identification and assignment of personnel responsible for the maintenance of systems and equipment installed to prevent or control fires.

- ✓ The identification and assignment of personnel responsible for the maintenance, housekeeping, and controlling of fuel hazard sources.

Instructions should include not using the elevator when the fire alarm sounds or when there is a reported fire. Using the elevator can be dangerous, as the elevator doors could open up on the floor involved in fire resulting in injury or death.

Floor plans should also identify the routes, both conventional and accessible routes, of evacuation for each room or portion of the occupancy and the locations of interior areas of refuge. Post the floor plan throughout the building. Mark YOU ARE HERE relative to the floor and the location on the posted plan. In hotels, motels, and shelters, a diagram depicting two evacuation routes shall be posted on or immediately adjacent to every required exit door from a sleeping room.

Posted site maps should show the locations of the exterior assembly area where occupants are requested to meet so a count can be taken.

A simple flyer can be created to distribute to the building occupants, or a more detailed document can be used outlining the roles of staff or tenants during a fire emergency. In residential buildings, a copy of the emergency guide should be given to each tenant prior to the time of initial occupancy. In employee environments, the emergency plan should be given to each new employee and a copy should be in an accessible location for all employees to review.

Required emergency plans shall be submitted to the fire department for review. Emergency plans shall be reviewed and updated annually. Additional reviews and updates shall be provided whenever changes are made in the occupancy or physical arrangement of the building.

5. *Specific Use and Occupancy Emergency Plan Requirements*

Hazardous Materials
Any occupancies listed in Section 2 that have a hazardous materials permit from the Fire Marshal's Office shall provide the following:

- ✓ MSDS's that are readily available.
- ✓ Rooms and buildings are identified with a hazard warning sign in compliance with the NFPA 704 diamond hazard placard.
- ✓ Persons responsible for areas where hazardous materials are stored, handled, or used shall be trained about the nature of the chemicals and know the mitigating actions to take to maintain a safe environment in the case of a release, spill, fire, or leak. They shall be trained as the liaison for the fire department.
- ✓ A Hazardous Materials Inventory Statement shall be available to fire officials.

Group A Occupancies (places of assembly)
Group A occupancies shall also provide a detailed seating plan(s) and occupant load(s) for rooms exceeding an occupant load of 50 with their emergency plan.

Group E occupancies (educational facilities, 1st - 12th grade)
Fire drills involve the following:
- ✓ The 1st drill of each school year shall be conducted within 10 days of the beginning of classes.
- ✓ In severe weather the Fire Code Official shall approve deferral requests of the drill frequency.
- ✓ The time of the drills shall be random to avoid the distinction between drills and actual fires.
- ✓ Persons evacuating from the facility shall meet at designated locations.

Semiconductor Fabrication Facilities (H-5)
Group H-5 occupancies shall also provide with their emergency plan:
- ✓ Plans and diagrams in approved locations indicating the amount and type of HPM stored, handled, and used, location of shutoff valves for HPM supply piping, emergency phone and exit locations.

Group I-1 Occupancies (buildings exceeding 16 persons providing 24 hour care: assisted living, half-way houses, group homes, etc)
Group I-1 occupancies shall also provide with their emergency plan:
- ✓ Special staff actions including fire protection procedures for residents.
- ✓ Residents capable of assisting in their own evacuation shall be trained in the proper actions in the event of an emergency.
- ✓ Emergency drills of the onsite emergency response team shall occur at least 6 times per year unless the facility is under licensure of the state and can evacuate within 3 minutes.
- ✓ Residents shall be involved in the evacuation drills by moving them to a selected assembly point. There are some exceptions so consult with your Fire Marshal's Office.

Group I-2 Occupancies (hospitals, nursing homes, outpatient clinics, etc)
Group I-2 occupancies shall also provide with their emergency plan:
- ✓ The movement of patients to safe areas or to the exterior of the building is not required during drills.
- ✓ When evacuation drills are conducted after visiting hours or when patients are expected to sleep, a code announcement is permitted instead of audible alarms.

Group I-3 Occupancies (jails, prisons, reformatories, etc)
Group I-3 occupancies shall also provide with their emergency plan:

- ✓ Employees shall be instructed in the use of portable extinguishers. Train new staff immediately and provide refresher training semiannually.
- ✓ 24 hour staffing is required to be within 3 floors or 300 ft. of the access door of each resident housing area. In Use Conditions 3, 4, and 5 as defined in Chapter 2 of the IFC, the staff must release locks necessary for evacuation or other emergency actions within 2 minutes of the alarm.

- ✓ Keys necessary for unlocking doors for exiting shall be identifiable by touch and sight.

R-1 Occupancies (transient boarding house, hotels and motels, etc)
R-1 occupancies shall also provide with their emergency plan:

- ✓ A diagram shall be posted on or adjacent to each guestroom interior exit door depicting 2 evacuation routes.
- ✓ Employees shall activate an available fire alarm, and call 9-1-1 when a fire or suspected fire is discovered.
- ✓ Guests shall be provided information to decide whether to evacuate, go to an area of refuge, or remain in place.

R-2 Occupancies (University dormitories, fraternities, and sororities)
R-2 occupancies shall also provide with their emergency plan:

- ✓ A fire emergency guide that describes the location, function and use of fire protection equipment including fire alarm systems, smoke alarms, and portable extinguishers. The guide shall be provided in each guest unit.

Fire drills involve the following:

- ✓ The 1st drill of each school year shall be conducted within 10 days of the beginning of classes.
- ✓ The time of the drills shall be random to avoid the distinction between drills and actual fires.
- ✓ One drill shall be before sunrise or after sunset.

R-4 Occupancies (24 hour assisted living from 5 to 16 persons, etc)
R-4 occupancies shall also provide with their emergency plan:

- ✓ Employees shall be periodically instructed in the proper actions to take in the event of a fire.
- ✓ Evacuation drills shall be done according to the licensing agency rules and not less than 6 times per year, 2 times per shift per year and 12 drills shall be conducted the first year.
- ✓ Drills will involve the evacuation of residents to selected assembly points and shall provide residents with experience in exiting the building. There are some exceptions so consult with your Fire Marshal's Office.

Covered Mall Buildings

Malls shall also provide with their emergency plan:

- ✓ A lease floor plan identifying each occupancy and tenant, fire protection features; fire department hose connections, fire command center, smoke management systems, hose valve outlets, elevator controls, sprinkler and standpipe control valves, firewalls and smoke detector zones.
- ✓ The lease plan shall be approved by the Fire Marshal's Office and be available to emergency responders.
- ✓ The second exit doors for each tenant shall be labeled on the exterior side with the tenant's name and number.
- ✓ Vacant tenant spaces are kept free of storage, combustible waste, and shall be kept clean.

6. *Emergency Drill Frequency*

Drills shall occur as follows:

2006 IFC
TABLE 405.2
FIRE AND EVACUATION DRILL FREQUENCY AND PARTICIPATION

GROUP OR OCCUPANCY	FREQUENCY	PARTICIPATION
Group A	Quarterly	Employees
Group B[c]	Annually	Employees
Group E	Monthly[a]	All occupants
Group I	Quarterly on each shift	Employees[b]
Group R-1	Quarterly on each shift	Employees
Group R-2[d]	Four annually	All occupants
Group R-4	Quarterly on each shift	Employees[b]
High-rise buildings	Annually	Employees

a. The frequency shall be allowed to be modified in accordance with Section 408.3.2.
b. Fire and evacuation drills in residential care assisted living facilities shall include complete evacuation of the premises in accordance with Section 408.10.5. Where occupants receive habilitation or rehabilitation training, fire prevention and fire safety practices shall be included as part of the training program.
c. Group B buildings having an occupant load of 500 or more persons or more than 100 persons above or below the lowest level of exit discharge.
d. Applicable to Group R-2 college and university buildings in accordance with Section 408.3.

Fire drills shall be conducted in a manner that carries out the procedures in the emergency plan. In addition, fire drills shall include a review of the emergency plan and assigned employee duties.

When fire drills are conducted, the orderly evacuation of the building shall receive priority over the speed of evacuation.

Records of fire drills shall be maintained on the premises for review by the fire department.

Always notify your fire alarm monitoring company and 9-1-1's nonemergency number before and after conducting a fire drill, so the fire department does not respond unnecessarily.

The following are suggestions for fire drills:

- ✓ Appoint someone to monitor the drill, activate and reset the fire alarm, and time the evacuation.
- ✓ Fire drills shall be conducted at varying times and under varying conditions to simulate conditions that could occur during a fire or other emergency. Make it realistic by requiring participants to use their second way out or to crawl low. This can be done by having someone hold up a sign reading "smoke" or "exit blocked by fire."
- ✓ After the evacuation, take a head count at the designated meeting place(s) to account for everyone's participation and safe evacuation.
- ✓ After the drill, gather everyone together to discuss questions or problems that occurred. Redesign the drill procedures as needed.

7. *Recordkeeping*

Records shall be maintained to record the events of the evacuation drill. The record shall include the following. A sample template is **provided** at the back of this document:

- ✓ Identify the person conducting the drill.
- ✓ Date and time of the drill.
- ✓ Notification method used.
- ✓ Staff members on duty and participating.
- ✓ Number of occupants evacuated.
- ✓ Special conditions simulated.
- ✓ Problems encountered.
- ✓ Weather conditions during the drill.
- ✓ Time required completing the evacuation.

8. *Surveying the Building*

Before starting to formulate an evacuation plan, take some time to conduct a walk-through of the building. During your walk-through, note the location of fire protection features, such as portable fire extinguishers, manual fire alarm boxes, and if your building is equipped with a fire alarm control system, the location of the fire alarm control unit and any annunciator panels.

For each area of the building, note the safest and shortest path for a primary evacuation route. Also note a secondary route if the primary path becomes unsafe or obstructed. In newer buildings, note "areas of refuge" which have been created as safe areas for individuals with wheelchairs or with walking disabilities.

Consider whether your building houses a changing population, as in a retail store or restaurant, or will the building occupants be familiar with the facility? Note any special needs that individuals may have, such as non-English-speaking or people with disabilities.

9. *Designating an Exterior Meeting Spot for All Occupants*

A location a safe distance away from the building shall be designated as a common meeting spot for people to gather after evacuating the building. The meeting spot should be located in a remote location clear of incoming fire equipment, away from traffic, and hazards such as falling glass. There could be several meeting locations depending on the size and configuration of the building.

A member of the Building Response Team should be assigned to verify that all building occupants are accounted for. That person should advise the fire department whether all occupants are accounted for or if some are still in the building.

To ensure the fastest, most accurate accountability of your employees, consider including the following items into your emergency plan:

- ✓ Designate assembly areas where employees should gather after evacuating. People should not re-enter the building for any reason.

- ✓ Take a head count after the evacuation. Identify the names and last known locations of anyone not accounted for and pass them to the Fire Official in charge.

- ✓ Establish a method for the accounting of nonemployees, such as suppliers and customers.

- ✓ Establish procedures for expanding the evacuation in case the incident expands. This may consist of sending employees home by normal means, or relocating them to a nearby building, or providing them with transportation to a more remote location.

10. *Employee Duties, Assignments and Training*

Employees should receive training in the emergency plans and their duties as part of new employee orientation and at least annually thereafter.

Employees should be assigned duties and be trained for emergencies. Training in fire safety, evacuation plans, and their duties shall occur annually (and semiannually for Group I-2 occupancies). Training should include familiarization with:

- ✓ **Fire Prevention:** Apprised of fire hazards of materials and processes in their work environment.
- ✓ **Evacuation:** Familiarization with the fire alarm and evacuation signals, evacuation procedures and routes, areas of refuge, exterior meeting areas, and their assigned duties in the event of an emergency.
- ✓ **Fire Safety:** Employees should be trained to know the locations and proper use of portable fire extinguishers.

11. Pick your Fire/Life Safety Team

The recommended structure of your team could involve one or more of the following roles.

Fire/life Safety Director
The person identified who will maintain and implement your Emergency Operations Plan. Pick a person who has not only the knowledge, but the authority to implement the procedures in the plan.

Building Response Team
Their primary role is to investigate the source of an alarm or emergency, communicate their findings to the other building occupants, and notify the Fire Department.

Floor Wardens
Floor wardens are volunteers selected from among the building staff and tenants to assist in the evacuation of occupants from specific building areas in the event of a fire alarm or emergency and conduct the accountability procedure for their area. They must be familiar with the building's evacuation plan.

Assistant Monitors
Assistant monitors are responsible for providing assistance to those individuals on a floor that require assistance to evacuate. These include people with disabilities or who have medical problems.

12. Reporting Emergencies

Building occupants should be instructed to call 9-1-1 whenever an emergency occurs. They should state:
- ✓ The nature of the emergency i.e., bomb threat, fire, or hazardous materials spill.
- ✓ The address of building.
- ✓ The nearest cross street.
- ✓ The extent of the fire or emergency, and specific information, if known.
- ✓ All callers should follow the 9-1-1 operator's instructions.

The caller should never hang up until told to do so by the emergency operator or unless his/her life is in danger.

Post the emergency number on all phones. In Eugene, call 9-1-1 for fire, police, or medical aid. The address of the building should be on the telephone. If the building security or manager wants to be notified, that number can be listed on the phone as well.

Controlling the fire and evacuating everyone safely depends on immediate notification of the emergency.

13. *Procedures for Persons Unable to Use Exit Stairs*

They should move to the exit or "Area of Refuge." If the stairwell is free of smoke, they should enter after all persons on the floor have evacuated, unless the stair landing is large enough for their presence without hindering the egress of others. After entering the exit stairway, make sure the stairwell door is securely shut. One person should wait with the disabled person while someone else goes to inform the arriving fire department of their location.

If stairwell traffic builds from the evacuation of upper floors, persons waiting should re-enter the floor to allow others to pass.

The disabled person and employee should stay at the stairwell landing as long as conditions are safe. It is preferred that appointed team members evacuate the person down the stairs to a safe area as determined by the plan. If the building does not provide a safe area for persons to wait, the emergency plan should incorporate a method and training to evacuate the disabled.

14. *If You Are Unable to Leave the Floor*

If you are unable leave the floor because exits become blocked or unsafe to enter due to smoke or fire, refuge should be sought on the floor in a totally enclosed room with a telephone and window.

Use towels or clothing to block openings around doors or vents where smoke might enter. Put a wet cloth over your mouth. Place a signal in the window. The signal can be anything that will call attention to your location. For instance, tie the curtains in a knot.

If smoke enters your unit, call 9-1-1 to report your location. Stay low to the floor to breathe the best air. It is not advisable to break the windows. Often smoke from outside of the building can enter through open windows and broken glass can injure those below.

15. *Sample Floor Plans*

Fire Plan Review and Inspection Guidelines

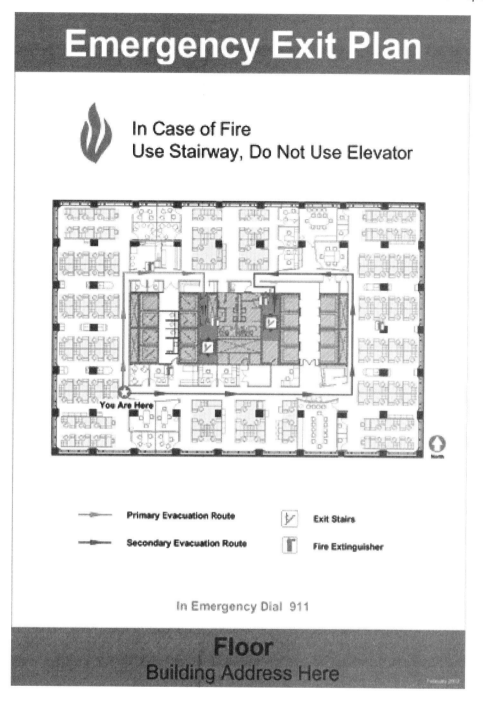

Reprinted with permission from The Environments Group, (312) 644-5080, Chicago, IL.

Fire Plan Review and Inspection Guidelines

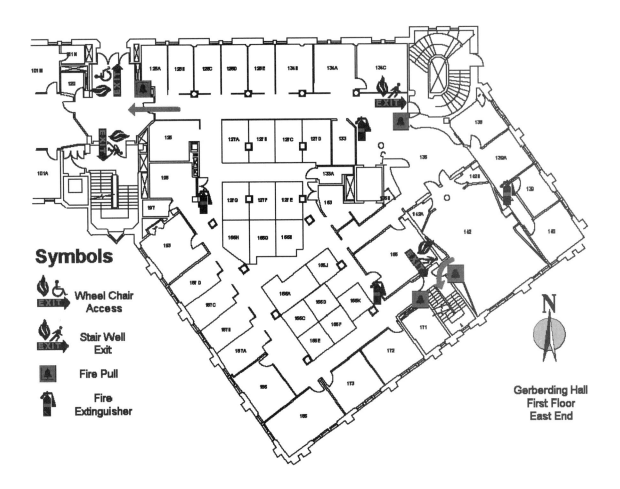

Reprinted with permission from the University of Washington, Planning and Budgeting, Seattle, WA.

Fire Plan Review and Inspection Guidelines

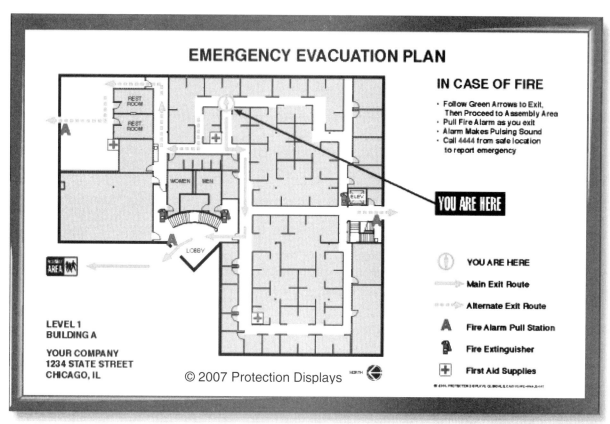

Reprinted with permission from Protection Displays, (818) 952-4044, Glendale, CA.

Fire Plan Review and Inspection Guidelines

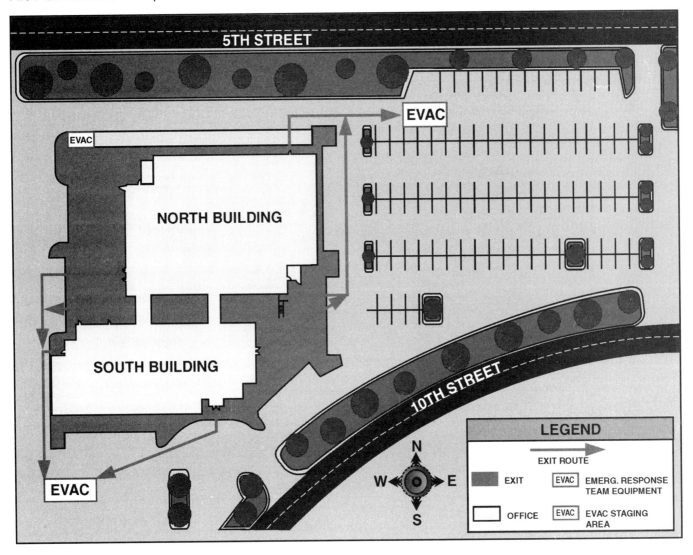

Fire Plan Review and Inspection Guidelines

Site Plan
Plan Review Worksheet
2006 IFC

Date of Review: _____ Permit Number: _____

Business/Bldg Name: _____ Address of Project: _____

Designer Name: _____ Designer's Phone: _____

The numbers that follow worksheet statements represent a IFC code section unless otherwise stated.

Appendix D and the references noted below are not mandatory unless the AHJ has incorporated the Appendix as a regulatory requirement.

Worksheet Legend: ✓ or **OK** = no problem,   **N** = need to provide,   **NA** = not applicable

Access:
1. _____ Drawings are provided.
2. _____ The required fire department access roads is a minimum unobstructed 20 ft. in width and 13 ft. 6 in. clear height, IFC 503.2.1. Check with local or state requirements that may have street planning regulations that supercede the IFC requirements.
3. _____ "No Parking Fire Lane" signs are provided at AHJ prescribed locations, IFC 503.3.
4. _____ Required fire department access roads are designed to support an apparatus with a gross axle weight of 75,000 lb, engineering specifications are provided, IFC App D102.1.
5. _____ Required fire department access roads are an all weather driving surface such as asphalt, concrete, chip seal (oil matting), or similar materials, IFC 503.2.3.
6. _____ The proposed building does have an emergency vehicle access road within 150 ft. of any exterior portion of the structure, if not, a fire department access road must be provided, IFC 503.1.1.
7. _____ The grade for required fire department access road does not exceed 10 percent unless approved by the Chief, Appendix D103.2.
8. _____ A local jurisdiction alternative to the 10 percent grade restriction could be the following: If the grade exceeds 10 percent, the first portion of the grade shall be limited to 15 percent for a length of 200 ft. and then 15 percent to 20 percent for a maximum of 200 ft., repeat the cycle as necessary unless the building is sprinklered.
9. _____ No access drive grades are greater than 10 percent if Appendix D is applicable at the local level, Appendix D 103.
10. _____ The access road design for a maximum grade conforms to specifications established by the fire code official, IFC 503.2.7.
11. _____ The dead-end fire department access roads (s) in excess of 150 ft. is provided with a turn-around, IFC 503.2.5.
12. _____ The turn-around cul-de-sac has an an approved inside and a outside radius, e.g. 30 ft. 50 ft. respectively, a hammerhead design is a minimum 70 ft. L x 20 ft. W, or another approved design may be used, IFC 503.2.4.
13. _____ The turning radius for emergency apparatus roads is 30 ft. inside and 50 ft. outside radius or as approved by the code official.
14. _____ Fire department access roads shall be constructed and maintained for all construction sites, IFC 1410.1.
15. _____ Dead-end streets in excess of 150 ft. resulting from a phased project are provided an approved temporary turnaround, IFC 503.2.5.

Water Flow and Hydrants: An in-depth plan review for private hydrants and private water mains will occur during the project plan review phase.
16. _____ A fire flow test and report is provided to verify that the fire flow requirement is available. Also, refer to the note at the bottom of the page.
17. _____ Water mains and pipe sizes are detailed on the site plan, IFC 508.1.
18. _____ All water mains and hydrants shall be installed and operate as soon as combustible materials arrive on a construction site, IFC 1412.1.
19. _____ The nearest hydrant(s) to the project structure and/or property road frontage are shown on the plan.

Fire Plan Review and Inspection Guidelines

20. _____ Prior to the installation of private water main systems, plans shall be submitted for a permit, review and approval.
21. _____ A hydrant is required within 400 ft. of any exterior portion of a nonsprinklered building or 600 ft. for an R-3 occupancy or sprinklered building, IFC 508.5.1.

Note: When a hydrant water flow report is required, the test should be performed by the local water purveyor or a company approved by the water purveyor. The report shall provide the water pressures measured and provide the available GPM at 20 PSI residual pressure. Existing reports may be used if not dated more than 3 years ago or as approved by the code official.

Additional Comments:

Review Date: _____ Approved or Disapproved FD Reviewer: _____

Review Date: _____ Approved or Disapproved FD Reviewer: _____

Review Date: _____ Approved or Disapproved FD Reviewer: _____

Fire Plan Review and Inspection Guidelines

Turnarounds and Secondary Emergency Access Roads

Office of the Fire Marshal

Turnarounds and Secondary Emergency Access Roads

References: 2006 IFC Section 503 and Appendix D

Access:
Every building shall be provided a fire apparatus access road. The road must extend to within 150 ft. of all portions of the exterior walls of the first story of a building as measured by an approved route.

Fire Apparatus Access Road: *A road that provides fire apparatus access from the fire station to a facility, building or portion thereof. This is a general term inclusive of all other terms such as fire lane, public street, private street, parking lot lane and access roadway.*

Turnarounds may be required due to the length of dead-end access roads:
Turnarounds for dead-end roads exceeding 150 ft. in length are necessary for fire operations and vehicle safety.

IFC Section 503.2.5 Dead-ends. Dead-end fire apparatus access roads in excess of 150 ft. in length shall be provided with approved area for turning around fire apparatus.

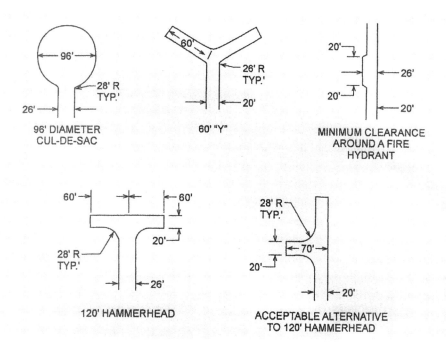

During an emergency response, numerous vehicles are dispatched to a location. These vehicles must be positioned in strategic locations. On dead-end access roads, emergency vehicles will have to turnaround and for time-crucial events the turnarounds help facilitate vehicle positioning quickly. A secondary vehicle access road does not serve the purpose of facilitating vehicle turnaround. Secondary access roads are usually obstructed by removable bollards or chains.

On a dead-end access road with a turnaround, there are times response personnel are finishing an emergency event and they have notified dispatch that they are available. They are then dispatched to another emergency and the turnaround will enable a quicker response than does a blocked secondary access. Backing large engines or ladder trucks for an extended distance is slow, impractical, and unsafe and that is why 150 ft. becomes the benchmark for dead-end access roads needing a turnaround.

Designers have the option to provide alternatives for turnarounds that cannot be provided due to location on property, topography, grade, or other valid reasons. To date, one alternative has been to sprinkler buildings beyond the 150 ft. distance, which is noted in the code, another alternative is using a secondary fire apparatus access road with "approved" break-away or collapsible barriers at each access road entry point. Check with the local road engineering division to see if these types of barriers are permissible.

The reason for secondary fire apparatus access roads is different than an access road or associated turnaround:
Secondary access roads are for areas or developments that can have the primary vehicle access road easily compromised thus obstructing access for emergency responders.

Each commercial or residential development design must be reviewed for secondary access needs but use the following assessment criteria:

> *IFC 503.1.2 Additional access: The fire official is authorized to require more than one fire apparatus access road based on the potential for impairment of a single road by vehicle congestion, condition of terrain, climatic conditions or other factors that could limit access.*

The code criteria along with site design factors create a unique set of variables making each review unique and making it impossible to draft standard and specific design parameters. The code's main purpose and intent is to give us another entry point if we determine, through a risk assessment, a secondary access is necessary. Its purpose is not related to the operational needs facilitated by access road turnarounds.

Surfacing:
In the past, there have been questions inquiring if a gravel surface is acceptable for access roads. The fire code states the following

> *IFC 503.2.3 Surface: Fire apparatus access roads shall be designed and maintained to support the imposed loads of fire apparatus (75,000 lbs.) and shall be surfaced so as to provide all-weather driving capabilities.*

A gravel surface is a less desirable surface material, but permitted. Gravel has been prone to problems resulting from extended rainy periods and/or due to the lack of a responsible party to maintain it. Other acceptable surfaces are interlocking pavers, chip seal, oil matting or some design that provides road integrity should be considered. Access road design must meet the community's Road Engineering specifications. We will request engineering data to validate that the road is designed to sustain the imposed load of 75,000 lbs. for roads not surfaced with asphalt or concrete.

PREPARED BY:		
EFFECTIVE DATE:	REVISED DATE:	BY:

Fire Access Road Turnarounds

Fire access roads in excess of 150 ft. should be provided an approved turnaround 503.2.5 of the IFC. The following designs from Appendix D, Figure D103.1 of the IFC can be consulted as a guide unless the Appendix is adopted at the local level. Consult Appendix D for additional information for turnarounds and signage.

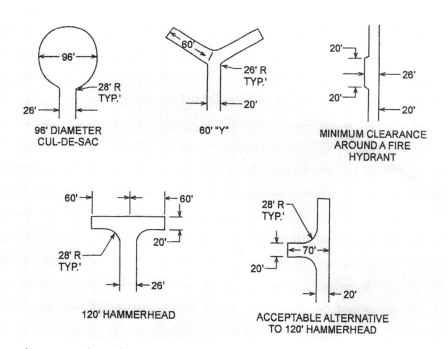

Following the generic examples pictured above, on the remaining pages, are sample engineered designs that can be used by a jurisdiction. Below are "No Parking" signs from Appendix D of IFC, Figure D103.6 that can be used for marking fire access roads or fire lanes in accordance with Section 503.3. Section D103.6 of the IFC can be consulted for additional requirements.

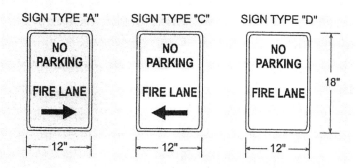

It is recommended that each fire jurisdiction verify that the generic designs do not conflict with local road engineering standards and determine if engineered designs may already be available as handouts.

Fire Plan Review and Inspection Guidelines

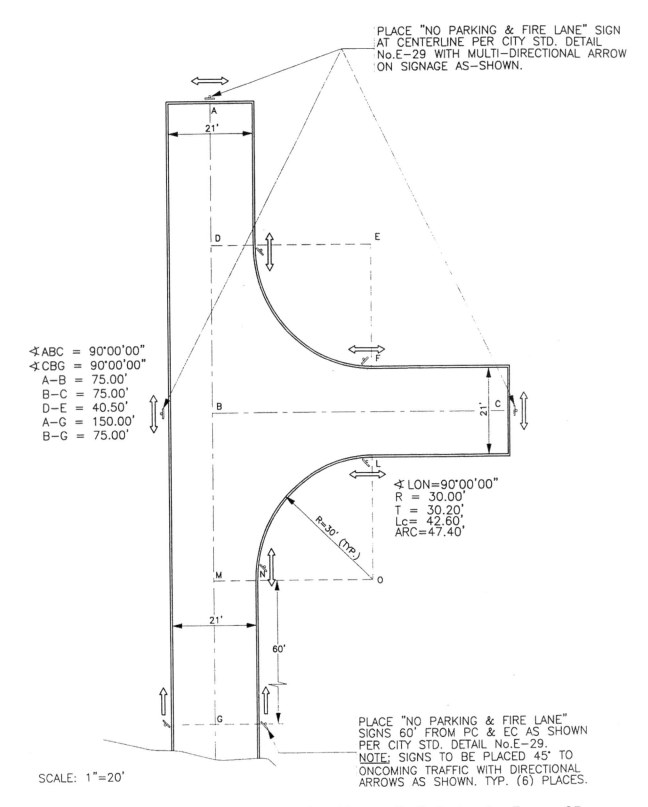

Reprinted with permission from City of Eugene, Traffic Engineering, Eugene, OR

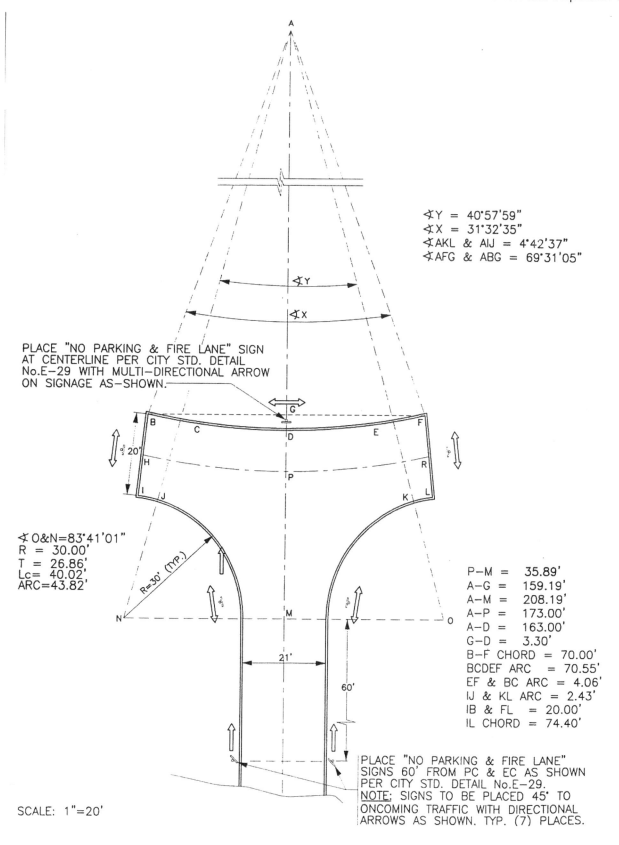

Reprinted with permission from City of Eugene, Traffic Engineering, Eugene, OR.

Fire Plan Review and Inspection Guidelines

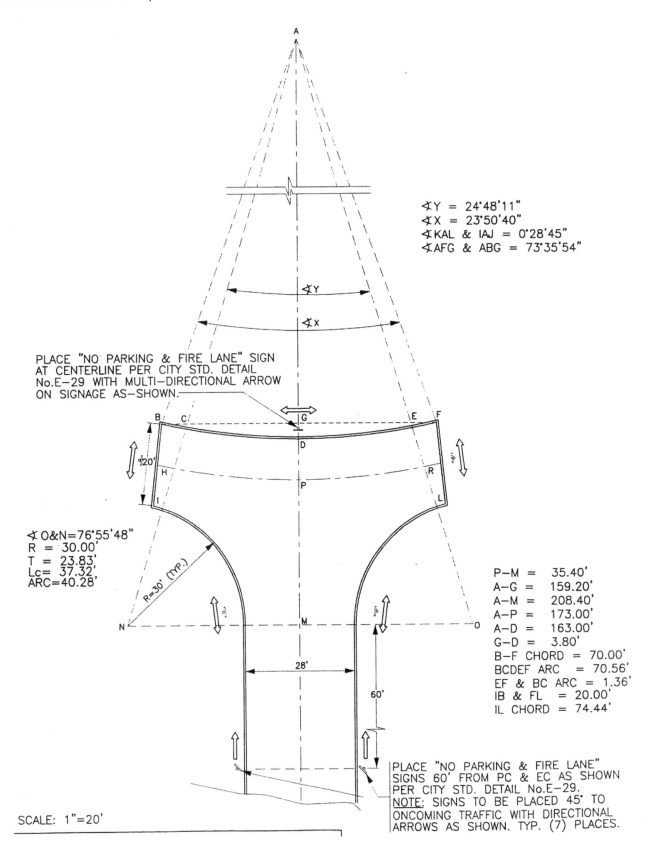

Reprinted with permission from City of Eugene, Traffic Engineering, Eugene, OR.

Fire Plan Review and Inspection Guidelines

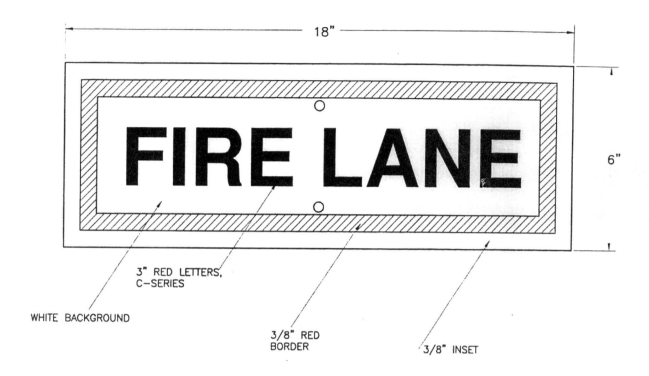

NOTE:
WHEN SPECIFIED "FIRE LANE" SIGNS TO BE MOUNTED DIRECTLY UNDER "NO PARKING"
SING WITH 5/16" X 2-1/2" LUG BOLTS.

Reprinted with permission from City of Eugene, Traffic Engineering, Eugene, OR.

Fire Plan Review and Inspection Guidelines

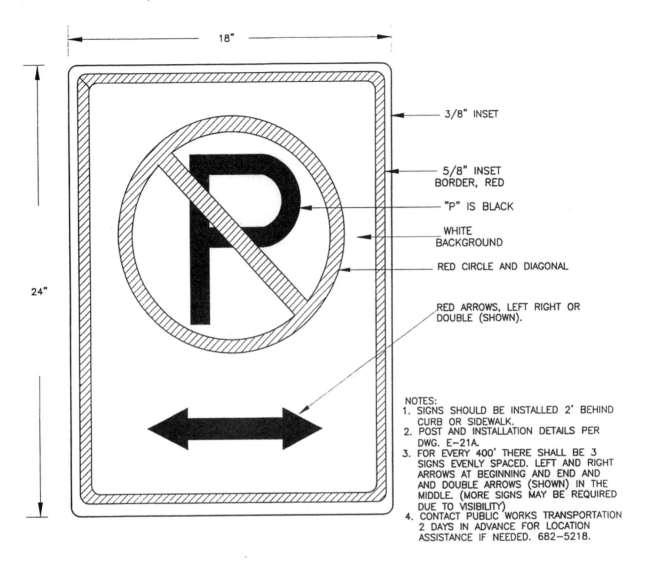

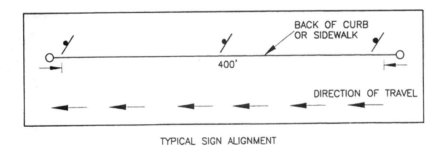

Reprinted with permission from City of Eugene, Traffic Engineering, Eugene, OR.

Emergency Access Lock Box
Sample Information Bulletin or Policy

Office of the Fire Marshal

Key Box

References: 2006 IFC Section 506

A key box allows the fire department to gain immediate access to a facility when a life-saving or fire-fighting emergency exists. The International Fire Code has a specific requirement for key boxes.

Key Box:

"Section 506.1. Where access to or within a structure or an area is restricted because of secured openings or where immediate access is necessary for life-saving or fire-fighting purposes, the fire code official is authorized to require a key box to be installed in an approved location. The key box shall be of an approved type and shall contain keys to gain necessary access as required by the fire code official."

"Section 506.1.1. Key box shall be installed on gates and similar barriers when required by the fire code official."

The key box chosen is manufactured by _____. The box may be purchased with or without a tamper alarm switch, which can be connected to a fire alarm system if a building is equipped with one. The benefit is if someone tampers with the key box the fire alarm system will activate.

Consult the local yellow pages for local locksmiths and fire protection companies that sell and install key boxes. Also, key boxes may be purchased from the company direct. Their number is _____. An authorization form from the Fire Marshal's Office is required if the lock box is purchased directly from the company.

After the key box is installed, please call the Fire Marshal's Office at (xxx) xxx-xxxx to schedule an appointment for fire personnel to come to your location to lock keys into the key box.

The Fire Department's key will accommodate the following models of key boxes _____. All fire engines are equipped with a key to allow our access into the key boxes.

Key boxes shall be mounted no greater than 6 ft. above ground level on the exterior wall to the left of the front entrance when you are facing the entrance. Key boxes are to be installed on buildings equipped with a fire alarm system or sprinkler system. If your business is open 24 hours a day, 365 days a year, or the building has a security guard stationed on premises, a key box is not required.

If you have any questions please call our office at xxx-xxxx.

Prepared By:		
Effective Date:	Revised Date:	By:

Private Fire Service Mains
Contents

1. Plan Review Worksheet. 53

2. Acceptance Test Checklist. 55

3. Installer Certification Form. 56

4. Pretest Information for Final Permit Inspection 57

5. Hydrant Flush Procedure 58

6. Hydrant Placement Requirements 59

7. Appendix C Table C105.1 61

8. Hydrant Orientation 62

9. Annual Maintenance Worksheet 63

10. Hydrant Flow Sample Test Procedure 65

Fire Plan Review and Inspection Guidelines

Private Water and Hydrant System
Plan Review Worksheet
2006 IFC and 2002 NFPA 24

Date of Review: _____ Permit Number: _____

Business/Building Name: _____ Address of Project: _____

Designer Name: _____ Designer's Phone: _____

Contractor: _____ Contractor's Phone: _____

Reference numbers following worksheet statements represent an NFPA code section unless otherwise specified.

Worksheet Legend: ✓ or **OK** = acceptable   **N** = need to provide   **NA** = not applicable

1. _____ Three sets of drawings are provided with component specification sheets, listed components shall be used.

Design Plans Show the Following, 4.1:

2. _____ The site plan shows pipe size and placement to the hydrants and the building from the point of connection at the city main or water supply source.
3. _____ Scale: a common scale is used and plan information shall be legible.
4. _____ An equipment symbol legend is provided.
5. _____ A current water flow test summary sheet and the results at 20 PSI residual from nearest public water supply is provided. Hydraulic calculations showing the available flow results for new system hydrants are provided, 5.1.2.
6. _____ Size, type, and the location of the system shut-off and isolation valves are shown.
7. _____ Listing data sheets are provided for system components required by NFPA 24 to be listed.
8. _____ If used, a thrust blocks size matrix with details or calculations is provided. Pipe system, thrustblocks, and fitting locations are detailed, 10.8.2.
9. _____ Method(s) of a restrained joint system is specified. If used, the rod size and number of rods is specified, apply Section 10.8.3 and Table 10.8.3.1.2.2. If used, the size of restraint straps for tees is specified, apply Table 10.8.3.2.3. If used, clamp and rod detail is specified (1 pair of rods for each clamp) and the clamp size is specified, 10.8.3.1, A10.8.2.

Valves:

10. _____ Connections to water supplies and supply pipes to sprinkler risers shall be controlled by valves meeting the requirements of Section 6.1.
11. _____ At least 1 indicating valve for each source of water supply is detailed, 6.2.1.
12. _____ For more than 1 source of water supply, a check valve is at each connection and is detailed, 6.2.3.
13. _____ Control valves required by 6.2.3 are provided on each side of the check valve, 6.2.5.
14. _____ Control valves for connections to pressure or gravity tanks are in compliance with Sections 6.2.6 through 6.2.8.
15. _____ All control valves are readily accessible and free of obstructions, 6.2.10.
16. _____ Water supply connections to the building shall be with a post indicating valve (PIV), except FDCs, 6.3.1. Indicating valves are not required if authorized by the AHJ and are incompliance with Sections 6.1 and 6.4.
17. _____ PIV installation and cross sectional elevation details are provided.
18. _____ PIVs are at least 40 ft. from the building unless authorized by the AHJ and the top of the posts are 36 in. above grade, 6.3.3.1 and are protected from mechanical damage, 6.3.4.2 and IFC 312.
19. _____ Valves in pits, used in lieu of PIVs, are detailed to show conformance with Section 6.4, e.g. large enough for equipment placement, maintenance, inspection, and testing, and constructed to protect equipment from damage and accumulation of water.
20. _____ Sectional valves are provided to isolate the system for repair and maintenance and where a supply main is near or under a building foundation, 6.5.
21. _____ Each valve shall have identification signs indicating its function and what it controls, signage requirement and locations are noted on the plans, 6.6.
22. _____ Valves shall be supervised in accordance with Section 6.6.2.
23. _____ Check valves are installed according to the listing data sheet, 6.7.

Hydrants:

24. _____ Hydrants are the type approved for the jurisdiction, 7.1.1.2.

Fire Plan Review and Inspection Guidelines

25. _____ Hydrants shall have a minimum 6 in. connection to the main, 7.1.1.
26. _____ Hydrants are to be at least 40 ft. from a building, 7.2.3, unless less distance is approved by the AHJ, 7.2.4.
27. _____ Hydrant placing is in accordance with IFC Appendices B and C.
28. _____ A cross section hydrant installation detail is provided, 7.3.1.
29. _____ Hydrant, pipe connection, support, restraint methods and locations are detailed, 7.3.
30. _____ Center of hose outlet not less than 18 in. above grade, 7.3.3.
31. _____ The method of hydrant protection from mechanical damage by curbs, bollards, etc. is detailed, 7.3.5, IFC 312.

Piping:
32. _____ Piping is not smaller than 6 in. when supplying a hydrant, 5.2.1, 13.1.
33. _____ Piping supplying a water-based fire protection system can have a diameter of less than 6 in. if it designed in accordance with Section 5.2.2, 13.2.
34. _____ The pipe is listed for fire protection service or complies with Table 10.1.1, is designed to withstand a system working pressure of at least 150 PSI, and a listing data sheet is provided, 10.1.1, 10.1.5.
35. _____ The type and class of pipe material is specified, 10.1.4.
36. _____ The method of joining pipe sections is specified and in compliance with Section 10.3 and the fittings are pressure compatible with the pipe, 10.2.5.
37. _____ The top of the pipe is detailed to be at least 1 ft. below the area's frost line, 10.4.2.
38. _____ The depth of pipe for areas where frost is not a concern is detailed with the minimum depth being at 2.5 ft., or 3 ft. when the pipe is located under vehicle traffic areas, or 4 ft. when the pipe is located under railroad tracks, 10.4.
39. _____ Above-ground pipe which is subject to freezing is protected by a means capable of maintaining at least a temperature of 40° F, 10.5.1, 12.2.
40. _____ Pipe laid in waterways or streams are designed in accordance with Section 10.5.3.
41. _____ Pipe does not run under a structure but it is allowed to enter the building adjacent the building foundation, 10.6.1, 10.6.3.
42. _____ For pipe that is run under a structure, the means to protect the pipe are detailed and conform to Section 10.6.2.
43. _____ The methods of restraining all tees, plugs, bends, reducers, valves, and hydrant branches are detailed and are designed in compliance with 10.8.2 and 10.8.3. Pipe with fused, threaded, grooved, or welded joints do not need strained if they pass the hydrostatic test of 10.10.2.2 without shifting or leaking excessively, 10.8.1.2.
44. _____ All bolted joint assemblies shall be coated for corrosion protection, the coating product and the application requirement is noted on the plan, 10.3.6.2, 10.8.3.5. Exceptions include soil conditions or if the local water purveyor does not require corrosion protection.
45. _____ Backfill material for tamping around the pipe is specified, 10.9.
46. _____ The flushing and hydrostatic test requirements are on the plans as specified in 10.10.2.
47. _____ The minimum flushing flow rate requirements are provided on the plan, Table 10.10.2.1.3.
48. _____ Above-ground piping is not located in hazardous areas unless the area is protected by an automatic sprinkler system. The location of the pipe is protected from damage or fire, 12.2.1.
49. _____ Above-ground pipe passing through areas subjecting it to freezing conditions protection is detailed and conform with Section 12.2.3.
50. _____ Above-ground piping is protected against corrosive conditions, 12.2.4.
51. _____ When above-ground piping is located an a seismic design area that requires bracing of piping, the piping shall be properly braced in accordance with NFPA 13, 12.2.5.
52. _____ If water supply piping connects to reservoirs, rivers, or lakes, the connections shall be designed in accordance with 5.8 to avoid accumulations of mud or sediment.

Additional Comments:

Review Date: _____ Approved or Disapproved FD Reviewer: _____

Review Date: _____ Approved or Disapproved FD Reviewer: _____

Fire Plan Review and Inspection Guidelines

Private Water and Hydrant System Acceptance Inspection
2006 IFC and 2002 NFPA 24

Date of Inspection: _____ Permit Number: _____

Business/Building Name: _____ Address of Project: _____

Contractor: _____ Contractor's Phone: _____

Reference numbers following worksheet statements represent an NFPA code section unless otherwise specified.

Pass | Fail | NA

1. ___ | ___ | ___ Received Private Fire Service Main Installation Certification from installer.
2. ___ | ___ | ___ Approved drawing is on the site.
3. ___ | ___ | ___ Pipe size and configuration and hydrant locations are in compliance with the approved set of plans.
4. ___ | ___ | ___ Water flow test results from public water supply are provided.
5. ___ | ___ | ___ Depth of the pipe is in compliance with the approved set of plans.
6. ___ | ___ | ___ Pipe is fully ground supported its entire length, it is especially important for plastic pipe.
7. ___ | ___ | ___ Backfill is tamped in layers at least 6 in. under and around the pipe and for trenches cut in rock the backfill is tamped for at least 2 ft. above the pipe.
8. ___ | ___ | ___ All bolted joint components are coated with asphalt or other corrosion-retarding product, if required by the AHJ or the local water purveyor.
9. ___ | ___ | ___ Size, type, and location of the valves are in compliance with the approved set of plans.
10. ___ | ___ | ___ All control valves are an indicating type.
11. ___ | ___ | ___ Size and location of the hydrants are in compliance with the approved set of plans.
12. ___ | ___ | ___ Hydrants have small stone fill around the base to allow for proper drainage.
13. ___ | ___ | ___ Hydrant outlets are not less than 18 in. above final grade and properly oriented for hose connection, 7.3.3.
14. ___ | ___ | ___ Hydrant is protected from mechanical damage, 7.3.5.
15. ___ | ___ | ___ All tees, plugs, caps, bends, and hydrant branches are restrained against movement, pipe clamps and tie rods, thrust blocks, locked mechanical or push on joints or other approved methods are used, 10.8.3.
16. ___ | ___ | ___ When used, thrust blocks are provided at each change of direction, including tees, plugs, caps, and bends, unless another restraining method is used such as rod and clamps, 10.8.1.
17. ___ | ___ | ___ Thrust blocks are concrete and sized according to the plans, and poured against compacted soil, 10.8.2, A.10.8.2.
18. ___ | ___ | ___ Other retraining methods match the approved plans.
19. ___ | ___ | ___ All steel bolted joint assemblies shall be thoroughly coated for corrosion protection if required by the local water purveyor, 10.8.3.5.
20. ___ | ___ | ___ Pipe flushing requirements shall be 880 GPM for 6 in. diameter pipe, 1,560 GPM for 8 inch diameter pipe, 2,440 GPM for 10 in. diameter pipe, 3,520 GPM for 12 in. diameter pipe, 10.10.2.1.
21. ___ | ___ | ___ The hydrostatic test is witnessed and performed at 200 PSI or 50 PSI greater than the system working pressure for 2 hours with no pressure drop, leakage is limited to 2 quarts per hour per 100 joints, 10.10.2 .2.
22. ___ | ___ | ___ The hydrant flow test was witnessed, graphed, and documented. The results document the static pressure, residual pressure, and available GPM at 20 PSI residual pressure.
23. ___ | ___ | ___ Each dry barrel hydrant and control valve is opened and closed to verify water flow and drainage, 14.1.

Additional Comments:

_____
_____

Inspection Date: _____   Approved or Disapproved   FD Inspector: _____

Inspection Date: _____   Approved or Disapproved   FD Inspector: _____

Fire Plan Review and Inspection Guidelines

Private Water and Hydrant System Installation Certification
The NFPA contractor certificate of compliance form is an acceptable substitution for this form.

Permit #: _____ Date: _____

	Property Protected	System Installer	System Supplier
Business Name:	_____	_____	_____
Address:	_____	_____	_____
	_____	_____	_____
Representative:	_____	_____	_____
Telephone:	_____	_____	_____

Type of System: _____

Location of Plans: _____

Location of Owner's Manual: _____

1. <u>Certification of System Installation:</u> Complete this section after the system is installed, but prior to conducting operational acceptance tests.

   This system installation was inspected and found to comply with the installation requirements of:
   - _____ NFPA 24
   - _____ IFC
   - _____ Manufacturer's Instructions
   - _____ Other (specify: FM, UL, etc.) _____

   Print Name: _____
   Signed: _____ Date: _____
   Organization: _____

2. <u>Certification of System Operation:</u> All operational features and functions of this system were tested and found to be operating properly in accordance with the requirements of:
   - _____ NFPA 24
   - _____ Design specifications
   - _____ IFC
   - _____ Manufacturer's Instructions
   - _____ Other (specify) _____

   Print Name: _____
   Signed: _____ Date: _____
   Organization: _____

Prefinal and Certificate of Occupancy Inspection Requirements For Contractors
Contractors Worksheet

Water Main and Hydrant System Test Requirements

1. All certification forms and documents are required to be on the site for review:
   - ____ Plans
   - ____ Permit
   - ____ A system hydrostatic test is required before calling for an inspection as well as the completion of the items on this pretest form. Use the Acceptance Inspection worksheet for the pretest.
   - ____ Installation certification is completed, use the form contained in this book.
2. ____ A person familiar with installation must be present to perform the test.
3. ____ Owner's representative approval is needed for the time and date of testing.
4. ____ All areas are accessible.
5. ____ Hydrostatic testing and flushing should been done during the same inspection.
6. ____ Only those contractors who have preapproval may backfill all straight pipe lays except bends and joint restraints will remain exposed.
7. ____ Hydrant placement is in accordance with the plan, thrust blocks are in place and exposed, verification has been done by contractor.
8. ____ If Items 1-7 are incomplete, the inspection will be cancelled and another inspection request is required. A reinspection fee may be assessed.

Prior to the next approval test:
9. ____ Items 1-7 are complete and ready.
10. ____ A reinspection fee may be assessed if the system and paper work are not ready.

Preapproved Contractors: They are contractors who have demonstrated an acceptable level of competency to the Fire Department by installing at least 5 underground water systems without installation or code deficiencies. The approval enables the contractor to install and backfill straight runs of pipe except bends and joint restraints will remain exposed without an inspection. Inspections for restraining devices, thrust blocks, hydrostatic testing and flushing are still required.

Fire Plan Review and Inspection Guidelines

Hydrant Main Flushing Procedures

Scope: This bulletin provides the minimum information to conduct flushing a main supply pipe serving a private hydrant or a sprinkler riser. Agencies are responsible to research other publications for additional information and to train their personnel in the correct testing procedures.

Purpose: Properly performed flushes provide results that help ensure that debris in the pipe is removed so as not to compromise the operation of the sprinkler system or damage a fire pump.

Equipment:
1. Burlap bag is desirable to collect debris that might cause damage to a building or vehicle.
2. Hydrant wrenches.
3. Elbow(s) to divert water.

Procedure Summary: (In some jurisdictions, the water purveyor may require notification before flushing)
1. The minimum flow rate shall not be less than one of the following methods;
   A. Hydraulically calculated water demand flow rate of the system, including any hose requirements.
   B. Flow water at a rate that is necessary to provide a velocity of 10 ft/sec in accordance with 2002 NFPA 24 Table 10.10.2.1.3.
   C. At a maximum flow rate available to the system under fire conditions.
2. Underground mains and lead-in connections to system risers should be flushed through hydrants at dead-ends of the system or through accessible aboveground flushing outlets.
3. Sprinkler systems can be configured according to 2002 NFPA 24 Figure A.10.10.2.1. A 4 in. fire department connection pipe can be modified to be an open butt or through a Y or Siamese connection with (2) 2½ in. hose outlets.
4. Flush the system until the water is clear.
5. Attach a burlap bag to the outlet port(s) to capture any debris and to minimize damage to the surrounding area.
6. The flow at a rate identified in Step 1, which is necessary for cleaning the pipe and lifting debris to an outlet.
7. After flushing the system, the water flow requirements for the project should be verified. When conducting a flow test confirm the street valve(s) or isolation valve(s) are open.

To achieve a velocity of 10 ft. per second $Q = 29.83 \times C_d \times D^2 \times \sqrt{P}$

Pipe Size (inches)	Flow Rate (GPM)	Velocity Pressure for 4½ in. outlet, if $C_d = .7$
4	390	1
6	880	4
8	1,603	14
10	2,460	33
12	3,560	69

Q = flow in gallons C_d = hydrant coefficient (0.7, 0.8, or 0.9)
D^2 = hydrant outlet diameter $\sqrt{P}$ = sq root of the pitot pressure

Fire Plan Review and Inspection Guidelines

Hydrant Placement
Plan Designer Information
This information is not inclusive of the criteria published in the code references.

References: Sections 503 and 508, Appendices B, C and Tables B105.1 and C105.1 of the 2006 IFC

1. Are *fire apparatus access roads within 150 ft. of any exterior portion of the first floor of the facility as measured by an **approved route?
   The 150 ft. criteria can be modified if, 1) The building is sprinklered throughout, or 2) Access roads cannot be constructed due to topography, location on property or grade, and alternative fire protection is provided, or 3) There are not more than two Group R-3 or U occupancies.

   *Fire apparatus access road is a general term inclusive of other terms such as fire lane, public street, private street, parking lot, and access roadway and it must be within 150 ft. of all portions of a building.
   **An approved route is one that is vehicle usable, not as a crow flies, or through a building or over difficult terrain.

2. What is the required fire flow for the building?
   Refer to Appendix B. Water supply for the fire flow can come from reservoirs, pressure tanks, elevated tanks, water mains, or other fixed systems. The fire flow be can be reduced 50 percent for sprinklered 1- and 2-family dwellings but not less than 1,000 GPM, up to 75 percent for all other buildings that have an approved sprinkler system but not less than 1,500 GPM. Also consult Table 105.1 footnotes.

3. How many hydrants are required?
   Refer to Appendix C, Table C105.1. In addition to public hydrants, on-site hydrants may be necessary to meet fire flow requirements and/or travel distance to a hydrant. Verify that the public water system has the capability to supply the fire flow demand with at least 20 PSI residual, otherwise additional hydrants may not serve a useful purpose. Private hydrants on other adjacent properties should not be considered to meet fire flow demands.

4. What is the required average spacing between the hydrants?
   Refer to Appendix C, Table C105.1. Hydrant spacing will range from 200 ft. to 500 ft. depending on the water flow requirements.

5. Does any portion of the non sprinklered building exceed 400 ft. from a hydrant on a fire apparatus access road?
   When any portion of a structure exceeds 400 ft. from a hydrant on a fire apparatus access road as measured by an approved route around the exterior of the building, on-site hydrants and mains shall be provided when required by the chief. Hydrants shall be capable of supplying the fire flow requirement for that structure. Group R-3 and U occupancies are permitted to be 600 ft. from a hydrant and buildings with an approved sprinkler system are permitted to be 600 ft.

   Remember, a fire apparatus access road must be within 150 ft. of any exterior portion of the building unless one of the three exceptions is applied, Section 503.

6. Site Scenarios, Figures 1 and 2:
   For Figure 1, a 100 ft. x 90 ft. building fronts a fire access road with hydrants. From the approved access route the hydrants are within 400 ft. of any portion of the exterior of the building. Also, any hydrant is within the maximum distance of 225 ft. from any point on the access road. Therefore, an on-site hydrant is not required in Figure 1.

Fire Plan Review and Inspection Guidelines

For Figure 2, the building requires a fire flow of 3,000 GPM, which requires 3 hydrants in accordance with Table C105.1. The spacing between the hydrants on the street should be no more than the average spacing of 400 ft. as stated in 3rd column of Table C105.1. Also, the 225 ft. maximum distance from any point on the street to a hydrant, as stated in the 4th column of Table C105.1, can not be exceeded.

Figure 1
Type V-B nonsprinklered building.
The access road is within 150 ft. of any exterior portion of the building in accordance with 503.1.1. A hydrant is within the 225 ft. maximum distance from any point on the street or road frontage in accordance with the 4th column of Table C105.1 for the building's required fire flow of 2,500 GPM. A hydrant is also within 400 ft. of any exterior portion of the building in accordance with Section 508.5.1. The hydrants are within the 450 ft. average spacing requirement. For this scenario, no on-site hydrant is required.

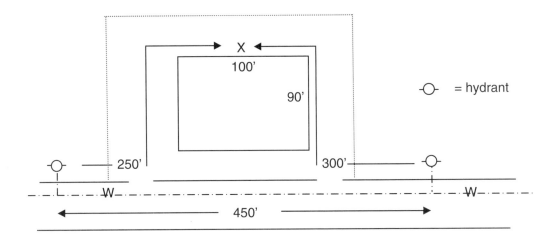

Figure 2
Type V-B nonsprinklered building, a 3,000 GPM fire flow is required.
It is 450 ft. to the nearest hydrant on the street from any exterior portion of the building, which exceeds the 400 ft. limit in Section 508.5.1. The average spacing between hydrants exceeds the 400 ft. required in the 3rd column of Table C105.1. The maximum distance from any point on the road to a hydrant exceeds 225 ft. stated in the 4th column of Table C105.1. The fire access road is not within 150 ft. of any exterior portion of the building. For this scenario, on-site hydrants would be required and fire access roads must be extended to within 150 ft. of all exterior portions of the building as measured by an approved route.

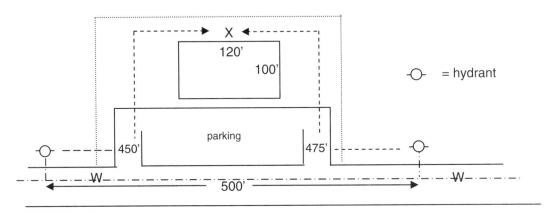

Fire Plan Review and Inspection Guidelines

2006 IFC Table C105.1
Number and Distribution of Fire Hydrants

Fire-Flow Requirement (GPM)	Minimum Number Of Hydrants	Average Spacing Between Hydrants[a,b,c] (Ft.)	Maximum Distance From Any Point on the Street or Road Frontage To A Hydrant[d]
X 3.785 for L/min.		X 304.8 for mm	
1,750 or less	1	500	250
2,000-2,250	2	450	225
2,500	3	450	225
3,000	3	400	225
3,500-4,000	4	350	210
4,500-5,000	5	300	180
5,500	6	300	180
6,000	6	250	150
6,500-7,000	7	250	150
7,500 or more	8 or more [e]	200	120

(a) Reduce by 100 ft. for dead-end streets or roads.

(b) Where streets are provided with median dividers which can be crossed by firefighters pulling hose lines, or where arterial streets are provided with four or more traffic lanes and have a traffic count of more than 30,000 vehicles per day, hydrant spacing shall average 500 ft. on each side of the street and be arranged on an alternating basis up to a fire-flow requirement of 7,000 gallons per minute and 400 ft. for higher fire-flow requirements.

(c) Where new water mains are extended along streets where hydrants are not needed for protection of structures or similar fire problems, fire hydrants shall be provided at spacing not to exceed 1,000 ft. to provide for transportation hazards.

(d) Reduce by 50 ft. for dead-end streets or roads.

(e) One hydrant for each 1,000 gallons per minute or fraction thereof.

Fire Plan Review and Inspection Guidelines

Hydrant Orientation
For Wet and Dry Barrel Hydrants

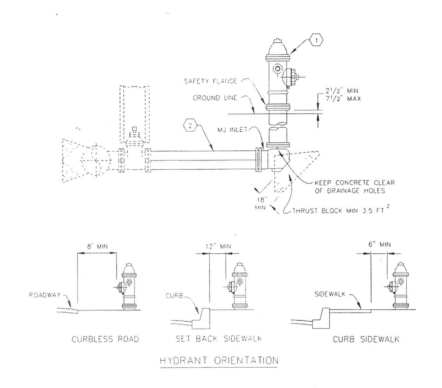

Note: Hydrant outlets shall be located a minimum of 18 inches above final grade.

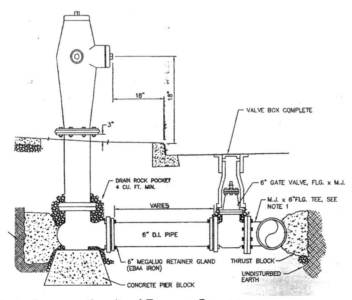

The source of the illustrations was the city of Eugene, Oregon.

Private Wet Hydrants
Annual Maintenance Worksheet
2006 IFC and 2002 NFPA 25

Date of Service: _____

Business/Bldg Name: _____ Address: _____

Contractor: _____ Contractor's Phone: _____

Hydrant No.: _____ Model: _____ Location: _____

Recommended: Water purveyor should be notified and their approval received before flowing any hydrants. Flows can only be conducted Monday through Thursday. Please provide the water purveyor the date, time, and location of the hydrant(s) to be tested.

<u>Yes | No | NA</u> Please use one form per hydrant.

General Maintenance

1. ___ | ___ | ___ Hydrant is accessible and visible; no plants or permanent items obstruct view, hinder operability, or prevent hose attachment. Generally 3 ft. side and back clearance and 10 ft. front clearance is required, IFC 508.5.4, 508.5.5.
2. ___ | ___ | ___ Outlets are tight and caps are lubricated, NFPA 25 7.4.3.1.
3. ___ | ___ | ___ Threads and port caps are in good condition, NFPA 25 7.4.3.1.
4. ___ | ___ | ___ When valve is open, there are no leaks from top of hydrant or ports, NFPA 25 7.2.2.5.
5. ___ | ___ | ___ Hydrant barrel is in good condition and not cracked, NFPA 25 7.2.2.5.
6. ___ | ___ | ___ Operating nut is in good condition and not worn, NFPA 25 7.2.2.5.
7. ___ | ___ | ___ Operating nut is lubricated, NFPA 25 7.2.2.5.
8. ___ | ___ | ___ Cap retaining chain or device is intact, NFPA 25 7.2.2.5.

Operational

9. ___ | ___ | ___ Valve opens fully and easily. Important; please open valve slowly, NFPA 25 7.2.2.5.
10. ___ | ___ | ___ Hydrant is fully opened (slowly) and flowed for at least 1 minute to ensure any foreign material has cleared and the water shows clear, NFPA 25 7.3.2.
11. ___ | ___ | ___ Valve seal will close and does not leak. Please close slowly, NFPA 25 7.2.2.5.
12. ___ | ___ | ___ Water properly drains from barrel, NFPA 25 7.2.2.5.
13. ___ | ___ | ___ The flow test is conducted (then at least annually, IFC 508.5.3) and the results are attached and provide the available GPM at 20 PSI residual pressure. If the results are less then 1,250 GPM at 20 PSI residual there may be an obstruction, a partially closed supply valve or it is the result of a small supply line design. Investigate to see if one of these issues may be influencing low flow test results from the system. It is recommended not to turn any water purveyor valves for any reason. Contact the water purveyor if you suspect a valve may be partially closed.

Repair Comments

Inspection Date: _____ Inspector: _____

Fire Plan Review and Inspection Guidelines

Private Dry Hydrants
Annual Maintenance Worksheet
2006 IFC and 2002 NFPA 25

Date of Service: _____

Business/Bldg Name: _____ Address: _____

Contractor: _____ Contractor's Phone: _____

Hydrant No.: _____ Model: _____ Location: _____

Recommended: Water purveyor should be notified and their approval received before flowing any hydrants. Flows can only be conducted Monday through Thursday. Please provide the water purveyor the date, time, and location of the hydrant(s) to be tested.

<u>Yes | No |NA</u>   Please use one form per hydrant.

General Maintenance
1. ____|____|____ Hydrant is accessible and visible; no plants or permanent items obstruct view, hinder operability, or prevent hose attachment. Generally 3 ft. side and back clearance and 10 ft. front clearance is required, IFC 508.5.4, 508.5.5.
2. ____|____|____ Outlets are tight and caps are lubricated, NFPA 25 7.4.3.1.
3. ____|____|____ Threads and port caps are in good condition, NFPA 25 7.4.3.1.
4. ____|____|____ When valve is open, there are no leaks from top of hydrant or ports, NFPA 25 7.2.2.4.
5. ____|____|____ Hydrant barrel is in good condition and not cracked, NFPA 25 7.2.2.4.
6. ____|____|____ Operating nut is in good condition and not worn, NFPA 25 7.2.2.4.
7. ____|____|____ Operating nut is lubricated, NFPA 25 7.2.2.4.
8. ____|____|____ Cap retaining chain or device is intact, NFPA 25 7.2.2.4.

Operational
9. ____|____|____ Valve opens fully and easily. Important; please open valve slowly, NFPA 25 7.2.2.4.
10. ____|____|____ Hydrant is fully opened (slowly) and flowed for at least 1 minute to ensure any foreign material has cleared and the water shows clear, NFPA 25 7.3.2.
11. ____|____|____ Dry barrel and wall hydrants drain within 60 minutes, NFPA 25 7.3.2.4.
12. ____|____|____ Dry barrel and wall hydrants with plugged drains located in a freezing environment are marked as requiring to be pump after use, NFPA 25 7.2.3.6.
13. ____|____|____ Valve seal will close and does not leak. Please close slowly, NFPA 25 7.2.2.4.
14. ____|____|____ Water properly drains from barrel, NFPA 25 7.2.2.4.
15. ____|____|____ The flow test is conducted (then at least annually, IFC 508.5.3) and the results are attached and provide the available GPM at 20 PSI residual pressure. If the results are less then 1,250 GPM at 20 PSI residual there may be an obstruction, a partially closed supply valve or it is the result of a small supply line design. Investigate to see if one of these issues may be influencing low flow test results from the system. It is recommended not to turn any water purveyor valves for any reason. Contact the water purveyor if you suspect a valve may be partially closed.

Repair Comments

Inspection Date: _____   Inspector: _____

Fire Plan Review and Inspection Guidelines

Hydrant Flow Testing

Conducting a flow test requires a minimum of two gauges. One gauge is mounted on a threaded cap for attachment to a hydrant outlet. The threaded cap gauge measures the static and residual pressure. The second gauge is mounted on a pitot tube, which is used to measure the force of a stream of water released when the hydrant is fully opened.

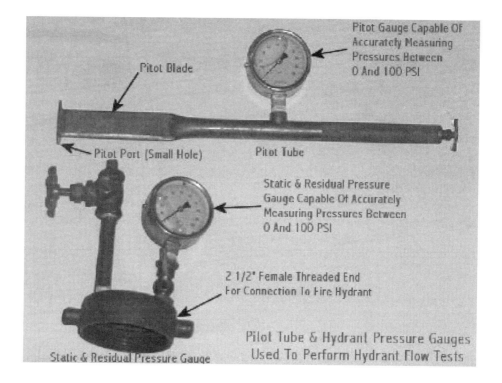

Figure 1

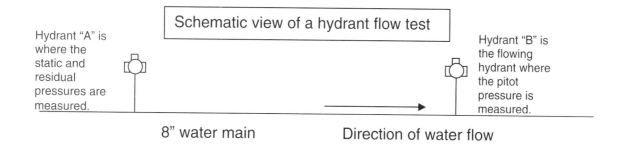

Fire Plan Review and Inspection Guidelines

Step 1 is the selection of the test and flow hydrants. If the water supply for a proposed sprinkler system is determined, the test hydrant will be closest to the building to be sprinklered. The "test hydrant" should be known has hydrant "A." The flow hydrant is the next downstream hydrant on the same system and for data recording purposes it should be known as hydrant "B."

Ensure damage will not occur to adjacent property, landscaping or nearby vehicles, and that traffic will not be obstructed when hydrant "B" flows water.

Step 2 is to obtain a static pressure from the actual hydrant that is being tested. Remove a port cap and install the cap and gauge unit to the hydrant port. Remove another cap from the hydrant and slowly and partially open hydrant "A" to expel the air in the barrel and hydrant. Close the hydrant, replace the cap on the open port, tighten all the port caps, and reopen fully the hydrant. Take a static pressure reading.

A static pressure reading is the one taken from the city water main without any water flowing. A high static pressure does not indicate the quality of the water supply.

To plot any water flow on the water flow test summary graph requires a minimum of two points. The first graph point will be the static pressure. This pressure reading is obtained from hydrant "A" as shown in Figure 1 on page 65.

In the photo above, the person reads a static pressure of 55 PSI. On some water systems it is not unusual to see fluctuations in the gauge pressure. When this happens the best result is obtained by taking an average of the fluctuations in pressure.

The residual pressure will be read from hydrant "A" after hydrant "B" has been flowing and the pressure drop has stabilized.

Figure 2

After the residual pressure reading is taken, remove a port cap from hydrant "B." Measure the port opening and record the inside diameter. Feel the inside of the port butt to determine if the butt is projecting, squared, or rounded so the correct discharge coefficient is applied later.

Fully open hydrant "B." Once hydrant "B" is fully open let it run for a minute or two to allow stabilization of pressure in the water system, to have clean water flow, and then take a pitot reading which can be converted into a rate of flow. The reading should be recorded on a Water Flow Test Summary Sheet similar to Figure 2 above.

The pitot tube measures the force of the water as it is discharged from the hydrant. The higher the velocity pressure (measured in PSI) the higher the GPM flowing from the hydrant.

As water is flowing, hold the pitot tube approximately half the diameter of the port opening away from the hydrant butt and insert the pitot tube so it is about in the middle of the stream. Once there, slightly move the pitot blade to obtain the most consistent pressure measurement. For this example flow test, the pitot pressure was 44 PSI.

For the entire example flow test, the following measurements were taken; a static pressure of 55 PSI, a residual pressure of 47 PSI from hydrant "A" while flowing a 2 1/2" outlet on hydrant "B." which provided a pitot reading of 44 PSI. Based on the discharge table on page 68 a preliminary flow rate of 1,114 GPM has been determined.

Fire Plan Review and Inspection Guidelines

How to determine the quantity of water that is actually flowing through an open hydrant butt involves using the table or using the flow formula noted below.

For example, the opening of the hydrant butt is 2 ½" and the pitot pressure was 44 PSI. Looking at the chart below the theoretical discharge is 1,237 GPM. Multiply the theoretical discharge by the appropriate discharge coefficient. The Flow formula can be used as an alternate method to achieve approximately the same results as shown below.

Since circular orifices on fire hydrants are imperfect, we have to apply a coefficient of discharge to compensate. Standard practice is to use a coefficient of 0.90 where the hydrant is well rounded (most new hydrants) while a coefficient of 0.80 or 0.70 would be used for older hydrants.

The example flow hydrant would have a coefficient of 0.90 so 0.90 x 1,237 = 1,114 GPM flowing.

Flow formula: $Q = (29.83) \times (C_d) \times (D^2) \times (\sqrt{P})$
Where Q = flow in gallons per minute (GPM)
  C_d = discharge coefficient
  D = diameter of orifice (inches)
  P = pitot pressure

$Q = (29.83) \times (0.90) \times (2.5^2) \times (\sqrt{44})$
$Q = \quad (26.84) \times (6.25) \times (6.6)$
$Q = \quad 1107.2$ GPM

Discharge (Pitot PSI and orifice size)

Pitot	1 ¼"	1 ¾"	2 ½"	4"	4 ½"	Pitot	1 ¼"	1 ¾"	2 ½"	4"	4 ½"	Pitot	1 ¼"	1 ¾"	2 ½"	4"	4 ½"
4	93	183	373	955	1210	20	209	409	834	2136	2710	50	330	646	1319	3377	4290
5	104	204	417	1068	1350	22	219	429	875	2240	2840	52	337	659	1345	3444	4375
6	114	224	457	1170	1450	24	229	448	914	2340	2970	54	343	672	1371	3510	4460
7	123	242	494	1263	1600	26	239	456	951	2435	3090	56	350	684	1396	3574	4540
8	132	259	526	1351	1710	28	247	484	987	2577	3210	58	356	696	1421	3637	4620
9	140	274	560	1435	1815	30	256	501	1022	2616	3320	60	362	708	1445	3700	4700
10	148	289	590	1510	1910	32	264	517	1053	2702	3430	62	368	720	1470	3761	4775
11	155	303	619	1584	2010	34	272	533	1088	2785	3540	64	374	731	1493	3821	4850
12	162	317	646	1655	2100	36	280	548	1119	2866	3640	66	379	742	1516	3880	4925
13	168	330	673	1722	2180	38	288	563	1130	2944	3740	68	385	754	1539	3938	5000
14	175	342	698	1767	2260	40	295	578	1180	3021	3840	70	391	765	1561	3990	5075
15	181	354	722	1849	2340	42	303	592	1219	3095	3935	72	395	775	1583	4053	5140
16	187	366	746	1910	2420	44	310	606	1237	3168	4030	74	402	785	1605	4109	5200
17	192	377	769	1969	2500	46	317	620	1265	3239	4120	76	407	797	1627	5154	5265
18	198	388	791	2026	2570	48	324	633	1293	3309	4205	78	412	807	1648	4218	5340
19	203	399	813	2052	2640							80	418	818	1669	4272	5405

Values given are the theoretical discharges through a variety of circular orifices and hydrant butts are 2 ½ in. and greater in size.
Coefficient of discharge through smoothly rounded of the internal portion of the hydrant butt, C_d =0.9.
Coefficient of discharge through square corners of a internal portion of the hydrant butt, C_d =0.8.
Coefficient of discharge through projection of corners of a internal portion of the hydrant butt, C_d =0.7.

Fire Plan Review and Inspection Guidelines

When water is flowing from hydrant "B," the pressure gauge will show a drop of pressure on test hydrant "A."

The photo here indicates the pressure dropped from 55 PSI to 47 PSI as hydrant "B" was opened and discharging 1,114 GPM.

The results of this test are as follows:

Static Pressure = 55 PSI

Residual Pressure = 47 PSI

Rate of Flow = 1,114 GPM

Graph the results.

Fire Plan Review and Inspection Guidelines

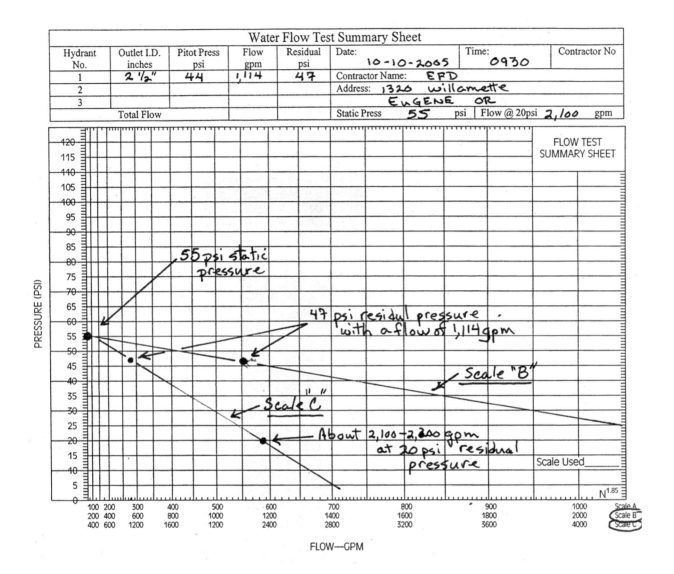

Note that this example plot the date using both Scale B and C. Notice that "Scale B" does not permit the water curve to show the quantity of available water at 20 PSI residual pressure. The same example provides a second plot using "Scale C," which permits the water curve to show the available water supply at 20 PSI residual pressure. Try to select a scale that will provide results for available water supply at 20 PSI.

FLOW TEST INFORMATION SHEET

1. **Reason for Test:** Bid Information ☐   Design Base ☐
   Other _____

2. **Location of Property** _____
   (Address)   (City)   (State)   (County)

3. **Date & Time of Test:**   Date: _____   Time: _____ (am) (pm)

4. **Test Conducted by:** _____
   Name   Title   Affiliation

5. **Test Witnessed by:** _____
   Name   Title   Affiliation

6. **Source of Water Supply:** Gravity ☐   Pump ☐   Other: _____

7. **Name of Water District** _____ Fire District _____

8. Is water supply provided with PRV STA's   Yes ☐   No ☐
   (If so what is PRV outlet setting? _____ PSIG

9. **Area Map:** (Draw Sketch showing property location; bounding streets and names, north arrow, hydrant locations and identification numbers, distances from hydrants to property elevations of hydrants and property floors or grade, all water mains and sizes and interconnection valves, etc.)

10. **Flow Test Data**

FLOW AT HYDR. NO.	STATIC AT HYDR NO.	STATIC PSIG	RESIDUAL PSIG	FLOW GPM	OUTLET COEFFICIENT	ADJUSTED GPM

11. See reverse side for graph

12. Signed _____

    Witness _____

Form No. 102
American Fire Sprinkler Association, Inc.
11325 Pegasus, Suite S-220
Dallas, Texas 75238
(214) 349-5965

Outlet Square and projecting into Barrel Coef. 0.70

Outlet Square and Sharp Coef. 0.80

Outlet Smooth and Rounded Coef. 0.90

Reprinted with permission from the American Fire Sprinkler Association, Dallas, Texas.

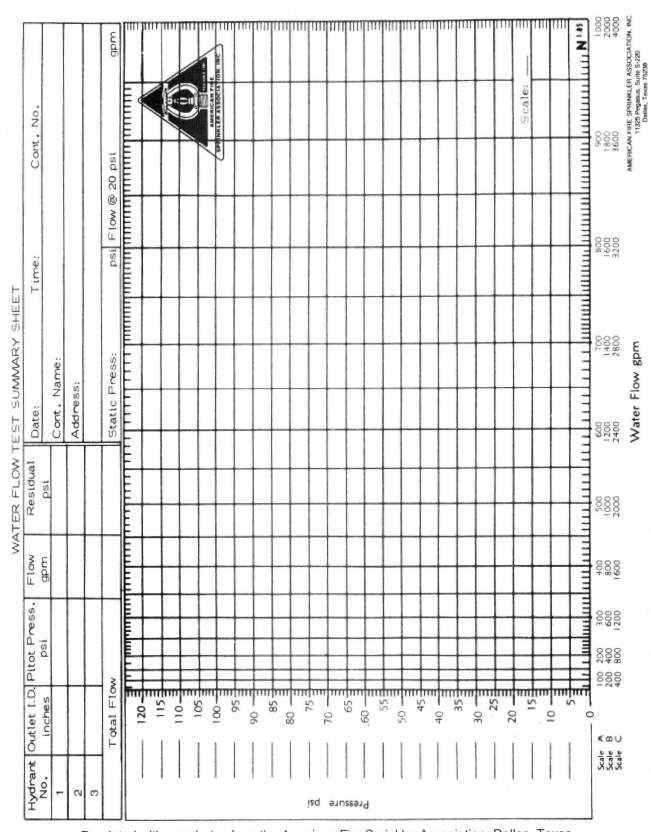

Reprinted with permission from the American Fire Sprinkler Association, Dallas, Texas.

Private Water Tank Systems
Contents

1. Plan Review Worksheet. 75

2. Acceptance Test Checklist. 78

3. Installer Certification Form. 80

4. Contractor Preinspection Information 81

Fire Plan Review and Inspection Guidelines

Private Water Tank System
Plan Review Worksheet
2006 IFC and 2003 NFPA 22

Date of Review: _____ Permit Number: _____

Business/Building Name: _____ Address of Project: _____

Designer Name: _____ Designer's Phone: _____

Contractor: _____ Contractor's Phone: _____

Manufacturer: _____ Model: _____ Capacity: _____

Reference numbers following worksheet statements represent an NFPA code section unless otherwise specified.

This worksheet will not include criteria for pressure tanks, wood tanks, embankment tanks and steel support towers. Unless otherwise noted the code references are NFPA.

Worksheet Legend: ✓ or OK = acceptable   N = need to provide   NA = not applicable

General Details and Information for all Tanks that Must be Shown on the Plans, 4.5:
1. _____ Three sets of drawings are provided, 4.5.
2. _____ Scale: a common scale is used and plan information shall be legible.
3. _____ An equipment symbol legend is provided.
4. _____ The tank material of construction is indicated on the shop drawings, 4.3. Materials are limited to steel, wood, concrete and coated fabric.
5. _____ Wood and steel tanks that are elevated are supported by steel or reinforced concrete towers, 4.3.1.1.
6. _____ The location of the tank in relation to the frost line shall be detailed, 4.3.2.1.1.
7. _____ The location of the tank does not subject it to fire exposure, 4.2.1.
8. _____ A minimum 2 hour fire-resistive assembly is provided when a tank and its steel supports are within 20 ft. of an exposure, 4.2.1.2 and 4.2.1.3 (interior timber is not permitted to support or brace tanks).
9. _____ Details and engineering calculations for anchoring the tank are provided, 4.2.3.
10. _____ When tanks are equipped with lightning protection, the protection system is detailed and designed in accordance with NFPA 780, 4.9.
11. _____ Dead, live, wind, earthquake, and equipment load calculations are provided, 4.12.1-.5.
12. _____ Provided are column and strut design details in accordance with AWWA D100, and columns that are in contact with water are a minimum of 0.25 in. thick, 4.12.6.
13. _____ A roof is provided for above-ground tanks and its maximum load design is noted on the plans, 4.12.8.
14. _____ OSHA 29 CFR 1910 compliant railings and ladders shall be provided when required by Section 4.14.1.1.
15. _____ Piping configuration, size, and materials used from the tank to the water destination point are detailed.
16. _____ Size and location of water supply pipe are provided on the plot or site plan.
17. _____ For an air tight roof, a roof vent designed to be at least 1/2 the area of the discharge pipe(s) or fill pipe, whichever is larger is detailed and designed in compliance with Section 4.15.

Fiberglass Tanks:
18. _____ Fiberglass tanks are for underground use only and comply with AWWA D120, 4.3.2.1.
19. _____ Fiberglass tanks are installed beneath the frost line; backfill material and install in relation to ground water level are in accordance with Section 4.3.2.1.

Welded Steel Gravity and Suction Tanks:
20. _____ The ASTM specification for the steel plates are provided, 5.2.
21. _____ Calculations for seismic anchor-bolt loading and seismic anchoring details are provided, 5.3.
22. _____ Shapes, bolts, rods, castings, and reinforcing steel grade specifications are noted on the plans, 5.2.2.
23. _____ Material plate thickness is detailed and meet minimum requirements of 5.5., Tables 5.5.1.6, 5.5.4(a) and (b).
24. _____ All openings on the shell, suspended bottom, steel plate riser, or tubular support that are greater than 4 in. are reinforced and detailed on the plans, 5.5.9.
25. _____ The underside of all bottom plates are provided corrosion protection and the method is noted on the plans, 5.6.7. The testing method for welded flat bottom tanks is on the plans, 5.6.8.
26. _____ Roof supports and roof plate anchorage are detailed, 5.5.10. and 5.7.2.

Fire Plan Review and Inspection Guidelines

27. _____ Exterior and interior ladder(s) shall comply with Sections 5.7.4, 5.7.5, 5.7.6 and 5.7.7.
28. _____ Interior and exterior painting requirements are specified on the plans, 5.7.9.

Factory-Coated, Bolted Steel Tanks:
29. _____ The ASTM specification for the steel plates are provided, 6.2.
30. _____ Shapes, bolts, rods, castings, and reinforcing steel grade specifications are noted on the plans, 6.2.
31. _____ Provided are calculations for seismic anchor-bolt loading and the method of anchorage is detailed, 6.3.
32. _____ Material thickness is detailed and meet minimum requirements of 6.4.
33. _____ The manufacturer's specifications, drawings, and the assembly instructions are provided 6.6.

Tank and Tower Foundations:
34. _____ The foundation and footings are detailed and constructed of concrete in accordance with Section 11.1.
35. _____ Foundation pier design and details for elevated tanks are in accordance with Section 11.3.
36. _____ Anchorage design and details for tanks and piers are in accordance with Section 11.4.

Pipe Connections and Fittings:
37. _____ Pipes intersecting roofs or floors shall be watertight, 13.1.1.
38. _____ When required, the heater thermometer location for a gravity circulating heating system is detailed, 13.1.8.
39. _____ Water level gauge is provided and it location is detailed, 13.1.11.
40. _____ Tank pipe risers shall be in frostproof casings unless the riser is 3 ft. or larger and is properly heated, 13.1.12, 13.1.13.
41. _____ The pipe specification shall be included with the plans, 13.1.15.
42. _____ When in contact with storage water, steel piping less than 2 in. is not used, 2 in. to 5 in. is extra-strong weight, equal or greater than 6 in. is standard weight and it is detailed and noted on plans, 13.1.15.
43. _____ The discharge pipe is a minimum 6 in. up to 25,000 gallon tanks, a minimum 8 in. for 30,000 - 100,000 gallon tanks, a minimum 10 in. for tanks greater than 100,000 gallons, 13.2.2.
44. _____ Discharge piping is erected using approved methods complying with Section 13.2.3.
45. _____ Pipe bracing is detailed and the threaded rod diameter meets Section 13.2.4.
46. _____ The discharge pipe is supported in accordance with Section 13.2.5.
47. _____ Discharge pipes that have offsets shall be supported at least every 12 ft., 13.2.6.
48. _____ An expansion joint for a discharge pipe that has a base elbow or support hanger less than 30 ft. from the tank bottom is detailed and design in accordance with Section 13.3 or 13.1.2 if rigidly connected, 13.2.8.
49. _____ A settling basin in the tank is detailed a minimum 4 in. for flat-bottom tanks and 18 in. for suspended-bottom tanks, 13.2.10.
50. _____ The location of a listed check valve is detailed and it is in a pit under the tower tank or if the tank is over a building, 13.2.11.
51. _____ A listed gate valve is located and detailed in the discharge pipe. Its location is on the yard side between the check valve and other pipe connections, 13.2.12.1.
52. _____ A listed indicating control valve is located and detailed in the discharge pipe. Its location is on the tank side of the check valve, 13.2.12.3. For tanks on an independent tower, the valve is located in the pit with the check valve, 13.2.12.3.1.
53. _____ A tank used as a suction source for fire pump(s) has the anti-vortex plate and location detailed, 13.2.13 and Annex Fig. B.1 (o-q).
54. _____ When required by Sections 13.2.7 and 13.2.8.1, the discharge pipe expansion joint is listed,13.3.
55. _____ When a tank is filled by a fire protection system or a special filling pump, all bypass and filling pump requirements are detailed and in accordance with Section 13.4.
56. _____ Calculations that verify filling of tank in 8 hours are provided, 13.4.
57. _____ Overflow and stub piping are detailed and in compliance with Section 13.5.
58. _____ Cleanouts, at least a 2 in. drain, control valve, sight glass or hose connection, etc. are designed and detailed in accordance with Section 13.6.
59. _____ Critical water level and temperature sensors are specified, monitored, and their location detailed, 13.8.
60. _____ Valve pit, if provided, is detailed and includes materials, size, load designs, access, drains, heater house, etc, 14.1.1-.10.

Additional Comments:

Review Date: _____   Approved or Disapproved   FD Reviewer: _____

Review Date: _____   Approved or Disapproved   FD Reviewer: _____

Review Date: _____   Approved or Disapproved   FD Reviewer: _____

Fire Plan Review and Inspection Guidelines

Private Water Tank System Acceptance Inspection
2006 IFC and 2003 NFPA 22

Date of Inspection: _____ Permit Number: _____

Business/Building Name: _____ Address of Project: _____

Contractor: _____ Contractor's Phone: _____

Reference numbers following worksheet statements represent an NFPA code section unless otherwise specified.

Pass | Fail | NA

1. ___|___|___ Approved drawing is on site.
2. ___|___|___ Received system certification from installer.
3. ___|___|___ Piping configuration and pipe size comply with the plans.
4. ___|___|___ Size and location of water supply piping comply with the plans.
5. ___|___|___ Lightning protection system is installed in accordance with NFPA 780 and the plans.
6. ___|___|___ Air tight roof has a roof vent the size that is in compliance with the approved set of plans.
7. ___|___|___ Fiberglass tank is underground, protected against freezing, has 12 in. minimum pea gravel or sand backfill around tank, and has anchors if below the water table level in accordance with the plan.
8. ___|___|___ Welded steel gravity and suction tanks: steel plates, shapes, bolts, rods, castings, and reinforcing steel comply with the plans.
9. ___|___|___ All openings greater than 4 in. are reinforced in accordance with the plans, 5.5.9.1.
10. ___|___|___ The underside of all bottom plates are coated for corrosion protection (paint or oiled sand).
11. ___|___|___ Roof anchorage is provided in accordance with the plans.
12. ___|___|___ Two roof hatches are provided in accordance with the plans.
13. ___|___|___ Ladder(s), cages, side bar and rungs are provided in accordance with the plans.
14. ___|___|___ Factory-coated, bolted steel tanks: steel plates, shapes, bolts, rods, castings, and reinforcing steel comply with the plans.

Pipe Connections and Fittings

15. ___|___|___ Water level gauge is provided and it works.
16. ___|___|___ Frostproof casings on pipe risers are in place.
17. ___|___|___ Fireproofing is in place for exposed steel supports and are within 20 ft. of exposures.
18. ___|___|___ Ladders and roofs hatches are properly installed in accordance with the plans.
19. ___|___|___ The size of the discharge pipe is in accordance with the plans, 13.2.2.
20. ___|___|___ Piping is flanged cast iron or steel pipe, welded steel or of listed corrosion-resistant with flanged or welded connections.
21. ___|___|___ Discharge pipe double-flanged base elbow is resting on a concrete or masonry foundation, 13.2.5.
22. ___|___|___ An expansion joint for a discharge pipe that has a base elbow or support hanger less than 30 ft. from the tank bottom is provided in accordance with the plans unless rigidly connected, 13.2.8.
23. ___|___|___ Settling basin in tank is minimum 4 in. for flat bottom and 18 in. for suspended bottom.
24. ___|___|___ If the tank is used as suction source for fire pump(s) it has a anti-vortex plate installed in accordance with the plans.
25. ___|___|___ The location of a listed check valve is in a pit under the tower tank or if the tank is over a building, 13.2.11.
26. ___|___|___ A listed gate valve is located in the discharge pipe in accordance with the plans. Its location is on the yard side between the check valve and other pipe connections, 13.2.12.1.
27. ___|___|___ A listed indicating control valve is located in the discharge pipe in accordance with the plans. Its location is on the tank side of the check valve, 13.2.12.3. For tanks on an independent tower, the valve is located in the pit with the check valve, 13.2.12.3.1.
28. ___|___|___ A fill by-pass is provided in accordance with the plans.
29. ___|___|___ A fill pump is provided in accordance with the plans.
30. ___|___|___ Overflow and stub pipe are installed in accordance with the plans.

Fire Plan Review and Inspection Guidelines

31. ____|____|____ Cleanouts, at least a 2 in. drain, control valve, sight glass or hose connection, etc. are installed in accordance with the plans 13.6.
32. ____|____|____ Tank heating is provided and system is provided a low temperature alarm, relief valve, and a thermometer, and its operation is verified as working.
33. ____|____|____ Size, type, and location of valves comply with the plans and all control valves are the indicating type.
34. ____|____|____ The tank is tested for being watertight by filling the tank; and flat bottom welds are tested.
35. ____|____|____ The signed tank inspection reports of inspection performed by the contractor and owner are provided, 4.7.

Additional Comments:

Inspection Date: _____   Approved or Disapproved   FD Inspector: _____

Inspection Date: _____   Approved or Disapproved   FD Inspector: _____

Inspection Date: _____   Approved or Disapproved   FD Inspector: _____

Fire Plan Review and Inspection Guidelines

Private Water Tank System Installation Certification

Permit #: _____   Date: _____

	Property Protected	System Installer	System Supplier
Business Name:	_____	_____	_____
Address:	_____	_____	_____
	_____	_____	_____
Representative:	_____	_____	_____
Telephone:	_____	_____	_____

Type of Tank(s): _____

Location of Plans: _____

Location of Owner's Manual: _____

1. <u>Certification of System Installation</u>: Complete this section after system is installed, but prior to conducting operational acceptance tests.

    This system installation was inspected and found to comply with the installation requirements of:
    _____ NFPA 22
    _____ Design Specifications
    _____ Manufacturer's Instructions
    _____ Other (specify: FM, UL, etc.) _____

    Print Name: _____

    Signed: _____ Date: _____

    Organization: _____

2. <u>Certification of System Operation</u>: All operational features and functions of this system were tested and found to be operating properly in accordance with the requirements of:
    _____ NFPA 22
    _____ Design Specifications
    _____ Manufacturer's Instructions
    _____ Other (specify) _____

    Print Name: _____

    Signed: _____ Date: _____

    Organization: _____

Prefinal and Certificate of Occupancy Inspection Requirements For Contractors
Contractors Worksheet

Private Water Tank System Preinspection Requirements

1. All certification forms and documents are required to be on the site for review:

    ____ Plans

    ____ Permit

    ____ Prefinal system equipment test and a device location inspection are required as well as the completion of the items on this pretest form. Use the Acceptance Inspection worksheet for the pretest.

    ____ Installation certification is completed, use the form contained in this book.

2. ____ Person familiar with installation must be present to perform any tests required.
3. ____ Owner's representative approval needed for time and date of testing.
4. ____ All tank areas are accessible.
5. ____ Equipment placement is in accordance with the plan, verification to have been done by contractor.
6. ____ If Items 1-5 are incomplete, the inspection will be cancelled and another inspection request is required. A reinspection fee may be assessed.

Prior to the next approval test:

7. ____ When there are device additions, contractor must provide:

    ____ As-builts and new calculations shall be submitted for review and approval.

    Note: New plan review will be submitted as "supplemental information" and proof of the additional review fee payment is required.

8. ____ A reinspection fee may be assessed if the system and paperwork are not ready.

Emergency and Standby Power
Contents

1. Plan Review Worksheet, Level 1 85

2. Plan Review Worksheet, Level 2. 87

3. Installer Certification Form 89

4. Guideline for Annual Service Tests 90

5. Contractor Prefinal and Certificate of Occupancy Inspection Requirements 92

Fire Plan Review and Inspection Guidelines

Emergency Power Supply System (EPSS), Level I (NFPA 110 4.4.1)
Plan Review
2006 IFC and 2002 NFPA 110

Date of Review: _____ Permit Number: _____

Business/Building Name: _____ Address of Project:: _____

Designer Name: _____ Designer's Phone: _____

Contractor: _____ Contractor's Phone: _____

System Manufacturer: _____ Model: _____

Occupancy Classification: _____

Reference numbers following worksheet statements represent an NFPA code section unless otherwise specified.

Worksheet Legend: ✓ or OK = acceptable   N = need to provide   NA = not applicable

1. _____ Three sets of drawings.
2. _____ Equipment is listed for intended use and specification listing sheets are provided.

Floor Plan Showing:

3. _____ Scale: a common scale is used and plan information is legible.
4. _____ Equipment and symbol legend.
5. _____ Overall floor plan showing room, room dimensions, and equipment placement.

Other Information to be Provided on the Plans in Accordance with NFPA 110:

6. _____ Prime mover manufacturer: _____ and model: _____, and generator manufacturer model: _____ are provided, also a copy of the equipment data sheets are required, (5.6.10.4).
7. _____ EPSS Type is provided, Table 4.1(b):   U   10   60   120   or M
8. _____ EPSS Class is provided, Table 4.1(a):   .083   .25   2   6   48   or X
9. _____ EPSS Level designation: 1
10. _____ In certain seismic design categories a Level 1 EPSS requires a fuel supply that can provide at least 96 hours of service, 5.1.2. IFC 604.2.15.1.1 requires a 2 hour supply.
11. _____ LPG and liquid petroleum fuels intended for the EPSS shall be dedicated only to the EPSS, 5.5.1.
12. _____ Low level sensing switch and alarm for fuel system is detailed, 5.5.2.
13. _____ Calculations are provided to verify main fuel reservoir capacity at 133 percent of the low-fuel sensor or In accordance with Table 4.1(a), 5.5.3.
14. _____ Fuel and water cooling line solenoid valves are battery powered with a manual override and are detailed on the plans, 5.6.3.2.
15. _____ Prime mover instrument panel and gauges are detailed, 5.6.3.3.
16. _____ Number, size, and type of batteries are noted on the plans, 5.6.4.3–5.6.4.5.
17. _____ Specification sheet and a location detail of the battery charger are provided, 5.6.4.6.
18. _____ Control panel and equipment are detailed, 5.6.5.1.
19. _____ When required the remote control and alarms are provided, 5.6.6.
20. _____ Prime mover cooling system is compliant with 5.6.7.
21. _____ When required the prime mover exhaust piping has a muffler sized for the unit, 5.6.8.
22. _____ The generator instrument panel has instrumentation complying with Section 5.6.9.9.
23. _____ When required the transfer switch shall be listed for emergency service, 6.1.4.
24. _____ Instructions for manual transfer process are provided for review and shall be provided at the transfer switch, this requirement is noted on plans, 6.2.4.
25. _____ Autotransfer switch is provided with source monitoring, interlocking, manual operation, time delay on starting, transfer and retransfer, and for generator exercising, a test switch, etc., 6.2.4 - 6.2.12.
26. _____ Overcurrent protective device rating is provided, 6.5.3.
27. _____ Level 1 systems are in a dedicated room of 2 hour construction and equipment has a minimum 30 in. working space around the unit, 7.2.1, 7.2.1.2 and 7.2.1.5.
28. _____ A level 1 EPSS room is provided with battery powered emergency lighting providing at least 3.0 footcandles (32.2 lux), 7.3.

Fire Plan Review and Inspection Guidelines

29. _____ Engine is located at least 6 inches above the floor, 7.4.1.
30. _____ When required the engine mounting foundation is isolated from the floor and is detailed in accordance with the manufacturer recommendations and installation information is provided, 7.4.3.
31. _____ Combustion and cooling air intake is provided and detailed, 5.6.7.2, 7.7.1, 7.7.2.
32. _____ Ventilation is provided from outside of the building. If the ventilation air is routed using a duct, the duct shall have the same fire resistance rating of the room housing the Level 1 EPSS, 7.7.3
33. _____ Fuel tank size is detailed and sized for its EPSS class, 7.9.
34. _____ Fuel is not gravity feed, exception: integral tanks, 7.9.2.
35. _____ Fuel piping materials are noted, 7.9.3.
36. _____ Diesel day tanks and return line are below engine fuel return elevation, 7.9.4.
37. _____ Fuel supply for gas and liquid fueled prime movers are designed in accordance with IFC Chapter 38 and NFPA 58 for LP-Gas and in accordance with IFC Chapter 34 for systems fueled with flammable or combustible liquids.
38. _____ The fuel gas supply for the EPSS unit shall be connected ahead of the building's main shutoff valve, 7.9.7.
39. _____ Main gas supply is marked to indicate it is the supply to an EPSS, 7.9.8.
40. _____ Gas-fuel supply is shown to meet the demands of the unit, 7.9.9.
41. _____ Exhaust system is properly vented and detailed, with flex connector at prime mover, 7.10.
42. _____ The system shall be tested in accordance with NFPA 110, 7.13, test procedure is noted on the plans, 7.13.
43. _____ Seismic anchoring and securing systems are detailed and calculations are provided for equipment and piping, 7.11.5.

Miscellaneous
44. _____ A remote manual stop is located outside the generator room or outdoors for exterior installations, 5.6.5.6. The manual stop shall be labeled to indicate its function, 5.6.5.6.1.

Additional Comments:

Review Date: _____   Approved or Disapproved   FD Reviewer: _____

Review Date: _____   Approved or Disapproved   FD Reviewer: _____

Review Date: _____   Approved or Disapproved   FD Reviewer: _____

Fire Plan Review and Inspection Guidelines

Emergency Power Supply System (EPSS), Level 2 (NFPA 110 4.4.2)
Plan Review
2006 IFC and 2002 NFPA 110

Date of Review: _____ Permit Number: _____

Business/Building Name: _____ Address of Project: _____

Designer Name: _____ Designer's Phone: _____

Contractor: _____ Contractor's Phone: _____

System Manufacturer: _____ Model: _____

Occupancy Classification: _____

Reference numbers following worksheet statements represent an NFPA code section unless otherwise specified.

Worksheet Legend: ✓ or OK = acceptable   N = need to provide,   NA = not applicable

1. ____ Three sets of drawings.
2. ____ Equipment is listed for intended use and specification listing sheets are provided.

Floor Plan Showing:

3. ____ Scale: a common scale is used and plan information is legible.
4. ____ Equipment and symbol legend.
5. ____ Overall floor plan showing room, room dimensions, and equipment placement.

Other Information to be Provided on the Plans in accordance with NFPA 110:

6. ____ Prime mover manufacturer: _____ and model:_____, and generator manufacturer and model: _____ are provided, also a copy of the equipment spec. sheets are required. (5.6.10.4)
7. ____ EPSS Type, Table 4.1(b):   U   10   60   120   or M
8. ____ EPSS Class, Table 4.1(a):   .083   .25   2   6   48   or X
9. ____ Low level sensing switch and alarm for fuel system is detailed, 5.5.2.
10. ____ Calculations are provided to verify main fuel reservoir capacity at 133 percent of the low-fuel sensor or in accordance with Table 4.1 (a), 5.5.3.
11. ____ Prime mover instrument panel and gauges are detailed, 5.6.3.3.
12. ____ Number, size, and type of batteries are noted on the plans, 5.6.4.3-.5.
13. ____ Specification sheet and the location detail of the battery charger are provided, 5.6.4.6.
14. ____ Control panel and equipment are detailed, 5.6.5.1.
15. ____ When required, the remote control and alarms are provided, 5.6.6.1.
16. ____ Prime mover cooling system is forced air, liquid cooled, or both, 5.6.7.
17. ____ Prime mover exhaust piping has a muffler sized for the unit, 5.6.8.
18. ____ When required the transfer switch is listed for emergency service, 6.1.6.
19. ____ The transfer switch is provided with source monitoring, interlocking, manual operation, time delay on starting, transfer and retransfer, and for generator exercising, a test switch, etc., 6.2.4.2 – 6.2.4.12.
20. ____ Overcurrent protective device rating is provided, 6.5.3.
21. ____ Level 2 systems are located in an area to minimize the possibility of damage, 7.2.3.
22. ____ Level 2 EPSS room is provided with battery powered emergency lighting, 7.3.1.
23. ____ Engine is elevated at least 6 inches above the floor, 7.4.1.1.
24. ____ Engine mounting foundation is isolated from the floor and is detailed per the manufacturer recommendations and installation information is provided, 7.4.3.
25. ____ Combustion and cooling air intake is provided and detailed, 5.6.7.2, 7.7.1, 7.7.2.
26. ____ Fuel is not gravity feed, 7.9.2.
27. ____ Fuel piping materials are noted, 7.9.3.
28. ____ Diesel day tanks and return line are below engine fuel return elevation, 7.9.4.
29. ____ Diesel integral tanks are limited to 660 gallons and gasoline fuel is limited to 25 gallons when inside or on the roof, 7.9.5.

Fire Plan Review and Inspection Guidelines

30. _____ Fuel supply for gas and liquid fueled prime movers are designed in accordance with IFC Chapter 38 NFPA 58 for LP-Gas and in accordance with IFC Chapter 34 for systems fueled with flammable or combustible liquids.
31. _____ The fuel gas supply for the EPSS unit shall be connected ahead of the building's main shutoff valve, 7.9.7.
32. _____ Main gas supply is marked to indicate it is the supply to an EPSS. 7.9.8.
33. _____ Gas-fuel supply is shown to meet the demands of the ESPS unit, 7.9.9.
34. _____ All manual fuel system valves shall be of the indicating type, 7.9.11.
35. _____ Exhaust system is properly vented and detailed, with flex connector at prime mover, 7.10.
36. _____ The system shall be tested in accordance with NFPA 110: 7.13, test procedure is noted on the plans.
37. _____ Seismic anchoring and securing systems are detailed and calculations are provided for equipment and piping, 7.11.5.

Miscellaneous
38. _____ A remote manual stop is located outside the generator room or outdoors for exterior installations, 5.6.5.6. The manual stop shall be labeled to indicate its function, 5.6.5.6.1

Additional Comments:

Review Date: _____	Approved or Disapproved	FD Reviewer: _____
Review Date: _____	Approved or Disapproved	FD Reviewer: _____
Review Date: _____	Approved or Disapproved	FD Reviewer: _____

Emergency and Standby Power Installation Certification

Permit #: _____   Date: _____

	Property Protected	System Installer	System Supplier
Business Name:	_____	_____	_____
Address:	_____	_____	_____
	_____	_____	_____
Representative:	_____	_____	_____
Telephone:	_____	_____	_____

Type of System: _____

Location of Plans: _____

Location of Owner's Manual: _____

1. <u>Certification of System Installation:</u> Complete this section after system is installed, but prior to conducting operational acceptance tests.

   This system installation was inspected and found to comply with the installation requirements of:
   - _____ NFPA 110
   - _____ Articles 700, 701, or 702 of the National Electrical Code
   - _____ Manufacturer's Instructions
   - _____ Other (specify: FM, UL, etc.) _____

   Print Name: _____

   Signed: _____   Date: _____

   Organization: _____

2. <u>Certification of System Operation:</u> All operational features and functions of this system were tested and found to be operating properly in accordance with the requirements of:
   - _____ NFPA 110
   - _____ Design Specifications
   - _____ Articles 700, 701, or 702 of the National Electrical Code
   - _____ Manufacturer's Instructions
   - _____ Other (specify) _____

   Print Name: _____

   Signed: _____   Date: _____

   Organization: _____

Guidelines for Conducting Acceptance or Annual Service Test: Emergency and Standby Power
These tests should be performed in the presence of an electrical inspector.

1. On-site acceptance test shall be required for the final Emergency Power Supply System (EPSS) approval. For test specifics consult NFPA 110:7.13.
2. Level 1 EPSS shall not be approved until tests are conducted.
3. Verify all required emergency systems are connected to the transfer switch branch circuits.
4. Tests: The contractor will observe and record every 15 minutes of 2 hour test the bulleted items.

   Prime mover: cold start and under a normal emergency load, an open switch or breaker power failure is initiated.
   - Observe and record start time delay.
   - Observe and record prime mover crank-to-start time.
   - Observe and record time to operating speed.
   - Observe and record voltage and frequency overshoot.
   - Observe and record time to achieve steady-state condition with switch(es) transferred.
   - Observe and record the voltage, frequency and amperage.
   - Observe and record prime mover oil pressure, water temperature, and the battery charge rate at 5 minute intervals for the first 15 minutes then every 15 minutes thereafter.
   - Load test (for intended load) for minimum time as per NFPA 110 Table 4.1(a) or 2 hour maximum.
   - Return building to normal power, observe and record time delays of retransfer switch(es) and cool-down and shutdown.
   - Prime mover shall be allowed 5 minutes to cool-down then immediately perform a full load test.
   - Two hour full load test of 100% of nameplate kW rating of the EPS less applicable derating factor.
   - For Otto or diesel prime movers perform cycle crank test required by the manufacturer and NFPA 110 5.6.4.2 and Table 5.6.4.2.
   - Test all safety features specified in NFPA 110 5.6.5, and 5.6.6 as recommended by the manufacturer.
   - Provide the following to the authority having jurisdiction, 7.13.11;
     - Evidence of the prototype test for Level 1 systems as detailed in 5.2.1.2.
     - Certification that a torsional vibration analysis for compatibility of prime mover and the generator as detailed in 5.6.10.2.
     - A letter of compliance in accordance with NFPA 110, 5.6.10.5.
     - A manufacturer's certification of a rated load test at the rated power factor in accordance with Section 7.13.11(4).

5. Verify that the fuel supply is adequate.
6. Annually verify records of maintenance, tests, and service.

Prefinal and Certification of Occupancy Inspection Requirements For Contractors
Contractors Worksheet

Emergency Power System Test Requirements

1. All certification forms and documents are required to be on the site for review:
   - ___ Plans
   - ___ Permit
   - ___ Prefinal inspection is required as well as the completion of the items on this pretest form. Use the Acceptance Guide and manufacture's requirements for the pretest.
   - ___ Installation certification is completed, use the form contained in this book.
2. ___ Person familiar with installation must be present to perform the test.
3. ___ Owner's representative approval needed for time and date of testing.
4. ___ All rooms or areas are unlocked and accessible.
5. ___ If Items 1-4 are incomplete, the inspection will be cancelled and another inspection request will be required. A reinspection fee may be assessed.

Prior to the next approval test:

6. ___ When installation deviates from the approved plans, the contractor must provide:
   - ___ As-builts for review and approval.
   - Note: A new plan review will be submitted as "supplemental information" and proof of the additional review fee payment is required.
7. ___ A reinspection fee may be assessed if the system and required documentation are not ready.

Automatic Sprinkler System
Contents

1. NFPA 13 Sprinkler System Plan Review Worksheet, 2006 IFC and 2002 NFPA 13 95

2. NFPA 13 Sprinkler System Acceptance Inspection, 2006 IFC and 2002 NFPA 1 101

3. NFPA 13 Sprinkler Installation Certification. 104

4. NFPA 13 Sprinkler System General Design Plan Review Worksheet, 2006 IFC and 2007 NFPA 13. 105

5. NFPA 13 Sprinkler System Acceptance Inspection, 2006 IFC and 2007 NFPA 13. 112

6. 13R Residential Sprinkler System Plan Review Worksheet, 2006 IFC, 2002 NFPA 13, and 2002 NFPA 13R 115

7. NFPA 13R Sprinkler System Acceptance Inspection, 2006 IFC and 2002 NFPA 13R..118

8. 13R Residential Sprinkler System Plan Review Worksheet, 2006 IFC, 2007 NFPA 13 and 2007 NFPA 13R 120

9. NFPA 13 R Sprinkler System Acceptance Inspection, 2006 IFC and 2007 NFPA 13R..124

10. NFPA 13R Sprinkler Installation Certification...126

11. 13D Residential Sprinkler System Plan Review Worksheet, 2006 IFC and 2002 NFPA 13D..127

12. NFPA 13D Sprinkler System Acceptance Inspection, 2006 IFC and 2002 NFPA 13D..130

13. 13D Residential Sprinkler System Plan Review Worksheet, 2006 IFC and 2007 NFPA 13D..131

14. NFPA 13D Sprinkler System Acceptance Inspection, 2006 IFC and 2007 NFPA 13D..134

15. NFPA 13D Sprinkler Installation Certification...135

16. Automatic Sprinkler Protection High-Piled Combustible Storage using 2002 NFPA 13, Chapter 12 and IFC Chapter 23 136

17. NFPA 13 Sprinkler System for Storage up to a Height of 25 Feet Plan Review Worksheet, 2006 IFC and 2002 NFPA 13 138

18. NFPA Sprinkler System for Storage up to 25 Feet Plan Review Worksheet, 2006 IFC and 2007 NFPA 13 143

19. NFPA 13 Sprinkler System for Storage Greater Than 25 Feet and Rubber Tires Plan Review Worksheet, 2006 IFC and 2002 NFPA 13 149

20. NFPA 13 Sprinkler System for Storage Greater Than 25 Feet and Rubber Tires Plan Review Worksheet, 2006 IFC and 2007 NFPA 13 154

21. Prefinal and Certificate of Occupancy Inspection Requirements for Contractors, Contractors Worksheet 160

Fire Plan Review and Inspection Guidelines

NFPA 13 Sprinkler System
Plan Review Worksheet
2006 IFC and 2002 NFPA 13

Date of Review: _____ Permit Number: _____

Business/Building Name: _____ Address of Project: _____

Designer Name: _____ Designer's Phone: _____

Contractor: _____ Contractor's Phone: _____

No. of Sprinklers: _____ Occupancy Classification: _____

Reference numbers following worksheet statements represent an NFPA code section unless otherwise specified.

Worksheet Legend: ✓ or OK = acceptable   N = need to provide   NA = not applicable

1. _____ A minimum of three sets of drawings are provided.
2. _____ Equipment is listed for intended use and compatible with the system; specification data sheets are provided.

Drawings shall detail the following (14.1.3.1-14.1.3.44):
General:

3. _____ Type of system is noted; __ hydraulic calc, __ pipe schedule, __ wet, __ dry, __ preaction, __ deluge, __ antifreeze. The plans declare the design standard is the 2002 edition year of NFPA 13.
4. _____ Scale: a common scale shall be used and plan information shall be legible.
5. _____ Plot plan illustrates fire protection water mains and pipe diameter(s) supplying the building.
6. _____ The location of smoke or fire partitions, fire walls, and building elevation views.
7. _____ Occupancy class and or use of each room or area, 5.1.1.
8. _____ Full height cross sectional drawing including ceiling construction.
9. _____ Total area protected by each system for each floor is provided.
10. _____ Dimensions for system piping, sprinkler spacing and branch line spacing and elevation changes.
11. _____ Equipment symbol legend and a north orientation arrow is provided.
12. _____ Area limitations for hazard classification; 52.000 sq. ft. for light and ordinary hazard, 25,000 sq. ft. for extra hazard pipe schedule, 40,000 sq. ft. for extra hazard-hydraulic calculations, and 40,000 for high-piled storage, 8.2.1.
13. _____ Hydrant flow test determining water supply capacity at 20 PSI residual pressure is provided.
14. _____ When used as a basis for design, hydraulic calculations are provided with summary, detail worksheets, and graph sheet, 14.3.
15. _____ Dry pipe system capacity in gallons is provided _____ gal., not to be greater than 750 gal. unless the requirements of 7.2.3.2 or 7.2.3.3 are met, 7.2.3.
16. _____ All water supply valves and water flow switches shall be electrically supervised, IFC 903.4.
17. _____ Exterior flow alarm location is detailed. Note: if electric, it shall be listed for outdoor use, IFC 904.3.2.
18. _____ When installed backflow prevention device pressure loss data is provided in the hydraulic calculations.

Sprinklers:

19. _____ Total number of each type of sprinkler is noted, 8.3.2.1.
20. _____ If the hazard classification of the occupancy is changed, the temperature of rating of sprinklers shall be evaluated in accordance with Section 8.3.2.6.
21. _____ Light hazard occupancies shall have quick-response or residential sprinklers, 8.3.3.1, IFC 903.3.2.
22. _____ Sprinkler are locat correctly for branch line spacing and area of protection limits, ceiling and roof cross sectional views are provided for clarification, 14.1.3.
23. _____ For each type of sprinkler the K-factor, temperature rating, and orifice size are provided, 14.1.3.(12).
24. _____ Each sprinkler coverage area is installed in accordance with its area limitations or its listing, 8.6.2.2, Table 8.6.2.2.1 (a-c).
25. _____ Specialty sprinklers, extra coverage, early suppression fast response, large drop, sidewall, etc. comply with the standard and listing limitations, 6.1.1 and 8.4.1–8.4.9.
26. _____ Maximum perpendicular distance to the walls is not greater than 1/2 of allowable distance between sprinklers, 8.6.3.2 and Tables 8.6.2.2.1(a through d), for sidewall sprinklers, 8.7.3.2. and Table 8.7.2.2.1.

Fire Plan Review and Inspection Guidelines

27. _____ Standard sprinkler spacing from vertical obstructions complies with Table 8.6.5.1.2 and for floor mounted obstructions, Table 8.6.5.2.2.
28. _____ Sidewalls sprinkler spacing for front obstructions refer to Table 8.7.5.1.3, for a side obstruction refer to Table 8.7.5.1.4, and for floor mounted obstructions refer to Table 8.7.5.2.2.
29. _____ Extended coverage uprights and pendent spacing for ceiling or wall obstructions refer to Table 8.8.5.1.2 and for floor mounted obstructions refer to Table 8.8.5.2.2.
30. _____ Extended coverage sidewall spacing for front obstructions refer to Table 8.9.5.1.3 and for floor mounted obstructions, Table 8.9.5.2.2.
31. _____ Residential upright and pendent sprinkler spacing from vertical obstructions complies with Table 8.10.6.1.2 and for floor mounted obstructions, Table 8.10.6.2.2.
32. _____ Residential sidewall sprinkler spacing from ceiling or hanging obstructions complies with Table 8.10.7.1.3 and for floor mounted obstructions, Table 8.10.7.2.2.
33. _____ Sprinkler coverage shall be provided beneath obstructions greater than 4 ft. wide, 8.5.5.3.1.
34. _____ Baffles are provided for sprinklers less than 6 ft. apart in accordance with Section 8.6.3.4.2.
35. _____ Locations or conditions requiring special consideration, 8.14.
36. _____ A. concealed spaces, for the 15 omissions, see 8.14.1.2.
37. _____ B. vertical shafts, 8.14.2.
38. _____ C. stairways, 8.14.3.
39. _____ D. vertical openings, 8.14.4.
40. _____ E. elevator hoistways and machine rooms, 8.14.5.
41. _____ F. spaces under ground floors, exterior docks, and platforms, 8.14.6.
42. _____ G. exterior roof and canopy, 8.14.7.
43. _____ H. dwelling unit, 8.14.8.
44. _____ I. library stack room, 8.14.9.
45. _____ J. electrical equipment, 8.14.10.
46. _____ K. ceilings: open-grid, drop-out, 8.14.12 and 8.14.13.
47. _____ L. stages, 8.14.15.
48. _____ Sprinkler is provided at top of shaft, refer to exceptions, shafts with combustible surfaces require coverage at alternate levels, accessible noncombustible shaft has sprinkler at bottom, 8.14.2.
49. _____ Vertical shaft has sprinklers at top opening, above bottom opening and alternate levels when it has combustible surfaces, 8.14.2.1, 8.14.2.2.
50. _____ Sprinklers are provided beneath combustible stairs, 8.14.3.1.
51. _____ Sprinklers are provided at the top of the stairway, under the first landing above the stairway shaft bottom when the shaft and stairs are noncombustible, 8.14.3.2.
52. _____ Closely spaced sprinklers with draft stops are provided around unenclosed floor openings except large openings like found in malls or atriums, and openings between floors of a common dwelling unit, 8.14.4.1 and 8.14.4.2.
53. _____ Elevator shaft has a sprinkler within 2 ft. of the shaft floor unless the shaft is noncombustible and there are no combustible hydraulic fluids, 8.14.5.
54. _____ Ordinary or intermediate temperature sprinklers are in the elevator machine room or at the top of the elevator shaft, 8.14.5.1 – 8.14.5.1.5.
55. _____ Sprinklers are provided under combustible ground floor, exterior dock, and platforms, 8.14.6.
56. _____ Sprinklers are provided under roofs and canopies unless constructed of noncombustible or limited combustible materials, less than 4 ft. wide, and no storage, refer to exceptions 8.14.7.1–8.14.7.4.
57. _____ Sprinklers are not required in noncombustible dwelling unit bathrooms, less than 55 sq. ft. or limited combustible with a 15 minute thermal barrier, except in nursing homes (I-1 and I-2) and in bathrooms that have direct access into corridors and exitways used by the public, 8.14.8.1.
58. _____ Sprinklers are not required in hotel or motel dwelling unit clothes closet, pantries, or linen closets less than 24 sq. ft. and the least dimension is not greater than 3 ft., 8.14.8.2.
59. _____ Sprinklers are provided in every aisle and at every tier stack, distance is not more than 12 ft. in library stack rooms, 8.14.9.
60. _____ Sprinklers are provided in electrical equipment rooms, exception: the room is dedicated use, has dry type equipment, 2 hour equipment enclosures, and no combustible storage, 8.14.10. Also consult the exceptions pertaining to spaces containing telecommunication equipment and associated power supplies as specified in IFC Section 903.2.
61. _____ Open grid ceilings shall not be installed under sprinklers, unless the grid opening and sprinkler placement criteria of Section 8.14.12 are met.

62. _____ Drop-out ceilings are installed under sprinklers in accordance with their listing, and sprinklers are not located below the ceilings, 8.14.13.
63. _____ Sprinklers for stages shall be provided in accordance with Section 8.14.15.
64. _____ Proscenium openings for stages shall be protected in accordance with Section 8.14.15.2.

Pipe Support and Hangers:
65. _____ Type and locations of hangers, sleeves, and braces are shown, 14.1.3. Non-listed hangers shall meet 5 performance criterion and the design shall be sealed by a registered professional engineer, 9.1.1.2
66. _____ If trapeze hangers are used, the locations are shown, a legend provided to specify span, size of pipe supported, angle and pipe used, and section modulus are provided and comply with Section 9.1.1.6.
67. _____ Pipe hanger spacing is in accordance with Table 9.2.2.1.
68. _____ Branch lines show one hanger for each section of pipe, exceptions are listed, 9.2.3.2.
69. _____ Cross mains show one hanger between each branch lines or in compliance with Table 9.2.2.1, and for additional spacing variations refer to Section 9.2.4.
70. _____ Supports can be on the horizontal pipe section if within 24 in. of the vertical pipe centerline, 9.2.5.1.
71. _____ Risers in multi-story buildings show supports at the lowest level, each alternate level, below offsets, and at the top, 9.2.5.3.
72. _____ The distance between supports for a riser does not exceed the limit specified in 9.2.5.4.

Pipe and Valves:
73. _____ Main drain pipe diameter is detailed and complies with Table 8.15.2.4.2, 8.15.2.4.
74. _____ Main drain routing is to the exterior or to an interior drain but ensure that the drain capacity is adequate, 8.15.2.4.4.
75. _____ Auxiliary drain location is detailed and its size is in accordance with Section 8.15.2.5.
76. _____ When required, the location of the listed backflow prevention device is detailed, 8.15.1.1.3.
77. _____ A listed control valve is provided on each side of the check valve, 8.15.1.1.4.1. Only one control valve on the system side of the check valve is necessary when the water supply is provided from the city connection, 8.15.1.1.4.3.
78. _____ The control valve(s) are accessible, 8.15.1.1.7.
79. _____ If a pressure reducing valve is used, its location and installation criteria are detailed in accordance with Section 8.15.1.2.
80. _____ If used, outside post-indicator control valve (PIV) locations and installation criteria are detailed in accordance with Section 8.15.1.3.
81. _____ If PIVs are approved to be located in a pit, the pit construction, location, and marking are designed and detailed in accordance with Section 8.15.1.4.2.

Seismic Bracing:
82. _____ Flexible couplings may be used for pipe 2½ in. or larger in accordance with Sections 9.3.2.2 and 9.3.2.3.
83. _____ A seismic separation assembly for piping is provided at building seismic joints, 9.3.3.
84. _____ Proper pipe clearance is noted on the plans for pipe penetrations in walls, floors, platforms or foundations, 9.3.4. Minimum clearance is in accordance with Section 9.3.4.2 - .5.
85. _____ Lateral sway bracing is required at a maximum spacing of 40 ft. for all feed and cross mains, and branch lines 2½ in. and larger, 9.3.5.3.1.
86. _____ Lateral sway bracing can be spaced up to 50 ft. if the design is in compliance with 9.3.5.3.3.
87. _____ Lateral sway bracing is within 20 ft. of the end of the pipe, 9.3.5.3.2.
88. _____ A lateral sway brace is provided on the last pipe of a feed or cross main, 9.3.5.3.4.
89. _____ Lateral sway bracing is required unless all the pipe is supported by rods less than 6 in. or by 30° wrap-around U-hooks for any size pipe, 9.3.5.3.7 and .8.
90. _____ Longitudinal sway bracing is a maximum of 80 ft. for mains and cross mains and within 40 ft. of the end of the pipe, 9.3.5.4.
91. _____ Four-way sway brace spacing on a riser does not exceed 25 ft. and a four-way sway brace is located at the top of the riser if the top of the riser exceeds 3 ft. in length, 9.3.5.5.
92. _____ Seismic bracing calculations are provided for each brace to be used as shown in Figure A.9.3.5(d).
93. _____ Longitudinal and lateral bracing is provided for each run of pipe between the change of direction unless the run is less than 12 ft. and supported by adjacent pipe run bracing, 9.3.5.11.
94. _____ Branch line method of restraint is detailed and in accordance with Section 9.3.6.1-.3.
95. _____ Restraints of branch lines shall be in accordance with Section 9.3.6.1.

Fire Department Connection (FDC):
96. _____ The FDC location is detailed on the street side or response side of building or as approved by the fire official, and when connected to the water supply it will not obstruct emergency vehicle access to the building, IFC 912.2.

Fire Plan Review and Inspection Guidelines

97. ____ Local water flow alarm is provided when the sprinkler system exceeds 20 sprinklers and its location is detailed, 8.16.1.1.
98. ____ FDCs for fire engine or fire boat are sized and arranged in accordance with, 8.16.2.3, and .4.
99. ____ The arrangement of the FDC piping supplying wet pipe, dry pipe, pre-action or deluge sprinklers shall be in accordance with Section 8.16.2.4.2.

Hydraulic Calculations, 11.2 and 14.1.3:
100. ____ Indicate the calculation method used: density area method or room design method, 11.2.3.2. and .3.
101. ____ Reference points in the calculation worksheet match with points on the plans, the occupancy hazard classifications are correct for the occupancy or use, 14.1.3.
102. ____ If design area adjustments are made, the selected shall be indicated, 11.2.3.2.7.
103. ____ Designs using QR sprinklers shall be in accordance with Section 11.2.3.2.3.
104. ____ Pipe size and length references in the calculation worksheet match the plans, 14.1.3(19).
105. ____ Sloped ceiling may require a 30 percent increase of design area, 11.2.3.2.4.
106. ____ Sprinkler data sheet information matches information on the plans.
107. ____ Water flow information is provided with static PSI, residual PSI, and available GPM at 20 PSI residual with graphed results.
108. ____ Density and design areas information are provided and comply with 12 conditions listed in Section 11.2.3.1.8, Figure 11.2.3.1.5.
109. ____ Calculations are correct: static PSI, pipe length, GPM, calculated K-factor values for drops or branch lines, elevation data, hose allowance, friction loss, and equivalent pipe and fitting lengths, 11.2.3.
110. ____ For the room design method the design area includes the most demanding room and if any, adjacent connecting compartments, 14.4.4.1.2.
111. ____ A minimum of two summary calculations are provided for a grid system, refer to the one exception, 14.4.4.2.
112. ____ Additional calculations may be required by the AHJ if the building design and room uses do not make the most demanding area obvious.
113. ____ Legend for calculation abbreviations is provided.
114. ____ Calculations are provided for extra hazard occupancies, deluge automatic sprinkler systems, and exposure protection systems.
115. ____ Dry pipe and double interlock preaction design areas are increased 30 percent but the density remains the same (11.2.3.2.5), use of high-temp sprinklers in extra hazard occupancies may reduce design area by 25 percent but not less than the area specified in 11.2.3.2.6.

Residential Sprinklers in a 13 System:
116. ____ The design area shall be in accordance with the requirements in Section 11.2.3.5.1.
117. ____ The calculation is based on the number of sprinklers and at the flow specified in 11.2.3.5.2.
118. ____ Hose streams and water duration requirements are for light hazard in accordance with Table 11.2.3.1.1, 11.2.3.5.5.

Special Design:
119. ____ Special design considerations for exposure protection, water curtain, and dry system are in accordance with 11.2.3.7–11.2.3.9.

Pipe Schedule:
Note: For systems less than 5000 sq. ft. the minimum water flow is proven to be available in accordance with Table 11.2.2.1. Systems less than 5,000 sq. ft. shall have 50 PSI residual pressure and meet the requirements of Table 11.2.2.1.
120. ____ Only ½ in. orifice sprinklers (nominal K-factor of 5.3 to 5.8) shall be used, 14.5.2.
121. ____ Light Hazard: 8 sprinklers maximum for each branch line, 14.5.2.1, 9 and 10 are permitted, see 14.5.2.
122. ____   A. pipe size, material and number of sprinklers are in accordance with Table 14.5.2.2.1.
123. ____   B. sprinklers above and below the ceiling are in accordance with Table 14.5.2.4.
124. ____ Ordinary Hazard: 8 sprinklers maximum for each branch line, 14.5.3.2.
125. ____   A. pipe size, material and number of sprinklers are in accordance with Table 14.5.3.4.
126. ____   B. sprinklersgreater than 12 ft. separation are in accordance with Table 14.5.3.5.
127. ____   C. sprinklers above and below the ceiling are in accordance with Table 14.5.3.7.
128. ____ Extra Hazard: the pipe schedule method is not allowed, 14.5.4.

Wet System:
129. ____ Relief valve not less than ¼ in. is detailed for gridded system, 7.1.2.1.
130. ____ An alarm test connection location for the waterflow alarm is provided and in compliance with 8.16.4.2.1–8.16.4.2.3.

Dry System, 7.2:

131. ____ Only upright, listed dry sprinklers are used, see exceptions for return bends and sidewall sprinklers, 7.2.2.
132. ____ System capacity is provided, 14.1.3(17).
133. ____ Only one dry pipe valve is permitted for each system that does not exceed 750 gallons unless the design complies with 7.32.3.2 or 7.2.3.3, 7.2.3.1.
134. ____ Water delivery calculations complying with 11.2.3.9 are provided for systems exceeding 750 gallons in order to confirm a water delivery time to be within 60 seconds, 7.2.3.3.
135. ____ A trip test connection sized according to 8.16.4.3.1 is equipped with a shutoff valve and the test connection is located in the upper story at the most remote sprinkler pipe, 8.16.4.3.
136. ____ Compressor capacity specification sheet is provided, restores system within 30 minutes, 7.2.6.
137. ____ Compressor piping system, air fill line not less than ½ in., and check-relief-shutoff valves are shown or noted, 7.2.6.3.
138. ____ Shown is the location for the quick opening device (QOD) for systems greater than 500 gallons, see exception in 7.2.4.
139. ____ Shown is the location of the check valve for QOD and the antiflooding device between the riser and the QOD, 7.2.4.6.

Preaction or Deluge:

140. ____ System capacity is provided, 14.1.3(17).
141. ____ Pressure gauge locations are above and below the preaction valve and on the air supply, 7.3.1.3.
142. ____ Location and spacing of the detection devices are detailed, 7.3.1.6.
143. ____ The preaction system is limited to 1,000 sprinklers, refer to the exception, 7.3.2.1., 750 gal. limit per valve.
144. ____ Only upright, listed dry sprinklers are used, see exceptions for return bends and sidewall sprinklers, 7.3.2.4.
145. ____ Double interlock systems are not gridded, 7.3.2.5, and valve room is heated, 7.3.1.8.

Combined Dry Pipe and Preaction:

146. ____ System capacity is provided, 14.1.3(17).
147. ____ Dry pipe riser location is shown.
148. ____ Two 6 in. dry pipe valves are provided for systems greater than 600 sprinklers or greater than 275 sprinklers in a fire area, 7.4.2.1.
149. ____ Multi-dry pipe valves are interconnected with 1 in. pipe with shut-off valve for simultaneous tripping, 7.4.2.4.
150. ____ QOD is provided at the dry pipe valves, 7.4.2.8.
151. ____ A minimum 2 in. exhaust valve is shown at the end of the common feed main, 7.4.3.1.
152. ____ Fire areas requiring greater than 275 sprinklers shall divide the system into sections of 275 sprinklers or less by the use of check valves, and a building with multi-fire areas shall limit 600 sprinklers per check valve, 7.4.4.2.
153. ____ The manual method of activating the detection system is within 200 ft. of travel, 7.4.1.3.

Valves:

154. ____ All water supply control valves, pressure switches and water flow switches are electrically supervised in accordance with IFC 903.4.
155. ____ Check valve is at/near connection to water supply, 8.15.1.
156. ____ Control valves are provided in accordance with 8.15.1.1.4.
157. ____ Water supply exceeding 175 PSI requires pressure reducing valves (PRVs), locations are detailed, 8.15.1.2.
158. ____ Gauges are provided on the inlet and outlet of the PRVs and an indicating valve is provided on the inlet side, 8.15.1.2.

Miscellaneous Storage:

159. ____ Class I-IV commodities, Group A plastics, and tires stored up to 12 ft. are protected as miscellaneous storage in accordance with Section 12.1.10.
160. ____ Rolled paper stored up to 10 ft. and idle pallets stored up to 4 ft. are protected as miscellaneous storage in accordance with Section 12.1.10.
161. ____ Hose stream demand has been added to the hydraulic calculation in accordance with 12.1.10.2.
162. ____ In-rack sprinkler location, operating pressure, and the design water flow demand is in compliance with 12.1.12
163. ____ Rack storage of Class I-IV commodities do not require hose connections, 12.3.1.3.
164. ____ The matching of the density design to the appropriate sprinkler K-factor is in compliance with 12.1.13.

Fire Plan Review and Inspection Guidelines

Flushing:

165. _____ Flushing instructions and criteria are on the plans. Flushing requirements shall be 880 GPM for 6 in. pipe, 1,560 GPM for 8 in., 2,440 GPM for 10 in., 3,520 GPM for 12 in., and the flush should be pitoted and calculated to ensure the flow and the velocity is at least 10 ft/sec.

Antifreeze System:	Refer to 7.5	**Protection against Exposure Fire:**	Refer to 7.7
Refrigerated Areas:	Refer to 7.8	**Commercial Cooking Equipment:**	Refer to 7.9
Storage:	Refer to Chapter 12	**Special Occupancy Requirements:**	Refer to Chapter 13

Private Fire Service Water Mains: Refer to Chapter 10, NFPA 24, and the Plan Review Worksheet contained in this book.

Additional Comments:

Review Date: _____   Approved or Disapproved   FD Reviewer: _____

Review Date: _____   Approved or Disapproved   FD Reviewer: _____

Review Date: _____   Approved or Disapproved   FD Reviewer: _____

Fire Plan Review and Inspection Guidelines

NFPA 13 Sprinkler System Acceptance Inspection
2006 IFC and 2002 NFPA 13

Date of Inspection: _____ Permit Number: _____

Business/Building Name: _____ Address of Project: _____

Contractor: _____ Contractor's Phone: _____

Reference numbers following worksheet statements represent an NFPA code section unless otherwise specified.

Pass | Fail | NA

1. ____ | ____ | ____ Approved drawing and above-ground piping certification documents are on site.
2. ____ | ____ | ____ Underground supply testing and flushing is witnessed and underground piping certification is provided. Flushing requirements shall be 880 GPM for 6 in., 1,560 GPM for 8 in., 2,440 GPM for 10 in., 3,520 for 12 in., have them pitot and calculate that flow and confirm the velocity is at least 10 ft/sec.
3. ____ | ____ | ____ Hydrostatic test: wet system, 200 PSI for 2 hours and it should include the FDC piping.
4. ____ | ____ | ____ Hydrostatic test: dry and double interlock system: 200 PSI for 2 hours and a 40 PSI air leak test for 24 hours with less than 1.5 PSI loss, 16.2.2.
5. ____ | ____ | ____ Backflow prevention device is installed in accordance with the approved set of plans and forward flow tested, 16.2.5.
6. ____ | ____ | ____ Systems subject to pressures greater than 150 PSI shall be hydrostatically tested at 50 PSI above system working pressure, 16.2.1.2.
7. ____ | ____ | ____ Operational test of the dry-pipe valve is performed and the quick opening device (500+ gallon systems) is tested, 750+ gallon system must trip within 60 seconds.
8. ____ | ____ | ____ PRVs are tested at maximum and normal inlet pressures or as specified be the manufacturer, the supply pressure is recorded on the certificate, a relief valve is on the discharge side and gauges on each side of the valve, 16.2.4.

Riser Room

9. ____ | ____ | ____ The main drain is routed to the exterior with a turned down elbow or an inside drain cabaple of handling the water flow. A flow test is performed. The main drain pipe is ¾ in. or greater for a riser up to 2 in., 1¼ in. or greater for a riser 2½ in. to 3½ in., 2 in. for a riser 4 in.or greater, 8.15.2.4, 16.2.3.4.
10. ____ | ____ | ____ Water control valves and flow switches are electronically supervised and tested, IFC 903.4 there are 7 exceptions: 13D systems, limited area systems, 13R systems where supply is common to the sprinkler and the domestic system.
11. ____ | ____ | ____ Paddle type water flow is not allowed for dry, preaction or deluge systems.
12. ____ | ____ | ____ 24 hour monitoring service agency or remote supervising station or proprietary supervising station received signals, IFC 903.4.1.
13. ____ | ____ | ____ Water flow alarm is tested and initiates an alarm within 5 minutes, located in accordance with the approved set of plans, and it is properly signed, 16.2.3.1.
14. ____ | ____ | ____ High-rise: each floor system shall have water flow device with a test connection and be connected to the fire alarm system.
15. ____ | ____ | ____ Permanent system identification signs for each control valve and what portion of the building each valve serves is provided, 6.7.4.
16. ____ | ____ | ____ Permanent hydraulic nameplate is attached to the riser, 16.5.1.
17. ____ | ____ | ____ Riser in a multistory structure is supported at the lowest level, each alternate level, above and below offsets, and at the top, 9.2.5.3.
18. ____ | ____ | ____ If flexible couplings are used, supports above the lowest level are designed in accordance with the approved plans to prevent an upward thrust of the piping, 9.5.3.2.
19. ____ | ____ | ____ Gauges are above and below riser check valve, 7.1.1.2.

FDC

20. ____ | ____ | ____ FDC capped and permanently signed with system type, the required pressure to support the system if the pressure demand is equal to or greater than 150 PSI, and area or building served, 8.16.2.4.7.
21. ____ | ____ | ____ FDC has check valve and drip valve, 8.16.2.5.

Fire Plan Review and Inspection Guidelines

22. ____|____|____ FDC for wet single riser system connects to the system side, 8.16.2.4.
23. ____|____|____ FDC for wet multi-riser system connects after the main system shutoff valve, 8.16.2.4.
24. ____|____|____ FDC for dry system connects between the indicating and dry-pipe valves.
25. ____|____|____ FDC pipe complies with the size indicated on the plans, FDC is 18 in. to 48 in. above grade and properly supported, 8.16.2.

Sprinklers

26. ____|____|____ Spare sprinklers – Provide at least 6 spare sprinklers for systems designed with 300 or less sprinklers; 12 spare sprinklers for system designed using 300 to 1000 sprinklers, and 24 spare sprinklers for systems designed using more than 1000 sprinklers, 6.2.9.
27. ____|____|____ Replacement wrench(s) are provided, 6.2.9.
28. ____|____|____ Sprinklers shall be a minimum of 4 inches from the wall and be properly spaced, 8.6.3.3.
29. ____|____|____ Sprinkler is equipped with a guard if it subject to damage.
30. ____|____|____ Sprinklers are not painted or covered.
31. ____|____|____ ESFR upright deflectors are a minimum 7 in. above the top of the pipe, 8.12.5.3.2.1.
32. ____|____|____ EFSR sprinklers are at least 1 ft. horizontally from the bottom edge of bar joist or open truss and at least 3 ft. above the top of the storage level, 8.12.6.
33. ____|____|____ The proper type and temperature sprinklers are used and match plans.
34. ____|____|____ Escutcheon plates are installed.
35. ____|____|____ Sprinklers are not obstructed, 8.5.5-8.12.5.

Pipe: Hangers, Seismic, and Penetrations

36. ____|____|____ Piping layout and pipe size are the same as the plans.
37. ____|____|____ Pipe penetrations have proper clearance 2 in. for pipe 1in. to 3½ in., 4 in. for pipe 4 in. and larger, 9.3.
38. ____|____|____ When flexible couplings are used in risers, above and below floor penetrations of multi-story structures, near penetrations of concrete or masonry walls, and near expansion joints, their location is in accordance with Section 9.3.2.3 (1)-(4).
39. ____|____|____ Minimum clearance around pipes penetrating construction element listed in 9.3.4.1 is in accordance with 9.3.4.2 unless the requirements of 9.3.4.3 – 9.3.4.5 are met.
40. ____|____|____ A seismic separation assembly is provided at building seismic joints, 9.3.3.
41. ____|____|____ Lateral sway bracing spacing is in compliance with the approved set of plans, 9.3.5.3.3.
42. ____|____|____ Lateral sway bracing is within 20 ft. of the end of the pipe, 9.3.5.3.2.
43. ____|____|____ Spacing does not exceed 80 ft. for longitudinal sway bracing, which is required for feed and cross mains, and the last brace is within 40 ft. of the end of the pipe, 9.3.5.4.1 and 9.3.5.4.3.
44. ____|____|____ A four-way sway brace spacing is not greater than 25 ft. and a four way brace is located at the top of the riser if the top of the riser exceeds 3 ft. in length, 9.3.5.5.
45. ____|____|____ Longitudinal and lateral bracing is provided for each run of pipe between the change of pipe direction unless the pipe run is less than 12 ft., 9.3.5.11.
46. ____|____|____ Sprigs greater than 4 ft. are restrained from lateral movement, 9.3.6.5.
47. ____|____|____ Splayed seismic bracing wire, wrap-around u-hooks, or lateral sway bracing shall not exceed 30 ft. spacing and are used to restrict sprinkler movement that could impact the building, equipment or finishing materials, 9.3.6.4.
48. ____|____|____ Restraining straps are on all C-clamps and the strap is bolted through if there is not a lip on the beam, 9.3.7.1.
49. ____|____|____ Branch lines have one hanger per section of pipe, 9.2.3.2.
50. ____|____|____ Mains and cross mains have one hanger between each branch line and at the end of the main.
51. ____|____|____ The maximum distance between the end sprinkler and hanger is 36 in. for 1in. pipe, 48 in. for 1 ¼ in., and 60 in. for 1½ in. pipe and greater, 9.2.4.
52. ____|____|____ Risers in multistory buildings have supports at the lowest level, at each alternate level, below offsets, and at the top, 9.2.5.3.
53. ____|____|____ Risers in vertical shafts or buildings with ceiling greater than 25 ft. have support for each pipe section.
54. ____|____|____ Hangers are not within 3 in. of upright sprinklers, 9.2.3.3.

Dry and Preaction Systems

55. ____|____|____ Dry system compressor, fill line, pressure gauges, check valve and shutoff valve and relief valve are installed in accordance with 7.2.6.2. The system fills the system within 30 minutes, 7.2.6.2.2.

Fire Plan Review and Inspection Guidelines

56. ___ |___ |___ Preaction and deluge systems are tripped by activation of the detection system.
57. ___ |___ |___ Riser room is heated, 7.2.5.
58. ___ |___ |___ Air pressure is set according to the manufacture instruction document or at least 20 PSI above the trip pressure, 7.2.6.7.1.
59. ___ |___ |___ Preaction systems exceeding 20 sprinklers shall be supervised in accordance with 7.3.2.3.1.
60. ___ |___ |___ Noninterlock and double interlock preaction systems supervise pipe pressure to maintain a minimum internal pressure of 7 PSI, 7.3.2.3.2.

Additional Comments:

Inspection Date: _____   Approved or Disapproved   FD Inspector: _____

Inspection Date: _____   Approved or Disapproved   FD Inspector: _____

Inspection Date: _____   Approved or Disapproved   FD Inspector: _____

Fire Plan Review and Inspection Guidelines

NFPA 13 Sprinkler Installation Certification

Permit #: _____   Date: _____

	Property Protected	System Installer	System Supplier
Business Name:	_____	_____	_____
Address:	_____	_____	_____
	_____	_____	_____
Representative:	_____	_____	_____
Telephone:	_____	_____	_____

Location of Plans: _____

Location of Owner's Manual: _____

1. <u>Certification of System Installation:</u> Complete this section after system is installed, but prior to conducting operational acceptance tests.

   This system installation was inspected and found to comply with the installation requirements of:
   - _____ NFPA 13
   - _____ IFC and IBC
   - _____ Manufacturer's Instructions
   - _____ Other (specify; FM, UL, etc) _____

   Print Name: _____

   Signed: _____ Date: _____

   Organization: _____

2. <u>Certification of System Operation:</u> All operational features and functions of this system were tested and found to be operating properly in accordance with the requirements of:
   - _____ NFPA 13
   - _____ IFC and IBC
   - _____ Manufacturer's Instructions
   - _____ Other (specify) _____

   Print Name: _____

   Signed: _____ Date: _____

   Organization: _____

Fire Plan Review and Inspection Guidelines

NFPA 13 Sprinkler System
General Design Plan Review Worksheet
2006 IFC and 2007 NFPA 13

This worksheet is for jurisdictions that permit the use of the 2007 NFPA 13 in lieu of IFC's referenced 2002 NFPA 13.

Date of Review: _____ Permit Number: _____

Business/Building Name: _____ Address of Project: _____

Designer Name: _____ Designer's Phone: _____

Contractor: _____ Contractor's Phone: _____

No. of Sprinklers: _____ Occupancy Classification: _____

Reference numbers following worksheet statements represent an NFPA code section unless otherwise specified.

Worksheet Legend: ✓ or **OK** = acceptable   **N** = need to provide   **NA** = not applicable

1. ____ A minimum of three sets of drawings are provided.
2. ____ Equipment is listed for intended use and compatible with the system; specification data sheets are provided.

Drawings shall detail the following (22.1.3.1-22.1.3.46):
General:

3. ____ Type of system is noted; __ hydraulic calc, __ pipe schedule, __ wet, __ dry, __preaction, __ deluge, __ antifreeze. The plans declare the design standard is the 2007 edition year of NFPA 13.
4. ____ Scale: a common scale shall be used and plan information shall be legible.
5. ____ Plot plan details illustrate the fire protection water supply piping and pipe diameter supplying the building.
6. ____ The location of smoke or fire partitions, fire walls and building elevation views.
7. ____ Occupancy class and or use of each room or area, 5.1.1.
8. ____ Full height cross sectionals and include ceiling construction as needed for clarification.
9. ____ Total area protected by each system for each floor is provided.
10. ____ Dimensions for system piping, sprinkler spacing and branch line spacing, and elevation changes.
11. ____ Equipment symbol legend and the compass point are provided.
12. ____ Area limitations for hazard classification; 52.000 sq. ft. for light and ordinary hazard, 25,000 sq. ft. for extra hazard pipe schedule, 40,000 sq. ft. for extra hazard-hydraulic calculations, and 40,000 for high-piled storage, 8.2.1.
13. ____ Hydrant flow test determining water supply capacity at 20 PSI residual pressure is provided.
14. ____ Hydraulic calculations are provided with summary, detail worksheets, and graph sheet, except for permissible pipe schedule systems, 22.3.
15. ____ Dry pipe system capacity in gallons is provided _____gal., not to be greater than 750 gal. unless the requirements of 7.2.3.2 or 7.2.3.3 are met, 7.2.3.
16. ____ All water supply valves and flow switches are supervised, IFC 903.4.
17. ____ Exterior flow alarm location is detailed and provided for systems exceeding 20 sprinklers, 8.17.1.1. Note: if electric, it shall be listed for outdoor use, IFC 904.3.2.
18. ____ If required, backflow prevention device pressure loss data is provided in the hydraulic calculations.

Sprinklers:

19. ____ Total number of each type of sprinkler is noted, ordinary temperature sprinklers are to be used, see other permitted temperature ratings from 8.3.2.2 to 8.3.2.5.
20. ____ If the hazard classification of the occupancy is changed, the temperature of rating of sprinklers shall be evaluated in accordance with Section 8.3.2.6.
21. ____ Light hazard occupancies shall have quick-response sprinklers unless residential sprinklers are required in accordance with, IFC 903.3.2 and NFPA 13: 8.3.3.1.
22. ____ Sprinkler locations are correct, ceiling and roof cross sectional views are provided for clarification, 22.1.3(45).
23. ____ For each type of sprinkler the K factor, temperature rating, and orifice size are provided, 22.1.3(12).
24. ____ Each sprinkler coverage area is within its area of protection limitations or its listing, 8.6.2.2, Table 8.6.2.2.1 (a-c).

Fire Plan Review and Inspection Guidelines

25. _____ Specialty sprinklers, extra coverage, early suppression fast response, large drop, sidewall, etc. comply with the standard and listing limitations, 6.1.1 and 8.4.1- 8.4.10.
26. _____ Maximum perpendicular distance to the walls is not greater than 1/2 of allowable distance between sprinklers, 8.6.3.2 and Tables 8.6.2.2.1(a through d), for sidewall sprinklers, 8.7.3.2 and Table 8.7.2.2.1. For irregular shaped or angled areas the sprinkler placement is in accordance with 8.6.3.2.3.
27. _____ Standard sprinkler spacing from vertical obstructions complies with Table 8.6.5.1.2 and for floor mounted obstructions, Table 8.6.5.2.2, 8.6.5.1.2 and 8.6.5.2.2.
28. _____ Sidewalls sprinkler spacing for a front obstruction refer to Table 8.7.5.1.3, for a side obstruction refer to Table 8.7.5.1.4, and for a floor mounted obstruction refer to Table 8.7.5.2.2.
29. _____ Extended coverage uprights and pendent spacing for ceiling or wall obstructions refer to Table 8.8.5.1.2 and for floor mounted obstructions refer to Table 8.8.5.2.2
30. _____ Extended coverage sidewall spacing for front obstructions refer to Table 8.9.5.1.3 and for floor mounted obstructions, Table 8.9.5.2.2.
31. _____ Residential upright and pendent sprinkler spacing from vertical obstructions complies with Table 8.10.6.1.2 and for floor mounted obstructions, Table 8.10.6.2.2.
32. _____ Residential sidewall sprinkler spacing from ceiling or hanging obstructions complies with Table 8.10.7.1.3 and for floor mounted obstructions, Table 8.10.7.2.2.
33. _____ Sprinkler coverage is provided under obstructions greater than 4 ft. wide, 8.5.5.3.1.
34. _____ Baffles are designed and provided for sprinklers less than 6 ft. apart in accordance with Section 8.6.3.4.2.
35. _____ Pilot line detector system design is in accordance with Section 8.14.
36. _____ Locations or conditions requiring special consideration, 8.15.
37. _____ A. concealed spaces, for the 15 omissions see 8.15.1.2.
38. _____ B. vertical shafts, 8.15.2.
39. _____ C. stairways, 8.15.3.
40. _____ D. vertical openings, 8.15.4.
41. _____ E. elevator hoistways and machine rooms, 8.15.5.
42. _____ F. spaces under ground floors, exterior docks, and platforms, 8.15.6.
43. _____ G. exterior roof and canopy, 8.15.7.
44. _____ H. dwelling unit, 8.15.8.
45. _____ I. library stack or medical record storage room, 8.15.9.
46. _____ J. electrical equipment, 8.15.10.
47. _____ K. duct protection, 8.15.12
48. _____ L. ceilings: open-grid, drop-out, 8.15.13 and 8.15.14.
49. _____ M. stages, 8.15.16.
50. _____ Sprinkler placement for the protection of a vertical shaft is in accordance with 8.15.2.1.
51. _____ Vertical shaft with combustible surfaces is protected in accordance with 8.15.2.2.
52. _____ Sprinklers are provided beneath combustible stairs, 8.15.3.1.
53. _____ Sprinklers are provided for stairways in accordance with 8.15.3. Refer to 8.15.3.2 for when there is storage use under the stair landing and 8.15.3.2.4 when a noncombustible construction exterior stair tower is 50 percent open.
54. _____ Closely spaced sprinklers with draft stops are provided around unenclosed floor openings except large openings like found in malls or atriums, and openings between floors of a common dwelling unit, 8.15.4.1 and 8.15.4.2.
55. _____ Elevator shaft has a sprinkler within 2 ft. of the shaft floor unless the shaft is noncombustible and there are no hydraulic fluids, 8.15.5.
56. _____ Ordinary or intermediate temperature sprinklers are in the elevator machine room or at the top of the elevator shaft, refer to exceptions, 8.15.5.1–8.15.5.5.
57. _____ Sprinklers are provided under combustible ground floor, exterior dock, and platforms, 8.15.6.
58. _____ Sprinklers are provided under roofs and canopies unless constructed of noncombustible or limited combustible materials, less than 4 ft. wide, and no storage, refer to exceptions 8.15.7.1 – 8.15.7.4.
59. _____ Sprinklers are not required in noncombustible dwelling unit bathrooms, less than 55 sq. ft. or limited combustible with a 15 minute thermal barrier, except in nursing homes, 8.15.8.1.
60. _____ Sprinklers are not required in hotel or motel dwelling unit clothes closet, pantries, or linen closets provided the closet area and its least dimension complies with 8.15.8.2.
61. _____ Sprinklers are provided in every aisle and at every tier stack, distance is not more than 12 ft. in library stack rooms, 8.15.9.

62. _____ Sprinklers are provided in electrical equipment rooms, exception: the room is dedicated use, has dry type equipment, 2 hour equipment enclosures, and no combustible storage, 8.14.10. Also consult the exceptions pertaining to spaces containing telecommunication equipment and associated power supplies as specified in IFC Section 903.2., 8.15.10.
63. _____ When required, ducts are protected in accordance with 8.15.12.1. Method of access for each sprinkler is detailed.
64. _____ Open grid ceilings shall not be installed under sprinklers, unless the grid opening and sprinkler placement criteria of Section 8.15.13 are met.
65. _____ Drop-out ceilings are installed under sprinklers in accordance with their listing, and sprinklers are not located below the ceilings, 8.15.14.
66. _____ Sprinklers for stages shall be provided in accordance with Section 8.15.16.
67. _____ Proscenium openings for stages shall be protected in accordance with Section 8.17.5.2.

Pipe Support and Hangers:
68. _____ Type and locations of hangers, sleeves, and braces are shown, 12.1.3(22). Nonlisted hangers shall meet 5 performance criterion and the design shall be sealed by a registered professional engineer, 9.1.1.2.
69. _____ If trapeze hangers are used, the locations are shown, a legend provides the span, size of pipe supported, angle and pipe used, and section modulus in accordance with Section 9.1.1.6.
70. _____ Pipe hanger spacing is in accordance with Table 9.2.2.1(a).
71. _____ Lightwall steel pipe hanger spacing is in accordance with Table 9.2.2.1(a).
72. _____ Branch lines show one hanger for each section of pipe, exceptions are listed, 9.2.3.2.
73. _____ Cross mains show one hanger between each branch lines or in compliance with Table 9.2.2.1(a), and for additional spacing variations refer to Section 9.2.4.
74. _____ Supports can be on the horizontal pipe section if within 24 in. of the vertical pipe centerline, 9.2.5.1.
75. _____ Risers in multi-story buildings show supports at the lowest level, each alternate level, below offsets, and at the top, 9.2.5.4.
76. _____ The distance between supports for a riser does not exceed the limit specified., 9.2.5.5.

Pipe and Valves:
77. _____ Main drain location and pipe diameter are detailed and complies with Section 8.16.2.4.
78. _____ Main drain routing is to the exterior or to an interior drain but ensure that the drain capacity is adequate, 8.16.2.4.4.
79. _____ Auxiliary drain location is detailed and its size is in accordance with Section 8.16.2.5.
80. _____ When required, the location of the listed backflow prevention device (can serve as a check valve) is detailed, 8.16.1.1.3.2.
81. _____ A listed control valve is provided on each side of the check valve, 8.16.1.1.4.1. Only one control valve on the system side of the check valve is necessary when the water supply is provided from the city connection, 8.16.1.1.4.3.
82. _____ The control valve locations are accessible, 8.16.1.1.7.
83. _____ If a pressure reducing valve is used, its location and installation criteria are detailed in accordance with Section 8.16.1.2.
84. _____ If used, outside post-indicator control valve (PIV) locations and installation criteria are detailed in accordance with Section 8.16.1.3.
85. _____ If PIVs are approved to be located in a pit, the pit construction, location, and marking are designed and detailed in accordance with Section 8.16.1.4.2.

Seismic Bracing:
86. _____ Flexible couplings may be used for pipe 2½ in. or larger in accordance with Sections 9.3.2.2 and 9.3.2.3.
87. _____ Flexible couplings are specified for drops to hose lines, rack sprinklers, and mezzanines, 9.3.2.4.
88. _____ A seismic separation assembly is provided and detailed at building seismic joints, 9.3.3.2 and 9.3.3.3.
89. _____ Proper pipe clearance is noted on the plans for pipe penetrations in walls, floors, platforms or foundations, 9.3.4. Minimum clearance is in accordance with Section 9.3.4.2 – 9.3.4.7.
90. _____ Lateral sway bracing is required at a maximum spacing of 40 ft. for all feed mains, cross mains, and branch lines 2½ in. and larger, 9.3.5.3.1.
91. _____ Lateral sway bracing is designed not to exceed the maximum zone of influence loading provided in Tables 9.3.5.3.2(a) and (b) for its spacing, 9.3.5.3.2.
92. _____ Bracing is provided for the last length of pipe of the end of a feed or cross main, 9.3.5.3.5.
93. _____ Bracing is required unless all the pipe is supported by rods less than 6 in. or by 30° wrap-around U-hooks for any size pipe, 9.3.5.3.8.

94. _____ Longitudinal sway bracing has a maximum span of 80 ft. for mains and cross mains and within 40 ft. of the end of the line, 9.3.5.4.1 and .3.
95. _____ A four-way sway brace spacing on a riser does not exceed 25 ft. and a four-way sway brace is located at the top of the riser if the top of the riser exceeds 3 ft. in length, 9.3.5.5.
96. _____ Seismic bracing calculations and the zones of influence are detailed and provided for each brace to be used as shown in NFPA Figure A.9.3.5.6(e), 9.3.5.6 through 9.3.5.8. The calculations shall include the basis for the selection of the seismic coefficient from Table 9.3.5.6.2.
97. _____ Longitudinal and lateral bracing is provided for each run of pipe between the changes of direction unless the run is less than 12 ft. and supported by adjacent pipe run bracing, 9.3.5.11.2.
98. _____ Branch lines are restrained at the end sprinkler of each line and restrained against vertical and lateral movement, 9.3.6.3.
99. _____ Branch line method of restraint is in accordance with Section 9.3.6.1.
100. _____ Restraints for branch lines shall be at intervals not greater than specified in Table 9.3.6.4 and justification for selection of the seismic coefficient is provided, 9.3.6.4.
101. _____ Detailed are restraints for sprigs 4 ft. long or greater against lateral movement, 9.3.6.6.

Fire Department Connection (FDC):
102. _____ The FDC location is detailed on the street side or response side of building or as approved by the fire official, and when connected to the water supply it will not obstruct emergency vehicle access to the building, IFC 912.
103. _____ Local water flow alarm is provided when the sprinkler system exceeds 20 sprinklers and its location is detailed, 8.17.1.1.
104. _____ FDCs for fire engine or fire boat are sized and arranged in accordance with A.8.17.2, 8.17.2.3, and 8.17.2.4.
105. _____ The arrangement of the FDC piping supplying wet pipe, dry pipe, preaction or deluge sprinklers shall be in accordance with Section 8.16.2.4.2.

Hydraulic Calculations, 22.3:
106. _____ Specify the calculation method used, density/area or room design, 22.3.
107. _____ The summary sheet, water supply graph sheet, supply analysis, node analysis, and worksheets are provided for computer generated calculations, 22.3.5. The summary sheet, water supply graph sheet, and work sheets are provided for hand calculations.
108. _____ Reference points in the calculation worksheet match with points on the plans, and the occupancy hazard classifications are correct for the occupancy or use, 22.3, 11.2.1.2.3.
109. _____ If design area adjustments are made, the adjustment methodology is provided, 22.32, 22.3.5.2.
110. _____ The use of quick response sprinklers in a design area shall meet the specific requirements in Section 11.2.3.2.3.
111. _____ Pipe size and length references in the calculation worksheet match the plans.
112. _____ Sloped ceiling may require a 30 percent increase of design area, 11.2.3.2.4.
113. _____ Sprinkler data shset matches information on the plans.
114. _____ Water flow information is provided with static PSI, residual PSI, and available GPM at 20 PSI residual with graphed results.
115. _____ Density and design areas information are provided and comply with the restrictions listed in Section 11.2.3.1.4, Fig 11.2.3.1.1.
116. _____ Calculations are correct: static PSI, pipe length, GPM, K factors for drops or branch lines, elevation data, hose allowance, friction loss, and equivalent pipe and fitting lengths, 22.3.
117. _____ For the room design method the design area includes the most demanding room and if any, adjacent communication compartments, 11.2.3.3, 22.4.4.1.2.
118. _____ A minimum of 2 summary calculations are provided for a grid system, refer to the one exception, 22.4.4.4.2.
119. _____ Additional calculations may be required by the fire code official if the building design and room uses do not make the most demanding area obvious.
120. _____ Legend for calculation abbreviations is provided.
121. _____ Calculations are also provided for extra hazard occupancies, deluge, and exposure systems.
122. _____ Dry pipe and double interlock preaction design areas are increased 30 percent but the density remains the same (11.2.3.2.5), use of high-temp sprinklers in extra hazard occupancies may reduce design area by 25 percent but not less than the area specified in 11.2.3.2.6.

Residential Sprinklers in a 13 System:
123. _____ Calculations for a single and for a multiple sprinkler discharge are provided, 11.3.1.1.

124. ____ The calculation design is based on the number of sprinklers and at the flow specified in 11.3.1.2.
125. ____ Hose streams and water duration requirements are based on a light-hazard occupancy classification in accordance with Table 11.2.3.1.2, 11.3.1.5.

Special Design:
126. ____ Special design considerations for exposure protection, water curtain, and dry system are in accordance with 11.2.3.7–11.2.3.9, 22.7.

Pipe Schedule:
Note: For systems less than 5000 sq. ft. the minimum water flow is proven to be available in accordance with Table 11.2.2.1. Systems less than 5,000 sq. ft. shall have 50 PSI residual pressure and meet the requirements of Table 11.2.2.1.

127. ____ Only ½ in. orifice sprinklers (nominal K-factor of 5.3 to 5.8) shall be used, 22.5.1.2.
128. ____ Light Hazard: 8 sprinklers maximum for each branch line, 22.5.2.1.1, 9 and 10 permitted see 22.5.2.
129. ____ A. pipe diameter, pipe material and number of sprinklers are in accordance with Table 22.5.2.2.1.
130. ____ B. sprinklers above and below the ceiling are in accordance with Table 22.5.2.4.
131. ____ Ordinary Hazard: 8 sprinklers maximum for each branch line, 22.5.3.1, 9 and 10 permitted see 22.5.3.
132. ____ A. pipe size, pipe material and number of sprinklers are in accordance with Table 22.5.3.4.
133. ____ B. sprinklers greater than 12 ft. separations are in accordance with Table 22.5.3.5.
134. ____ C. sprinklers above and below the ceiling are in accordance with Table 22.5.3.7.
135. ____ Extra Hazard: not allowed, 22.5.4.

Wet System:
136. ____ Relief valve not less than ¼ in. is detailed for gridded system, 7.1.2.1.
137. ____ An alarm test connection location for the waterflow alarm is provided and in compliance with 8.17.4.2.

Dry System, 7.2:
138. ____ Only upright, listed dry sprinklers are used, see exceptions for return bends and sidewall sprinklers, 7.2.2.
139. ____ System capacity is provided and a quick opening device is provided when required by 7.2.3.2.
140. ____ System is designed to meet the water delivery times for the hazard classification in accordance with Table 7.2.3.6.1 and calculations are provided, 7.2.3.6.
141. ____ A trip test connection sized according to 8.16.4.3.1 is equipped with a shutoff valve and the test connection is located in the upper story at the most remote sprinkler, 8.17.4.3.
142. ____ Compressor capacity specification sheet is provided, restores system within 30 minutes, 7.2.6.2.2.
143. ____ Compressor piping system, air fill line not less than ½ in., and check-relief-shutoff valves are shown or noted, 7.2.6.3.
144. ____ Shown is the location for the quick opening device (QOD) for systems greater than 500 gallons, see exception in 7.2.34.3.
145. ____ Shown is the location of the check valve for QOD and the antiflooding device between the riser and the QOD, 7.2.4.5, 7.2.4.8.

Preaction or Deluge:
146. ____ System capacity is provided, 14.1.3(17).
147. ____ Pressure gauge locations are above and below the preaction valve and on the air supply, 7.3.1.3.
148. ____ Location and spacing of the detection devices are detailed, 7.3.1.7.
149. ____ The single and non-interlock preaction system is limited to 1,000 sprinklers, 7.3.2.2., 750 gallon limit for each valve.
150. ____ The double-interlock preaction system is based on water delivery of not exceeding 60 seconds but the water delivery time is also based on Table 7.2.3.6.1, 7.3.2.2.
151. ____ Preaction system is supervised in accordance with 7.3.2.4.
152. ____ Only upright, listed dry sprinklers are used, see exceptions for return bends and sidewall sprinklers, 7.3.2.5.
153. ____ Double interlock systems shall not be gridded, 7.3.2.6, and valve room is heated, 7.3.1.8.2.

Combined Dry Pipe and Preaction:
154. ____ System capacity is provided, 14.1.3(17).
155. ____ Dry pipe riser location is shown.
156. ____ Two 6 in. dry pipe valves are provided for systems greater than 600 sprinklers or greater than 275 sprinklers in a fire area, 7.4.3.1.
157. ____ Multidry pipe valves are interconnected with 1 in. pipe with shutoff valve for simultaneous tripping, 7.4.3.4.
158. ____ QOD is provided at the dry pipe valves, 7.4.3.8.
159. ____ A minimum 2 in. exhaust valve is shown at the end of the common feed main, 7.4.4.1.

160. ____ Fire areas requiring greater than 275 sprinklers shall divide the system into sections of 275 sprinklers or less by the use of check valves, and a building with multifire areas shall limit 600 sprinklers per check valve, 7.4.5.
161. ____ The manual method of activating the detection system is within 200 ft. of travel, 7.4.2.3.
162. ____ Only upright, listed dry sprinklers are used, see exceptions for return bends and sidewall sprinklers, 7.4.2.5.

Valves:
163. ____ A check valve is at/near connection to water supply, 8.16.1, 8.16.1.1.3.5.
164. ____ All water supply control valves and water flow switches are electrically supervised in accordance with, IFC 903.4.
165. ____ Control valves are provided in accordance with 8.16.1.1.4.
166. ____ Water supply exceeding 175 PSI requires pressure reducing valves (PRVs), locations are detailed, 8.16.1.2.
167. ____ Gauges are on the inlet and outlet sides of PRVs and an indicating valve on the inlet side, 8.16.1.2.

General Storage Requirements:
168. ____ Ceiling slope is detailed, cross sectional view provided, and does not exceed a 2 in 12 slope, 12.1.2.
169. ____ Storage design requirements are based on the absence of draft curtains and roof vents, 12.1.
170. ____ If the building has two or more storage hazard areas nonseparated and an extended design area is provided in accordance with 12.3.
171. ____ Dry pipe and preaction system design areas are increased 30 percent but not to exceed the area specified in 12.5.2, 12.5.1.
172. ____ When adjustments are made to the design area, the designer provided a calculations explaining and showing how the adjustments were made, 12.7.7.
173. ____ Design for idle pallets is in accordance with 12.12.
174. ____ Densities up to 0.20 GPM/ft^2 are sprinklers with the minimum K-factor specified in 12.6.1.
175. ____ Densities 0.21 GPM/ft^2 to 0.34 GPM/ft^2 protecting rack storage, tire, roll paper, and baled cotton storage are protected with sprinklers with the minimum k-factor specified in 12.6.2.
176. ____ Densities greater than 0.34 GPM/ft^2 protecting rack storage, tire, roll paper, and baled cotton storage are protected with sprinklers with the minimum K-factors as specified in 12.6.3.
177. ____ If the design area is adjacent a combustible concealed space then the minimum design area is in compliance with 12.9.1 unless the concealed space meets the criteria of Sections 12.9.2 (1) – (9), 12.9.1.
178. ____ A general information sign that indicates the design capabilities and limitations of the automatic sprinkler system shall be provided at each system riser, antifreeze loop and auxiliary system control valve. The sign shall contain the required information specified in Section 24.6.2, 24.6.1

Miscellaneous Storage:
179. ____ Miscellaneous and Class I through IV and Group A plastic, tires storage up to 12 ft., and rolled paper up to 10 ft. high are designed in accordance with density curve Figure 13.2.1 and Table 13.2.1.
180. ____ Hazard classification when using the design/area method complies with Figure 13.2.1 and commodity protection complies with Table 13.2.1.
181. ____ In-rack sprinklers K-factors and the minimum design pressure complies with 13.3.2.
182. ____ In-rack sprinkler water demand is based on the number and the location of sprinklers specified in 13.3.3.

Miscellaneous:
183. ____ Flushing instructions and criteria are on the plans. Flushing requirements shall be 880 GPM for 6 in. pipe, 1,560 GPM for 8 in., 2,440 GPM for 10 in., 3,520 GPM for 12 in. pipe. The water flow should be measured to ensure the velocity is at least 10 ft/sec.

Antifreeze System:	Refer to 7.6	**Protection against Exposure Fire:**	Refer to 7.8
Refrigerated Areas:	Refer to 7.9	**Commercial Cooking Equipment:**	Refer to 7.10
Special Occupancy Requirements:	Refer to Chapter 21		

Private Fire Service Water Mains: Refer to Chapter 10, NFPA 24, and the Plan Review Worksheet contained in this book.

Additional Comments:

Review Date: _____ Approved or Disapproved FD Reviewer: _____
Review Date: _____ Approved or Disapproved FD Reviewer: _____
Review Date: _____ Approved or Disapproved FD Reviewer: _____

Fire Plan Review and Inspection Guidelines

NFPA 13 Sprinkler System Acceptance Inspection
2006 IFC and 2007 NFPA 13
This worksheet is for jurisdictions that permit the use of the 2007 NFPA 13 in lieu of IFC's referenced 2002 NFPA 13.

Date of Inspection: _____ Permit Number: _____

Business/Building Name: _____ Address of Project: _____

Contractor: _____ Contractor's Phone: _____

Reference numbers following worksheet statements represent an NFPA code section unless otherwise specified.

Pass | Fail | NA

1. ___ | ___ | ___ Approved drawing and above-ground piping certification documents are on site.
2. ___ | ___ | ___ Underground supply testing and flushing is witnessed and underground piping certification is provided. Flushing requirements shall be 880 GPM for 6 in., 1,560 GPM for 8 in., 2,440 GPM for 10 in., 3,520 for 12 in., have them pitot and calculate that flow and confirm the velocity is at least 10 ft/sec.
3. ___ | ___ | ___ Hydrostatic test: wet system, 200 PSI for 2 hours and it should include the FDC piping.
4. ___ | ___ | ___ Hydrostatic test: dry and double interlock system: 200 PSI for 2 hours and a 40 PSI air leak test for 24 hours with less than 1.5 PSI loss, 24.2.2.
5. ___ | ___ | ___ Back flow prevention device is installed in accordance with the approved set of plans and forward flow tested, 24.2.5.
6. ___ | ___ | ___ Systems subject to pressures greater than 150 PSI shall be hydrostatically tested at 50 PSI above system working pressure, 24.2.1.2.
7. ___ | ___ | ___ Operational test of the dry-pipe valve is performed and the quick opening device (500+ gallon systems) is tested, 750+ gallon system must trip within 60 seconds.
8. ___ | ___ | ___ Pressure reducing valves are tested at maximum and normal inlet pressures or as specified by the manufacturer, the supply pressure is recorded on the certificate, a relief valve is on the discharge side and gauges on each side of the valve, 24.2.4.

Riser

9. ___ | ___ | ___ The main drain is routed to the exterior with a turned down elbow or an inside drain cabaple of handling the water flow. A flow test is performed. The main drain pipe is ¾ in. or greater for a riser up to 2 in., 1¼ in. or greater for a riser 2½ in. to 3½ in., 2 in. for a riser 4 in. or greater, 8.16.2.4.2, 24.2.3.4.
10. ___ | ___ | ___ Water control valves and flow switches are electronically supervised and tested, IFC 903.4.
11. ___ | ___ | ___ Paddle type water flow is not allowed for dry, preaction or deluge systems.
12. ___ | ___ | ___ 24 hour monitoring service agency or remote supervising station or proprietary supervising station received signals, 903.4.1.
13. ___ | ___ | ___ Water flow alarm is tested and initiates an alarm within 5 minutes, located in accordance with the approved set of plans, and it is properly signed, 24.2.3.1.
14. ___ | ___ | ___ High-rise: each floor system shall have water flow device with a test connection and be connected to the fire alarm system.
15. ___ | ___ | ___ Permanent system identification signs for each control valve and what portion of the building each valve serves is provided, 6.7.4.
16. ___ | ___ | ___ A permanent hydraulic nameplate is attached to the riser, 24.5.1. A general information sign that indicates the design capabilities and limitations of the automatic sprinkler system shall be provided at each system riser, antifreeze loop and auxiliary system control valve. The sign shall contain the required information specified in Sections 24.6.2, 24.6.1
17. ___ | ___ | ___ Riser is supported by hanger or attachment, for multistory at the lowest level, each alternate level, above and below offsets, and at the top, 9.2.5.4.
18. ___ | ___ | ___ Gauges are above and below riser check valve, 7.1.1.2.

FDC

19. ___ | ___ | ___ FDC capped and permanently signed with system type, the required pressure to support the system if the pressure demand is equal to or greater than 150 PSI, and area or building served, 8.17.2.4.7.
20. ___ | ___ | ___ FDC has check valve and drip valve, 8.17.2.5.
21. ___ | ___ | ___ FDC for wet single riser system connects to the system side, 8.17.2.4.

Fire Plan Review and Inspection Guidelines

22. ____|____|____ FDC for wet multiriser system connects after the main system shut off valve, 8.17.2.4.
23. ____|____|____ FDC for dry system connects between the indicating and dry-pipe valves.
24. ____|____|____ FDC pipe complies with the size indicated on the plans, 18 in. to 48 in. above grade, and properly supported, 8.17.2, A.8.17.2.

Sprinklers
25. ____|____|____ Spare sprinklers – Provide at least 6 spare sprinklers for systems designed with 300 or less sprinklers; 12 spare sprinklers for system designed using 300 to 1000 sprinklers, and 24 spare sprinklers for systems designed using more than 1000 sprinklers, 6.2.9.
26. ____|____|____ Replacement wrench(s) are provided, 6.2.9.
27. ____|____|____ Sprinklers shall be a minimum of 4 in. from the wall and be properly spaced, 8.6.3.3.
28. ____|____|____ Sprinkler heads have a guard if subject to damage.
29. ____|____|____ Sprinkler heads are not painted or covered.
30. ____|____|____ ESFR upright deflectors are a minimum 7 in. above the top of the pipe, 8.12.5.3.2.1.
31. ____|____|____ EFSR sprinklers are at least 1 ft. horizontally from the bottom edge of bar joist or open truss and at least 3 ft. above the top of the storage level, 8.12.6.
32. ____|____|____ Proper type and temperature sprinklers are used and match plans.
33. ____|____|____ Escutcheon plates are installed.
34. ____|____|____ Sprinklers are not obstructed, 8.5.5-8.12.5.

Pipe: Hangers, Seismic, and Penetrations
35. ____|____|____ Piping layout and size are the same as on the approved set of plans.
36. ____|____|____ Pipe penetrations have proper clearance 2 in. for pipe 1 in. to 3½ in., 4 in. for pipe 4 in. and larger, 9.3.
37. ____|____|____ When flexible couplings are used in risers, above and below floor penetrations of multi-story structures, near penetrations of concrete or masonry walls, and near expansion joints, their location is in accordance with Section 9.3.2.1-.4
38. ____|____|____ Minimum clearance around pipes penetrating construction elements listed in 9.3.4.1 is in accordance with 9.3.4.2 unless the requirements of 9.3.4.3 – 9.3.4.7 are met.
39. ____|____|____ A seismic separation assembly is provided at building seismic joints, 9.3.3.
40. ____|____|____ Lateral sway bracing is provided and spaced is in accordance with the approved set of plans for all mains, cross mains, and branch lines 2½ in. and larger.
41. ____|____|____ Longitudinal sway bracing is provided and spaced is in accordance with the approved set of plans for feed mains and cross mains, 9.3.5.4.
42. ____|____|____ A 4-way sway brace is provided at least every 25 ft. and at the top of each riser, 9.3.5.5.
43. ____|____|____ Longitudinal and lateral bracing is provided for each run of pipe between the change of pipe direction unless the pipe run is less than 12 ft., 9.3.5.11.3.
44. ____|____|____ Riser nipples greater than 4 ft. are restrained from lateral movement, 9.3.6.6.
45. ____|____|____ Seismic bracing wire, wrap-around u-hooks, or lateral sway bracing shall not exceed 30 ft. spacing and are used to restrict sprinkler movement that could impact the building, equipment or finishing materials, 9.3.6.
46. ____|____|____ Restraining straps are on all C-clamps and the strap is bolted through if there is not a lip on the beam, 9.3.7.1.
47. ____|____|____ Branch lines have one hanger per section of pipe, 9.2.3.2.
48. ____|____|____ Mains and cross mains have one hanger between each branch line and at the end of the main, 9.2.4.
49. ____|____|____ The maximum distance between the end sprinkler and hanger is 36 in. for 1in. pipe, 48 in. for 1¼ in., and 60 in. for 1½ in. pipe and greater, 9.2.3.4.
50. ____|____|____ Risers in multistory buildings have supports at the lowest level, at each alternate level, below offsets, and at the top, 9.2.5.4.
51. ____|____|____ Hangers are not within 3 in. of upright sprinklers, 9.2.3.3.

Dry and Preaction Systems
52. ____|____|____ Dry system compressor, fill line, pressure gauges, check valve and shutoff valve and relief valve are installed in accordance with the approved set of plans and 7.2.6.2. The system fills the system within 30 minutes, 7.2.6.2.2.
53. ____|____|____ Preaction and deluge systems are tripped by activation of the detection system.
54. ____|____|____ Riser room is heated, 7.2.5.
55. ____|____|____ Air pressure is set according to the manufacture instruction document or at least 20 PSI above the trip pressure, 16.2.2.

Fire Plan Review and Inspection Guidelines

56. ____|____|____ Dry and preaction systems are supervised and water reaches furthest point within the time period provided on the plans or water delivery calculations in accordance with Table 7.2.3.6.1.

57. ____|____|____ Preaction systems exceeding 20 sprinklers automatically supervise (constant monitoring) pipe pressure (maintain at least 7 PSI) and detection devices, 7.3.2.4.

Additional Comments:

Inspection Date: _____ Approved or Disapproved FD Inspector: _____

Inspection Date: _____ Approved or Disapproved FD Inspector: _____

Inspection Date: _____ Approved or Disapproved FD Inspector: _____

Fire Plan Review and Inspection Guidelines

13R Residential Sprinkler System
Plan Review Worksheet
2006 IFC, 2002 NFPA 13, and 2002 NFPA 13R

Date of Review: _____ Permit Number: _____

Business/Building Name: _____ Address of Project: _____

Designer Name: _____ Designer's Phone: _____

Contractor: _____ Contractor's Phone: _____

No. of Sprinklers: _____ Occupancy Classification: _____

Reference numbers following worksheet statements represent an NFPA code section unless otherwise specified.

Worksheet Legend: ✓ or OK = acceptable   N = need to provide,   NA = not applicable

1. _____ Three sets of drawings are provided. The plans declare the design standard is the 2002 edition year of NFPA 13R.
2. _____ System components are listed for intended use and compatible with the system, and equipment data sheets are provided.

Drawings shall detail the following:
General:

3. _____ The type of system is noted: __ wet, __ dry, __ antifreeze not exceeding 40 gals., __ preaction, and type of sprinklers are noted: ___ pendent, ___ upright, ___ sidewall, 5.3.2.
4. _____ Scale: a common scale shall be used and plan information is legible, 6.1.
5. _____ Plot plan showing supply piping and pipe size from the water source to the building, 6.1.
6. _____ Building dimensions, location of partitions, and fire walls, 6.1.
7. _____ Room dimensions, labeled rooms, occupancy class of each room, 6.1.
8. _____ Full height cross elevation views and include ceiling construction, 6.1.
9. _____ Type of protection for nonmetallic pipe, 6.1.
10. _____ Dimensions for system piping, type of pipe, and component spacing, 6.1.
11. _____ Equipment symbol legend and the North orientation arrow, 6.1.
12. _____ A water flow alarm and test connection are provided, 6.4.3 and 6.6.8.
13. _____ All water supply valves and flow switches are supervised, IFC 903.4.
14. _____ Exterior flow alarm location is shown and the type identified, if electric, it is listed for outdoor use, IFC 903.4.2., and it is connected to the building fire alarm, if provided, 6.6.8.
15. _____ Backflow prevention device, when required, is shown in the pipe schematic, listed specification sheet and pressure loss data is provided, IFC 903.3.5.
16. _____ Antifreeze systems are detailed and designed in accordance with NFPA 13: 7.5.
17. _____ The system demand has at least 30 minutes of water supply, 6.5.2.
18. _____ If a fire pump is required it is designed and detailed in accordance with NFPA 20 and this book's worksheet, 6.5.4.
19. _____ Pressure gauges are provided and detailed for supply and system pressure, 6.6.5.

Sprinklers:

20. _____ Total number of each type of sprinkler is noted and the number of sprinklers per floor are noted, 6.1.
21. _____ Sprinkler location is correct, ceiling and roof sectionals are provided for clarification.
22. _____ Type of sprinklers: sprinkler K-factors, temperature rating, and orifice size, 6.1.
23. _____ Residential sprinklers are limited for use for wet pipe automatic sprinkler systems unless specifically listed for another use, 6.6.7.
24. _____ When listed quick-response sprinklers are used in dwelling units, the dwelling unit shall meet the definition of a compartment and a maximum of four sprinklers are used. The sprinkler density complies with 6.6.7.1.3.
25. _____ Sprinklers are rated for ordinary temperature (135°F-175°F) when ceiling temperature does not exceed 100°F, 6.6.7.1.5.
26. _____ Sprinklers in areas with a ceiling temperature of 101°F-150°F are equipped with intermediate temperature sprinklers (175°F-225°F), 6.6.7.1.5.

27. _____ Distance of sprinklers from heat sources complies with Table 6.6.7.1.5.3.
28. _____ Quick-response sprinklers are used when protection is on the outside a dwelling unit, 6.6.7.2.
29. _____ Each sprinkler coverage area is within its listing limitations, 6.6.7.
30. _____ Residential sprinklers without a listed coverage criteria: Sprinkler separation is a maximum of 12 ft. and a maximum of 6 ft. from the wall unless the listing states otherwise, 6.7.1.3.1.2 and 6.7.1.3.1.3.
31. _____ Residential sprinklers without a listed coverage criteria: Sprinkler separation is a minimum of 8 ft. within a compartment unless the listing states otherwise, 6.7.1.3.1.4.
32. _____ Sidewall sprinklers distance from the ceiling complies with 6.7.1.5.2.1.
33. _____ A single sprinkler at the highest ceiling level can provide coverage for closets and storage areas not exceeding 300 cu. ft. and the lowest point of the ceiling height is 5 ft., 6.7.1.5.4.
34. _____ Sprinklers are not required in noncombustible dwelling unit bathrooms where the area and the walls and ceiling meet the construction requirements of 6.8.2.
35. _____ Sprinklers are not required in dwelling unit clothes closets, pantries, or linen closets, provided the closet area, its least dimension, and its method of construction complies with 6.8.3.
36. _____ Sprinkler protection for open and attached porches, balconies, corridors, and stairs are not required, 6.8.4. If the building construction is of Type V balconies and decks require sprinkler protection in accordance with IFC Section 903.3.1.2.1.
37. _____ Sprinklers are not required for areas not used for living purposes or used for storage as listed in 6.8.5.

Pipe Support and Hangers are in Accordance with NFPA 13, 13R 6.6.6.:
38. _____ Type and locations of hangers, sleeves, braces, and methods of securing pipe are shown, 6.1.7.
39. _____ Pipe hanger spacing is in compliance with NFPA 13 Table 9.2.2.1.
40. _____ Branch lines show one hanger per section of pipe, exceptions are listed, NFPA 13 9.2.3.2.
41. _____ Mains show one hanger between each branch line unless the requirements in NFPA 13 9.2.4.2 through 9.2.4.5. are met, 9.2.4.
42. _____ Cross mains show one hanger between each two branch lines, exceptions are listed, NFPA 13 9.2.4.
43. _____ Risers in multistory buildings show supports at the lowest level, each alternate level, below offsets, and at the top, NFPA 13 9.2.5.3.
44. _____ Risers have a distance between supports not to exceed 25 ft., NFPA 13 9.2.5.4.

Drains and Test Connection:
45. _____ At least a 1 in. nominal diameter drain with a valve is detailed as being on the system side of the control valve, 6.6.2.1 and 6.6.2.2.
46. _____ Each portion of trapped dry system piping that is subject to freezing is provided a ½ in. drain, 6.6.2.4.
47. _____ The location and size of a test connection with a valve is detailed and complies with 6.6.3.1.

Pipe and Valves:
48. _____ One control valve is provided for both the domestic water and sprinkler, unless a separate control valve is provided for the sprinkler system, 6.6.1.1 and it is electronically supervised, IFC 903.4.

Seismic Bracing in Accordance with NFPA 13 Chapter 9, 13R 6.6.6:
49. _____ Flexible couplings may be used for pipe 2½ in. or larger in accordance with NFPA 13 Sections 9.3.2.2 and 9.3.2.3.
50. _____ A seismic separation assembly for piping is provided at building seismic joints, NFPA 13 9.3.3.
51. _____ Proper pipe clearance is noted on the plans for pipe penetrations in walls, floors, platforms or foundations, 9.3.4. Minimum clearance is in accordance with section NFPA 13 9.3.4.2 – 9.3.4.5.
52. _____ Lateral sway bracing is required at a maximum spacing of 40 ft. for all feed and cross mains, and branch lines 2½ in. and larger, NFPA 13 9.3.5.3.1.
53. _____ Lateral sway bracing can be spaced up to 50 ft. if the design is in compliance with NFPA 13 9.3.5.3.3.
54. _____ Lateral sway bracing is within 20 ft. of the end of the pipe, NFPA 13 9.3.5.3.2.
55. _____ A lateral sway brace is provided on the last pipe of a feed or cross main, NFPA 13 9.3.5.3.4.
56. _____ Lateral sway bracing is required unless all the pipes are supported by rods less than 6 in. or by 30^0 wrap-around U-hooks for any size pipe, NFPA 13 9.3.5.3.7 and 9.3.5.3.8.
57. _____ Longitudinal sway bracing is a maximum of 80 ft. for mains and cross mains and within 40 ft. of the end of the line, NFPA 13 9.3.5.4.
58. _____ A four-way sway brace spacing on a riser does not exceed 25 ft. and a four-way sway brace is located at the top of the riser if the top of the riser exceeds 3 ft. in length, NFPA 13 9.3.5.5.
59. _____ Seismic bracing calculations are detailed and provided for each brace to be used as shown in NFPA 13 Figure A.9.3.5.6(e).
60. _____ Longitudinal and lateral bracing is provided for each run of pipe between the change of direction unless the run is less than 12 ft. and supported by adjacent pipe run bracing, NFPA 13 9.3.5.11.
61. _____ Branch line method of restraint is detailed and in accordance with NFPA 13 Sections 9.3.6.1-9.3.6.3.

62. _____ Restraints for branch lines shall be at intervals not greater than 30 ft. if line movement will impact equipment or structural elements, NFPA 13 9.3.6.4, and restrain riser nipples 4 ft. long or greater against lateral movement, NFPA 13 9.3.6.5.
63. _____ Calculations for sway bracing zone of influence may be required, NFPA 13 9.3.5.6 – 9.3.5.11.

Fire Department Connection:
64. _____ For buildings whose area and height exceed the values specified in 6.6.4.1 a FDC is required.
65. _____ The FDC location is detailed on the street side or response side of building or as approved by the fire official, and when connected to the water supply it will not obstruct emergency vehicle access to the building, IFC 912.2.
66. _____ FDC is provided a connection that is at least a 1½ in., 6.6.4.2.

Design Criteria and Hydraulic Calculations:
67. _____ Hydraulic reference points match the plans.
68. _____ Pipe diameters match the plans.
69. _____ Sprinkler information matches the plans.
70. _____ Water flow information is provided; static PSI, residual PSI, GPM at 20 PSI residual with graphed results.
71. _____ The domestic water design demand is added to the sprinkler design when there is a single water supply, 6.5.5.
72. _____ Calculations are correct: static PSI, pipe length, GPM, calculated K-for for riser nipples or drop nipples, elevation data, hose allowance, friction loss, and equivalent pipe length, 6.7.1.4.
73. _____ Sprinklers without a listed discharge criteria are assigned a discharge criteria in accordance with, 6.7.1.1.1 and 6.7.1.1.2.
74. _____ Sprinklers with a listing discharge criteria: sprinklers comply with the discharge criteria for multiple and single sprinkler operation as required by their listing, 6.7.1.1.2.1, and at the discharge flow complies with 6.7.1.1.2.2.
75. _____ Sprinkler design for flat, smooth ceilings are calculated in accordance with Section 6.7.1.2 for the greatest hydraulic demand, 6.7.1.2.
76. _____ Sprinkler design for sloped, beamed, and pitched ceilings could require special design features such as larger flows or a design of 5 or more sprinklers to operate in the compartment, A.6.7.1.2.
77. _____ Sprinklers without a listed coverage criteria shall not exceed the area limits for sprinkler coverage area, 6.7.1.3.
78. _____ Hydraulic calculations are provided for single sprinkler and multisprinkler design.
79. _____ Areas outside dwelling unit shall have the design discharge, number of design sprinklers, coverage area, and sprinkler positions designed in accordance with Section 6.7.2.1.
80. _____ Areas outside dwelling unit: Residential sprinklers can protect building areas with flat smooth ceilings not exceeding 10 ft. as listed in Section 6.7.2.3.
81. _____ A garage separated from the residential building by fire-resistive construction that qualifies the garage as a separate building is sprinklered in accordance with NFPA 13 criteria, 6.7.3.1.
82. _____ Garage areas accessible by people from more than 1 dwelling unit and where the area is not constructed like 6.7.3.1 is a part of the building and is protected in accordance with 6.7.2, 6.7.3.2.
83. _____ A garage that is only accessible from 1 dwelling unit is a part of that dwelling and is sprinklered with residential sprinklers in accordance with NFPA 13R 6.7.1 or quick-response in accordance with, 6.7.3.3.
84. _____ A legend for calculation abbreviations is provided.
85. _____ A single combination water supply shall be allowed provided that the domestic demand is to the sprinkler demand as required by NFPA 13, IFC 903.3.5.1.2.

Additional Comments:

Review Date: _____   Approved or Disapproved   FD Reviewer: _____

Review Date: _____   Approved or Disapproved   FD Reviewer: _____

Review Date: _____   Approved or Disapproved   FD Reviewer: _____

Fire Plan Review and Inspection Guidelines

NFPA 13R Sprinkler System Acceptance Inspection
2006 IFC and 2002 NFPA 13R

Date of Inspection: _____ Permit Number: _____

Business/Building Name: _____ Address of Project: _____

Contractor: _____ Contractor's Phone: _____

Reference numbers following worksheet statements represent an NFPA code section unless otherwise specified.

Pass | Fail | NA

1. ___|___|___ Approved plans and above ground piping certification documents are on site, 6.2.2.
2. ___|___|___ Underground water supply testing and flushing are witnessed and underground piping certification is provided, 6.3.1.1.
3. ___|___|___ Hydro test for a wet system is 200 PSI for 2 hrs. and should include the FDC piping, 6.3.2.1.
4. ___|___|___ Hydro test for a dry system is 200 PSI for 2 hrs, 6.3.2.1.
5. ___|___|___ Hydro test of 50 PSI above the maximum design pressure of the system is permitted for a system without an FDC and less than 20 heads, 6.3.2.2.
6. ___|___|___ Backflow prevention device is installed in accordance with its listing and approved plans, IFC 903.3.5.

Riser Room

7. ___|___|___ Water flow drain location is in compliance with the approved set of plans. Flow test is performed, 6.6.2.
8. ___|___|___ Test valve and flow switch are monitored and tested.
9. ___|___|___ Paddle type water flow device is not allowed for dry systems.
10. ___|___|___ The monitoring signals are received by the 24 hour monitoring location.
11. ___|___|___ Water flow alarm is located according to the approved set of plans, is properly signed, and connected to the fire alarm system, if a fire alarm system is provided.
12. ___|___|___ Water supply valves are indicating type and electrically supervised, IFC 903.4.
13. ___|___|___ Valves on the riser are labeled with signage and pressure gauges are on the supply and system sides of the check valve, NFPA 13: 6.7.4.
14. ___|___|___ A permanent label with hydraulic calculations is attached to the riser.
15. ___|___|___ The riser supports for a multistory building are in compliance with the approved set of plans.
16. ___|___|___ There are a minimum of 3 spare sprinklers for each sprinkler type listed on the legend of the approved set pf plans, 6.4.1.

FDC if provided

17. ___|___|___ FDC is capped and permanently signed, 6.8.4.
18. ___|___|___ FDC has check valve and drip valve.
19. ___|___|___ FDC for wet single riser system connects to the system side of any water supply control valves.
20. ___|___|___ FDC is a minimum 1½ in. connection and easily accessible.

Sprinklers

21. ___|___|___ Sprinkler locations are installed in accordance with the approved set of plans.
22. ___|___|___ Pendent deflectors are 1 in. to 4 in. from the ceiling unless listing permits otherwise, 6.7.1.5.1.
23. ___|___|___ Sidewall deflectors are 4 in. to 6 in. from the ceiling unless listing permits otherwise, 6.7.1.5.2.
24. ___|___|___ Sprinklers have guards if subject to damage.
25. ___|___|___ Sprinkler heads are not painted or covered.
26. ___|___|___ The type and temperature rating of the sprinklers installed are in accordance with the approved set of plans.
27. ___|___|___ Escutcheon plates are installed.

Pipe: Hangers and Seismic Bracing

28. ___|___|___ Piping size and layout are in accordance with the approved set of plans.

29. ____ |____| ____ Lateral sway bracing is installed in accordance with the approved set of plans, NFPA 13: 9.3.5.3.
30. ____ |____| ____ If provided, seismic separation assemblies are installed in accordance with the approved plans.
31. ____ |____| ____ Longitudinal sway bracing is installed in accordance with the approved set of plans, NFPA 13: 9.3.5.4
32. ____ |____| ____ 4-way sway bracing is installed in accordance with the approved set of plans, NFPA 13: 9.3.5.5.
33. ____ |____| ____ Restraining straps are on all C-clamps and the strap is bolted through if there is not a lip on the beam, NFPA 13: 9.3.7.
34. ____ |____| ____ Branch lines have one hanger per section of pipe, NFPA 13: 9.2.3.2.
35. ____ |____| ____ Hangers are installed in accordance with the approved set of plans, NFPA 13: 9.2.4.
36. ____ |____| ____ Risers in multistory buildings have supports installed in accordance with the approved set of plans, NFPA 13: 9.2.5.3.

Additional Comments:

Inspection Date: _____ Approved or Disapproved FD Inspector: _____

Inspection Date: _____ Approved or Disapproved FD Inspector: _____

Inspection Date: _____ Approved or Disapproved FD Inspector: _____

Fire Plan Review and Inspection Guidelines

13R Residential Sprinkler System
Plan Review Worksheet
2006 IFC, 2007 NFPA 13 and 2007 NFPA 13R

This worksheet is for jurisdictions that permit the use of the 2007 NFPA 13R in lieu of IFC's referenced 2002 NFPA 13R.

Date of Review: _____ Permit Number: _____

Business/Building Name: _____ Address of Project: _____

Designer Name: _____ Designer's Phone: _____

Contractor: _____ Contractor's Phone: _____

No. of Sprinklers: _____ Occupancy Classification: _____

Reference numbers following worksheet statements represent an NFPA code section unless otherwise specified.

Worksheet Legend: ✓ or **OK** = acceptable   **N** = need to provide,   **NA** = not applicable

1. _____ Three sets of drawings are provided.
2. _____ System components are listed for intended use and compatible with the system, and specification data product sheets are provided.

Drawings shall detail the following and items listed in 6.2.7.:
General:
3. _____ The type of system is noted: __ wet, __ dry, __antifreeze not exceeding 40 gals., __ preaction, __ dry and type of sprinklers are noted: ___pendent, ____upright, _____sidewall, 5.3.2.
4. _____ Scale: a common scale shall be used and plan information is legible, 6.2.
5. _____ Plot plan showing supply piping and pipe size from the water source to the building, 6.2.
6. _____ Building dimensions, location of partitions, and fire walls, 6.2.
7. _____ Room dimensions, labeled rooms, occupancy class of each room. If the room label is not descriptive provide the room's type of use, 6.2.
8. _____ Full height cross elevation views and include ceiling construction, 6.2.
9. _____ Type of protection for nonmetallic pipe, 6.2.
10. _____ Dimensions for system piping, type of pipe, and component spacing, 6.2.
11. _____ Equipment symbol legend and the compass point, 6.2.
12. _____ A water flow alarm test connection is provided, 6.5.4.
13. _____ All water supply valves and flow switches are supervised, IFC 903.4.
14. _____ Exterior flow alarm location is shown and the type identified, if electric, it is listed for outdoor use, IFC 903.4.2., and it is connected to the building fire alarm, if provided, 6.7.8.
15. _____ Backflow prevention device, when required by state or local regulations, is shown in the pipe schematic, listed specification sheet and pressure loss data are provided, IFC 903.3.5.
16. _____ Antifreeze systems are detailed and designed in accordance with NFPA 13: 7.6.
17. _____ Water supply provides the system demand for at least 30 minutes, 6.6.2.
18. _____ If a fire pump is required it is designed and detailed per NFPA 20 and this book's worksheet, 6.6.4.
19. _____ Pressure gauges for the riser are provided and detailed for supply and system pressure, 6.7.5.
20. _____ The area of coverage does not exceed 52,000 sq. ft., 6.1.1.1.
21. _____ Above-ground water supply pipe is protected against freezing conditions, 6.1.4.1.

Sprinklers:
22. _____ Total number of each type of sprinkler is noted and the number of sprinklers per floor are noted, 6.2.
23. _____ Sprinkler location is correct, ceiling and roof sectionals are provided for clarification.
24. _____ Type of sprinklers: sprinkler K factors, temperature rating, and orifice size, 6.2.
25. _____ Residential sprinklers are limited for use for wet pipe automatic sprinkler systems unless specifically listed for another use, 6.7.7.1.2.
26. _____ When listed quick-response sprinklers are used in dwelling units, the dwelling unit shall meet the definition of a compartment and a maximum of four sprinklers are used, The sprinkler density complies with 6.6.7.1.3. Refer to 2002 NFPA Handbook Exhibit 6.5 for clarification., 6.7.7.1.3.
27. _____ Sprinklers are rated for ordinary temperature (135°F-175°F) when ceiling temperature does not exceed 100°F, 6.7.7.1.5.1.
28. _____ Sprinklers in areas with a ceiling temperature of 101°F-150°F are equipped with intermediate temperature sprinklers (175°F-225°F), 6.7.7.1.5.2.

29. _____ Distance of sprinklers from heat sources complies with Table 6.7.7.1.5.3.
30. _____ Quick-response sprinklers are used when protection is on the outside a dwelling unit, 6.7.7.2.
31. _____ Each sprinkler coverage area is within its listing limitations, 6.8.1.3.1.
32. _____ Sprinkler coverage not required for some architectural areas that meet the area and dimensions as specified in 6.7.7.1.5.5.
33. _____ Sprinkler coverage is not required for shadowed areas in a compartment when the area and dimensions are in compliance with 6.7.7.1.5.7.
34. _____ Sloped ceiling spacing is in accordance with Figure 6.8.1.3.1.1(A) and Section 6.8.1.3.1.
35. _____ Sidewall sprinklers distance from the ceiling complies with 6.8.1.5.2.1.
36. _____ Closets and storage areas limited to 400 cu. ft., a single sprinkler provides is located at the highest ceiling height, 6.8.1.5.3.1.
37. _____ Pendent sprinkler are distanced from obstructions e.g. light fixtures, ceiling fans, etc. in accordance with 6.8.1.5.3.2. Sprinkler locations for continuous obstructions are in compliance with 6.8.1.5.3.4.
38. _____ Sidewall sprinklers are distanced from obstructions e.g. light fixtures, ceiling fans, etc. in accordance with 6.8.1.5.3.3. Sprinkler locations for continuous obstructions are in compliance with 6.8.1.5.3.5.
39. _____ Soffits and cabinets are provided sprinkler coverage in accordance with 6.8.1.5.3.6.
40. _____ Ceiling pockets are sprinklered unless the pocket volume, its depth, the floor protection, its separation from other pockets and the construction finish material complies with 6.8.1.5.4.
41. _____ Sprinklers are not required in dwelling unit bathrooms less than 55 sq. ft. and constructed of materials providing a 15-minute thermal barrier, 6.9.2.
42. _____ Sprinklers are not required in dwelling unit clothes closets, pantries, or linen closets, provided the closet area, its least dimension, and its method of construction complies with 6.9.3.
43. _____ Sprinklers are provided in closets containing heating or air-conditioning equipment, 6.9.4.
44. _____ Sprinklers are not required for porches, balconies, corridors, and stairs that are open and attached, 6.9.5. If the building construction is Type V then balconies, decks and ground floor sprinkler coverage is required and sprinkler placement is in accordance with Section IFC 903.3.1.2.1.
45. _____ Sprinklers are not required for areas not used for living purposes or used for storage as listed in, 6.9.6.
46. _____ Areas with flat smooth ceilings outside dwelling unit: residential sprinklers can protect the spaces listed in 6.8.2.3.
47. _____ Garage separated from the residential building by fire-resistive construction, which qualifies the garage as a separate building is sprinklered in accordance with NFPA 13, 6.8.3.1.
48. _____ A garage accessible by people from more than 1 dwelling unit and not constructed like 6.8.3.1 is part of the building and is protected in accordance with 6.8.2, 6.8.3.2 and 6.8.3.3.

Drains and Test Connection:
49. _____ At least a 1 in. nominal diameter drain with a valve is detailed on the system side of the control valve, 6.7.2.1 and 6.7.2.2.
50. _____ Each portion of trapped dry system piping that is subject to freezing is provided a ½ in. drain, 6.7.2.4.
51. _____ The location and size of a test connection with a valve is detailed and complies with 6.7.3.1.

Pipe and Valves:
52. _____ One control valve is provided for both the domestic water and sprinkler, unless a separate control valve is provided for the sprinkler system, 6.6.1.1 and it is electronically supervised, IFC 903.4.

Pipe Support and Hangers are in Accordance with NFPA 13, 13R 6.7.6:
53. _____ Type and locations of hangers, sleeves, braces, and methods of securing pipe are shown and the manufacture's installation manual for plastic pipe is provided, NFPA 13R 6.2.7(21).
54. _____ Steel pipe hanger spacing is in compliance with Table 9.2.2.1(a) and for copper and plastic see Table 9.2.2.1(a) and the pipe listing information.
55. _____ Light wall steel pipe hanger spacing is in compliance with Table 9.2.2.1(a).
56. _____ Branch lines are provided with one hanger per section of pipe, exceptions are listed, 9.2.3.2.
57. _____ Mains are provided with one hanger between each branch line unless 9.2.4.2 through 9.2.4.5. are met, 9.2.4.
58. _____ Cross mains are provided with one hanger between each two branch lines or in compliance with Table 9.2.2.1(a), and for additional spacing variations refer to Section 9.2.4.
59. _____ Risers in multi-story buildings are provided with supports at the locations listed in 9.2.5.4.
60. _____ The maximum span between riser supports does not exceed 25 ft., 9.2.5.5.

Seismic Protection in Accordance with NFPA 13, 13R 6.7.6:
61. _____ Flexible couplings may be used for pipe 2½ in. or larger in accordance with Section 9.3.2.

Fire Plan Review and Inspection Guidelines

62. _____ A seismic separation assembly for piping is provided at building seismic joints in accordance with 9.3.3.2 and 9.3.3.3 and a detail is provided.
63. _____ Proper pipe clearance is noted on the plans for pipe penetrations, 9.3.4. also refer to the 3 spacing variations, 9.3.4.2 – 9.3.4.7.
64. _____ Lateral sway bracing is required at a maximum spacing of 40 ft. for all mains, cross mains, and branch lines 2½ in. and larger, 9.3.5.3.1 and 9.3.5.3.2.
65. _____ Lateral sway bracing is designed not to exceed the maximum zone of influence loading provided in Tables 9.3.5.3.2(a) and (b) for its spacing, 9.3.5.3.2.
66. _____ Lateral bracing is provided for the last length of pipe of the end of a feed or cross main, 9.3.5.3.5.
67. _____ Lateral bracing is required unless all the pipe is supported by rods less than 6 in. or by 30^0 wrap-around U-hooks for any size pipe, 9.3.5.3.8 and 9.3.5.3.9.
68. _____ Longitudinal sway bracing is a maximum of 80 ft. for mains and cross mains and within 40 ft. of the end of the line, 9.3.5.4.1 and 9.3.5.4.3.
69. _____ A four-way sway brace is provided at least every 25 ft. and at the top of the riser if the top of the riser exceeds 3 ft. in length, 9.3.5.5.
70. _____ Seismic bracing calculations and the zones of influence are detailed and provided for each brace to be used as shown in NFPA Figure A.9.3.5.6(e).
71. _____ Longitudinal and lateral bracing is provided for each run of pipe between the changes of direction unless the run is less than 12 ft. and supported by adjacent pipe run bracing, 9.3.5.11.2.
72. _____ Branch lines are restrained at the end sprinkler of each line and restrained against vertical and lateral movement, 9.3.6.3.
73. _____ Branch line method of restraint is in accordance with Section 9.3.6.1.
74. _____ Restraints for branch lines shall be at intervals not greater than specified in Table 9.3.6.4 and justification for selection of the seismic coefficient is provided, 9.3.6.4.
75. _____ Detailed are restraints for sprig 4 ft. long or greater against lateral movement, 9.3.6.6.

Fire Department Connection:
76. _____ For buildings whose area and height exceed the values specified in 6.7.4.1 an FDC is provided.
77. _____ The FDC location is detailed on the street side or response side of building or as approved by the fire official, and when connected to the water supply it will not obstruct emergency vehicle access to the building, IFC 912.2.
78. _____ FDC is at least a 1½ in. connection and 18 in. to 48 in. above grade, 6.7.4.2.

Design Criteria and Hydraulic Calculations:
79. _____ Reference points match with plans.
80. _____ Pipe size references match the plans.
81. _____ Sprinkler information matches the plans.
82. _____ Water flow information is provided; static PSI, residual PSI, GPM at 20 PSI residual with graphed results.
83. _____ Calculations are correct: static PSI, pipe length, GPM, K for drops or branch, elevation data, hose allowance, friction loss, and equivalent pipe length. Minimum sprinkler pressure is that specified by the listing or 7 PSI, whichever is greater and at least a density of 0.05 GPM/sq. ft. are provided, 6.8.1.1.1.2.
84. _____ A single combination water supply shall be allowed provided that the domestic demand is to the sprinkler demand as required by NFPA 13, IFC 903.3.5.1.2.
85. _____ The system provides at least the flow required for multiple and single sprinkler operation as specified by the listing, 6.8.1.1.1.1., and at flow, sprinklers must produce a minimum density specified in 6.8.1.1.1.2.
86. _____ Sprinkler design for flat, smooth ceilings consists of up to 4 sprinklers within the same compartment with the largest flow and pressure demand, 6.8.1.1.1.3.
87. _____ Areas outside dwelling unit are designed in accordance with NFPA 13 or as a compartmented area with a demand of no greater than 4 sprinklers when all the following are met: 1) Is compartmented into areas 500 sq. ft. or less by 30 minute fire construction, 2) Area is protected by quick-response or residential sprinklers nor exceeding 130 sq. ft. for ordinary hazard occupancies, 225 sq. ft. for light hazard occupancies or in accordance with the sprinkler listing, 3) Openings have a lintel of at least 8 in. in depth, 4) Total area of openings does not exceed 50 sq. ft. per compartment, 5) Design densities arein accordance with NFPA 13R 6.8.2.2.
88. _____ Garage that are only accessible from 1 dwelling unit are part of that dwelling and are sprinklered with residential sprinklers in accordance with NFPA 13R 6.8.1 or quick-response sprinklers with a density and number of sprinklers as specified in 6.8.3.3.
89. _____ A legend for calculation abbreviations is provided.

Fire Plan Review and Inspection Guidelines

Additional Comments:

Review Date: _____   Approved or Disapproved   FD Reviewer: _____
Review Date: _____   Approved or Disapproved   FD Reviewer: _____
Review Date: _____   Approved or Disapproved   FD Reviewer: _____

Fire Plan Review and Inspection Guidelines

NFPA 13R Sprinkler System Acceptance Inspection
2006 IFC and 2007 NFPA 13R
This worksheet is for jurisdictions that permit the use of the 2007 NFPA 13R in lieu of IFC's referenced 2002 NFPA 13R.

Date of Inspection: _____ Permit Number: _____

Business/Building Name: _____ Address of Project: _____

Contractor: _____ Contractor's Phone: _____

Reference numbers following worksheet statements represent an NFPA code section unless otherwise specified.

Pass | Fail | NA

1. ___ | ___ | ___ Approved plans and above-ground piping certification documents are on site, 6.3.2.
2. ___ | ___ | ___ Underground supply testing and flushing are witnessed and underground piping certification is provided, 6.4.1.1.
3. ___ | ___ | ___ Hydro test for a wet system is 200 PSI for 2 hrs. and should include the FDC piping 6.4.2.1.
4. ___ | ___ | ___ Hydro test for a dry system is 200 PSI for 2 hrs, 6.4.2.1.
5. ___ | ___ | ___ Hydro test of 50 PSI above the maximum design pressure of the system is permitted for a system without an FDC and less than 20 heads, 6.4.2.2.
6. ___ | ___ | ___ Backflow prevention device is installed in accordance with its listing and approved plans, IFC 903.3.5.

Riser Room

7. ___ | ___ | ___ Water flow drain location is in compliance with the approved set of plans. Flow test is performed, 6.3 and Figure 6.3.2.
8. ___ | ___ | ___ Test valve and flow switch are monitored and tested.
9. ___ | ___ | ___ Paddle type water flow is not allowed for dry systems.
10. ___ | ___ | ___ The monitoring signals are received by the 24 hour monitoring location.
11. ___ | ___ | ___ Water flow alarm is located according to the approved set of plans, is properly signed, and connected to the fire alarm system, if a fire alarm system is provided.
12. ___ | ___ | ___ Water supply valves are indicating type and electrically supervised, IFC 903.4.
13. ___ | ___ | ___ Valves on the riser are labeled with signage and pressure gauges are on the supply and system sides of the check valve, NFPA 13: 6.7.4.
14. ___ | ___ | ___ A permanent label with hydraulic calculations is attached to the riser.
15. ___ | ___ | ___ The riser supports for a multistory building are in compliance with the approved set of plans.
16. ___ | ___ | ___ There are a minimum of 3 spare sprinklers for each sprinkler type listed on the legend of the approved set pf plans, 6.5.1.

FDC if provided

17. ___ | ___ | ___ FDC is capped and permanently signed, 6.8.2.
18. ___ | ___ | ___ FDC has check valve and drip valve.
19. ___ | ___ | ___ FDC for wet single riser system connects to the system side of any water supply control valves.
20. ___ | ___ | ___ FDC is a minimum 1½ in. connection and easily accessible

Sprinklers

21. ___ | ___ | ___ Sprinkler locations are installed in accordance with the approved set of plans.
22. ___ | ___ | ___ Pendent deflectors are 1 in. to 4 in. from the ceiling unless listing permits otherwise, 6.8.1.5.1.
23. ___ | ___ | ___ Sidewall deflectors are 4 in. to 6 in. from the ceiling unless listing permits otherwise, 6.8.1.5.2.
24. ___ | ___ | ___ Sprinkler heads have guards if subject to damage.
25. ___ | ___ | ___ Sprinkler heads are not painted or covered.
26. ___ | ___ | ___ The type and temperature rating of the sprinklers installed are in accordance with the approved set of plans.
27. ___ | ___ | ___ Escutcheon plates are installed.

Pipe: Hangers, Seismic, and Penetrations

28. ____ |____| ____ Piping size and layout are in accordance with the approved set of plans.
29. ____ |____| ____ Minimum clearance around pipes: Holes are 2 in. larger than pipe 1 in. to 3½ in., and 4 in. larger than pipe 4 in. or larger., NFPA 13R 6.6.6, NFPA 13: 9.3.4. Refer to NFPA 13: 9.3.4 for exceptions.
30. ____ |____| ____ Lateral sway bracing is installed in accordance with the approved set of plans, 13: 9.3.5.3.
31. ____ |____| ____ If provided, seismic separation assemblies are installed in accordance with the approved plans.
32. ____ |____| ____ Longitudinal sway bracing is installed in accordance with the approved set of plans, NFPA 13: 9.3.5.4.
33. ____ |____| ____ 4-way sway bracing is installed in accordance with the approved set of plans, NFPA 13: 9.3.5.5.
34. ____ |____| ____ Restraining straps are on all C-clamps and the strap is bolted through if there is not a lip on the beam, NFPA 13: 9.3.7.1.
35. ____ |____| ____ Branch lines have one hanger per section of pipe, NFPA 13: 9.2.3.2.
36. ____ |____| ____ Hangers are installed in accordance with the approved set of plans, NFPA 13: 9.2.4.
37. ____ |____| ____ Risers in multi-story buildings have supports installed in accordance with the approved set of plans, NFPA 13: 9.2.5.4.
38. ____ |____| ____ Risers in multi-story buildings have supports installed in accordance with the approved set of plans, NFPA 13: 9.2.5.5.

Additional Comments:

Inspection Date: _____   Approved or Disapproved   FD Inspector: _____
Inspection Date: _____   Approved or Disapproved   FD Inspector: _____
Inspection Date: _____   Approved or Disapproved   FD Inspector: _____

Fire Plan Review and Inspection Guidelines

NFPA 13R Sprinkler Installation Certification

Permit #: _____   Date: _____

	Property Protected	System Installer	System Supplier
Business Name:	_____	_____	_____
Address:	_____	_____	_____
	_____	_____	_____
Representative:	_____	_____	_____
Telephone:	_____	_____	_____

Location of Plans: _____

Location of Owner's Manual: _____

1. <u>Certification of System Installation:</u> Complete this section after system is installed, but prior to conducting operational acceptance tests.

    This system installation was inspected and found to comply with the installation requirements of:
    _____ NFPA 13R
    _____ IFC and IBC
    _____ Manufacturer's Instructions
    _____ Other (specify; FM, UL, etc.) _____

    Print Name: _____
    Signed: _____ Date: _____
    Organization: _____

2. <u>Certification of System Operation:</u> All operational features and functions of this system were tested and found to be operating properly in accordance with the requirements of:
    _____ NFPA 13R
    _____ IFC and IBC
    _____ Manufacturer's Instructions
    _____ Other (specify) _____

    Print Name: _____
    Signed: _____ Date: _____
    Organization: _____

Fire Plan Review and Inspection Guidelines

13D Residential Sprinkler System
Plan Review Worksheet
2006 IFC and 2002 NFPA 13D

Date of Review: _____ Permit Number: _____

Business/Building Name: _____ Address of Project: _____

Designer Name: _____ Designer's Phone: _____

Contractor: _____ Contractor's Phone: _____

No. of Sprinklers: _____ Occupancy Classification: _____

Reference numbers following worksheet statements represent an NFPA code section unless otherwise specified.

Worksheet Legend: ✓ or **OK** = acceptable, **N** = need to provide, **NA** = not applicable

1. _____ A minimum of three sets of drawings are provided. The plans declare the design is based on the 2002 edition of NFPA 13D.
2. _____ System components are listed for intended use, specification information sheets are provided, 5.1.2. Nonlisted items can be tanks, pumps, hangers, waterflow detection devices, and waterflow valves, 5.1.3.

Drawings shall show the following:
General:
3. _____ Scale: a common scale shall be used and information shall be legible, IFC 901.2.
4. _____ Plot plan details illustrate the water supply connection, pipe diameters, lengths, and fittings to the building, IFC 901.2.
5. _____ Building dimensions, cross sectional views, and the location of partitions are provided, IFC 901.2.
6. _____ Type of protection for nonmetallic pipe is provided, IFC 901.2.
7. _____ Dimensions for system piping, type of pipe, and component spacing, IFC 901.2.
8. _____ Equipment symbol legend is detailed, IFC 901.2.
9. _____ Total number of each type of sprinkler is noted on the plans, IFC 901.2.
10. _____ Type of sprinklers, K-factors, temperature rating, coverage area, minimum operating pressure, and orifice size are provided.
11. _____ Dry systems are not permitted unless all components are approved and listed and it serves unheated areas, 8.3.2.
12. _____ For a dry system, or a system using a pressurized tank as a water supply source, a pressure gauge is detailed, 7.3.
13. _____ When required, antifreeze system is designed in accordance with, 8.3.3 and local plumbing codes, IFC 903.3.5.
14. _____ Wet pipe system is used when not subject to freezing, 8.3.1.
15. _____ Antifreeze systems are detailed and usually limited to 40 gals, 8.3.3.1.
16. _____ Type of antifreeze solution and percentage is noted on the plans, 8.3.3.2.
17. _____ Is the supply riser in a heated environment?
18. _____ If a stored water supply is used, it shall provide the water demand rate in accordance with 6.1.2 and 6.1.3.
19. _____ A reliable water supply is provided in accordance with Section 6.2.

Multipurpose Piping Systems:
20. _____ Multipurpose system, without an FDC, that uses nonmetallic fittings, the fittings are designed to an operating pressure of 130 PSI or greater, 5.2.5.3.
21. _____ The piping system serving both sprinkler and domestic needs is acceptable if: 1) The common water supply is serving more than 1 dwelling unit, 5 GPM is added to the sprinkler demand, 2) Smoke alarms are provided, 3) All pipe used is listed, 4) Pipe connected to the system serving plumbing fixtures need not be listed, 5) Permitted by the plumbing code official, 6) A sign adjacent to the main shutoff indicates it serves the fire sprinkler system with verbiage per the code section, 7) Devices that restrict the flow shall not be added and water treatment and filtering systems shall be bypassed, 6.3.

Sprinklers:
22. _____ Sprinkler location is accordance with its listing criteria and the requirements in Sections 8.1.3 and 8.2.
23. _____ Residential sprinklers are limited to wet systems unless listed for other uses, 7.5.2.
24. _____ Dry pendent or sidewall sprinklers are permitted to be used in unheated areas not used for living, 7.5.3.

Fire Plan Review and Inspection Guidelines

25. _____ Sprinklers are ordinary temperature (135°F-175°F) when the ceiling temperature does not exceed 100°F, 7.5.5.1.
26. _____ Sprinklers that are in areas with ceiling temperatures of 101°F-150°F are intermediate temperature (175°F-225°F), 7.5.5.2.
27. _____ For skylights exposed to direct sun, unvented concealed spaces under uninsulated roofs or unvented attics, sprinklers when required are provided in accordance with 7.5.5.3.
28. _____ Sprinklers near heat sources are located in accordance with Table 7.5.5.3, 7.5.5.3.
29. _____ Each sprinkler coverage area is within its listing limitation, 8.1.1.
30. _____ Spacing between sprinklers and distance from walls does not exceed the limits specified in 8.1.3.
31. _____ Sidewall deflectors are 4 ft. to 6 ft. from the ceiling unless listing allows otherwise, 8.2.2.1.
32. _____ A single sprinkler at the highest ceiling level can provide coverage for closets and storage areas not exceeding 300 cu. ft. and the lowest point of the ceiling height is 5 ft., 8.2.6.
33. _____ Sprinklers are required in dwelling unit bathrooms exceeding 55 sq. ft., 8.6.2.
34. _____ Sprinklers are not required in dwelling unit clothes closets, pantries, or linen closets, provided the closet area, its least dimension, and its method of construction complies with 8.6.3.
35. _____ Sprinkler protection for unused attics, crawl spaces, and concealed spaces is not required, 8.6.5.
36. _____ Sprinkler protection for covered unheated projections from buildings at entrances is not required if another exit is available from the dwelling unit, 8.6.6.

Alarms:
37. _____ Local flow alarm location and inspector's test connection are provided and detailed. The exception is if the dwelling has smoke alarms in compliance with the building code, 7.2.4, 7.6.

Hydraulic Calculations or Design Discharge:
38. _____ Reference points match with plans.
39. _____ Pipe diameters match the plans and size is determined by hydraulic calculations, 8.4.4.
40. _____ Hydraulic calculations are also required when a system is gridded, looped, or connected to a city main less than 4 in., 8.4.7-8.4.9.
41. _____ Legend for calculation abbreviations are provided.
42. _____ Sprinkler data sheet matches what is on the plans and hydraulic calculations.
43. _____ Water flow information such as static PSI, residual PSI, and available GPM at 20 PSI residual is provided.
44. _____ Hydraulic calculations can be provided using one of three methods described in Section 8.4.4 when the system is connected to a municipal water main of at least 4 in. in diameter and typical calculations include static PSI, pipe length, discharge GPM, the calculated K factor for riser or drop nipples, elevation data, friction loss, friction loss data for gate valve and backflow prevention device and equivalent pipe length, 8.4.4.
45. _____ Sprinklers without a listed discharge criteria: a single sprinkler discharge is not less than 18 GPM and a multisprinkler discharge design is not less than 13 GPM, 8.1.1.1.1.
46. _____ Sprinkler with a listed discharge criteria: the system provides at least the flow required for multiple and single sprinkler operation as specified by the listing, 8.1.1.2.1, and must produce a minimum discharge density of .05 GPM/ft^2, 8.1.1.2.2.
47. _____ Sprinkler design for flat, smooth ceilings consists of up to 2 sprinklers within the same compartment with the largest water demand, A.8.1.2.
48. _____ Sprinkler design for sloped, beamed, and pitched ceilings could require special design features such as larger flows or a design of 3 or more sprinklers to operate in the compartment, A.8.1.2.
49. _____ Sprinklers without a listed coverage criteria shall not exceed 144 sq. ft. coverage area, 8.1.3.1.

Pipe Support and Hangers:
50. _____ Piping support shall comply with submitted manufacturers instructions and/or listing criteria. The plumbing code will be consulted for piping that does not have support criteria provided. Lateral movement is prevented for pipe laid on joists or rafters and in general, pipe movement is to be supported to restrain movement, 7.4.

Pipe and Valves:
51. _____ One control valve is provided for both the domestic water and sprinkler, unless a separate control valve is provided for the sprinkler system, 6.6.1.1.
52. _____ A drain and test connection is provided in accordance with 7.2.1.
53. _____ Each portion of trapped dry system piping that is subject to freezing is provided a drain, 7.2.3.
54. _____ A waterflow test connection is provided if a waterflow alarm is provided, 7.2.4.
55. _____ Type and size of pipe is provided. Pipe to be used complies with the pipe listed in Table 5.2.1 and nonmetal pipe is listed for fire protection service, 5.2.1.

Fire Plan Review and Inspection Guidelines

56._____ Minimum 1 in. diameter steel pipe or minimum ¾ in. diameter metallic or nonmetallic piping shall be used in the sprinkler system, 8.4.3.

57._____ Network systems are allowed when ½ in. nonmetallic pipe or copper pipe with listed special fittings is used in conjunction with the 10 conditions specified in 8.4.3.3.

Additional Comments:

Review Date: _____ Approved or Disapproved FD Reviewer: _____

Review Date: _____ Approved or Disapproved FD Reviewer: _____

Review Date: _____ Approved or Disapproved FD Reviewer: _____

Fire Plan Review and Inspection Guidelines

NFPA 13D Sprinkler System Acceptance Inspection
2006 FC and 2002 NFPA 13D

Date of Inspection: _____ Permit Number: _____

Business/Building Name: _____ Address of Project: _____

Contractor: _____ Contractor's Phone: _____

Reference numbers following worksheet statements represent an NFPA code section unless otherwise specified.

Pass | Fail | NA
1. ___ | ___ | ___ Approved drawing and certification documents on site.
2. ___ | ___ | ___ System is leak tested at normal operating pressure when an FDC is not provided, 4.4.1.
3. ___ | ___ | ___ System control valve for both the sprinkler and domestic systems is on.
4. ___ | ___ | ___ Signage is adjacent to the main water shut-off valve: Warning, the water system for this house supplies a fire sprinkler system that depends on certain flows and pressures being available to fight a fire…Don't remove this sign., 6.3.

Riser Room
5. ___ | ___ | ___ Operate the drain valve on the system side of the control valve.

Sprinklers
6. ___ | ___ | ___ Spacing between sprinklers does not exceed 12 ft., sprinklers are not greater than 6 ft. from a wall, and sprinklers are not within 8 ft. of each other unless listing allows it, 8.1.3.
7. ___ | ___ | ___ Sprinkler heads are not painted or covered or blocked.
8. ___ | ___ | ___ Proper type and temperature sprinklers are used.
9. ___ | ___ | ___ Escutcheon plates are installed and pendent/upright deflectors are within 1 in. to 4 in. from the ceiling, sidewalls are within 4 in. to 6 in. from the ceiling or all are per their listing.
10. ___ | ___ | ___ Pendent and upright deflectors in closets can be installed within 12 in. of the ceiling, 8.2.1.3.
11. ___ | ___ | ___ Sprinklers are in all areas as shown on the approved set of plans, 8.6.

Pipe and Support
12. ___ | ___ | ___ Piping layout and pipe size are the same as the plans.
13. ___ | ___ | ___ Pipe hangers and supports are per the manufacturer's requirements.
14. ___ | ___ | ___ Pipe laid on open joists is secured to prevent lateral movement and other piping is secured to restrict movement.
15. ___ | ___ | ___ Pipes in attics are adequately insulated, 7.7.
16. ___ | ___ | ___ Antifreeze system is installed in accordance with the approved set of plans.

Additional Comments:

Inspection Date: _____ Approved or Disapproved FD Inspector: _____

Inspection Date: _____ Approved or Disapproved FD Inspector: _____

Fire Plan Review and Inspection Guidelines

13D Residential Sprinkler System
Plan Review Worksheet
2006 IFC and 2007 NFPA 13D

This worksheet is for jurisdictions that permit the use of the 2007 NFPA 13D in lieu of IFC's referenced 2002 NFPA 13D.

Date of Review: _____ Permit Number: _____

Business/Building Name: _____ Address of Project: _____

Designer Name: _____ Designer's Phone: _____

Contractor: _____ Contractor's Phone: _____

No. of Sprinklers: _____ Occupancy Classification: _____

Reference numbers following worksheet statements represent an NFPA code section unless otherwise specified.

Worksheet Legend: ✓ or **OK** = acceptable, **N** = need to provide, **NA** = not applicable

1. _____ A minimum of three sets of drawings are provided. The plans declare the design is based on the 2007 edition of NFPA 13D.
2. _____ System components are listed for intended use, specification data sheets are provided, 5.1.2. Nonlisted items that are permitted by the standard can be tanks, pumps, hangers, waterflow detection devices, and waterflow valves, 5.1.3.

Drawings shall show the following:

General:
3. _____ Scale: a common scale shall be used and information shall be legible, IFC 901.2.
4. _____ Plot plan details illustrate the water supply connection, pipe diameters, lengths, and fittings to the building, IFC 901.2.
5. _____ Building dimensions, cross sectional views, and the location of partitions are provided, IFC 901.2.
6. _____ Type of protection for nonmetallic pipe is provided, IFC 901.2.
7. _____ Dimensions for system piping, type of pipe, and component spacing, IFC 901.2.
8. _____ Equipment symbol legend is detailed, IFC 901.2.
9. _____ Total number of each type of sprinkler is noted on the plans, IFC 901.2.
10. _____ Type of sprinklers, K factors, temperature rating, coverage area, minimum operating pressure, and orifice size are provided, 8.1.1.
11. _____ Dry systems are not permitted unless all components are approved and listed and it serves unheated areas, 8.3.2.
12. _____ For a dry system, or a system using a pressurized tank as a water supply source, a pressure gauge is detailed, 7.3.
13. _____ Wet pipe system is used when not subject to freezing, 8.3.1.
14. _____ Type of antifreeze solution and percentage is noted on the plans, 8.3.3.2.
15. _____ Systems in areas subject to freezing shall be well insulated or shall be a dry pipe or antifreeze system, 8.3.1 and 8.3.2.
16. _____ When required, the antifreeze system is designed in accordance with Figure 8.3.3.3.1.1, and local plumbing codes, 8.3.3 and 8.3.3.1, and IFC 903.3.5.
17. _____ If a stored water supply is used it shall provide the water demand rate in accordance with 6.1.2 and .3.
18. _____ Is the supply riser in a heated environment?
19. _____ A reliable water supply is provided in accordance with Section 6.2.

Multipurpose Piping Systems:
20. _____ Multipurpose system, without an FDC, that uses nonmetallic fittings, the fittings are designed to an operating pressure in accordance with 5.2.5.3.
21. _____ The piping system serving both sprinkler and domestic needs is acceptable if: 1) The common water supply is serving more than 1 dwelling unit, 5 GPM is added to the sprinkler demand, 2) All pipe used is listed, 3) Pipe connected to the system serving plumbing fixtures need not be listed, 4) Permitted by the plumbing code official, 5) A sign adjacent to the main shutoff indicates it serves the fire sprinkler system with verbiage per the code section, 6) Devices that restrict the flow shall not be added and water treatment and filtering systems shall be bypassed, 6.3.

Sprinklers:
22. _____ Sprinkler location is correct according to listing criteria and Sections 8.1.3 and 8.2.

Fire Plan Review and Inspection Guidelines

23. _____ Only residential sprinklers are specified for wet systems unless listed for other uses, 7.5.2.
24. _____ Dry pendent or sidewall sprinklers are permitted to be used in unheated areas not used for living, 7.5.3.
25. _____ Sprinklers are ordinary temperature when the ceiling temperature does not exceed the threshold specified in 7.5.5.1.
26. _____ Sprinklers that are in areas with ceiling temperatures of 101°F-150°F are intermediate temperature (175°F-225°F), 7.5.5.2.
27. _____ For skylights exposed to direct sun, unvented concealed spaces under uninsulated roofs or unvented attics, sprinklers when required are provided in accordance with 7.5.5.3.
28. _____ Ceiling pockets are sprinklered unless the pocket volume is 100 sq. ft. or less, its depth is 1 ft. or less, the floor below is protected, it is separated from other pockets by at least 10 ft., and the finish material is non-combustible or limited-combustible, 8.6.7.
29. _____ Each sprinkler coverage area is within its listing limitation, IFC 901.2.
30. _____ Sloped ceiling sprinkler spacing is in accordance with Figure 8.1.3.1.3.1 and Section 8.1.3.1.3.
31. _____ Closets, which may include mechanical equipment, that is limited to 400 cu. ft., a single sprinkler is provided and is located at the highest ceiling height, 8.2.5.1.
32. _____ Pendent sprinklers are distanced from obstructions e.g. light fixtures, ceiling fans, etc. in accordance with 8.2.5.2. Sprinkler locations for continuous obstructions are in compliance with 8.2.5.4.
33. _____ Sidewall sprinklers are distanced from obstructions e.g. light fixtures, ceiling fans, etc. in accordance with 8.2.5.3. Sprinkler locations for continuous obstructions are in compliance with 8.2.5.5.
34. _____ Soffits and cabinets are provided sprinkler coverage in accordance with 8.2.5.6.
35. _____ Dry pipe and preaction systems can use only listed sprinklers which are installed in accordance with 8.3.4.1.1.
36. _____ Dry pipe and preaction systems can K-factors with corrosion resistant or galvanized coated pipe as specified in 8.3.4.1.2.
37. _____ Dry pipe and preaction systems can use K-factors with other pipe as specified in 5.2.
38. _____ Dry pipe and double interlock preaction systems have calculations showing water delivery at the most remote sprinkler is within 15 seconds or within 15 seconds from an inspectors test outlet that is provided at the furtherest end of the system piping. The test outlet will flow at least the amount of water the system's smallest sprinkler will flow, 8.3.4.3.1 and 8.3.4.3.2.
39. _____ Dry pipe and preaction systems riser is in a location that is protected from freezing conditions, 8.3.4.4.
40. _____ Dry pipe and preaction systems detection is provided in all sprinkler protected compartments and the detection system plans are provided, 8.3.4.5.
41. _____ Pipe is sloped at least ¼ in. for each 10 ft. in order to drain dry pipe and preaction systems, 8.3.4.7.
42. _____ Dry pipe and preaction systems air maintenance system is detailed and equipment data sheets are provided, 8.3.4.9.
43. _____ Sprinklers are in all areas except bathrooms, clothes closets where the areas, least dimension, and construction methods comply with 8.6. Other building areas not protected are in compliance with the areas listed in 8.6.4 and 8.6.5.

Alarms:
44. _____ Local flow alarm location and inspector's test connection are provided and detailed, except if the dwelling has smoke detectors in compliance with the building code, 7.6.

Hydraulic Calculations or Design Discharge:
45. _____ Reference points match with plans.
46. _____ Pipe size references match the plans and size is determined by hydraulic calculations based on one of the following methods in Section 8.4.4 or 8.4.5, or using the calculation methods in NFPA 13.
47. _____ Hydraulic calculations are also required when a system is gridded, looped, or connected to a city main less than 4 in., 8.4.7-8.4.9.
48. _____ Legend for calculation abbreviations is provided.
49. _____ Sprinkler specification matches what is on the plans and hydraulic calculations.
50. _____ Water flow information such as static PSI, residual PSI, and available GPM at 20 PSI residual is provided.
51. _____ Hydraulic calculations are provided using one of three methods described in Section 8.4.4 when the system is connected to a city main of at least 4 in. in size. Calculations include information as specified in 8.4.4.
52. _____ Sprinklers without a listed discharge criteria meet the discharge criteria specified in 8.1.1.1.1, and .2.
53. _____ Sprinkler with a listing discharge criteria: the system provides at least the flow required for multiple and single sprinkler operation as specified by the listing, 8.1.1.2.1, and the minimum density complies with 8.1.1.2.2.

Fire Plan Review and Inspection Guidelines

54._____ Sprinkler design for flat, smooth ceilings include the number of sprinklers having the largest water demand as specified in 8.1.2.

Pipe Support and Hangers:

55._____ Piping support shall comply with submitted manufacturers instructions and/or listing criteria. The plumbing code will be consulted for piping that does not have support criteria provided. Lateral movement is prevented for pipe laid on joists or rafters and in general, pipe movement is to be supported to restrain movement, 7.4.

Pipe and Valves:

56._____ One control valve is provided for both the domestic water and sprinkler, unless a separate control valve is provided for the sprinkler system, 7.1.1.

57._____ A drain and test connection is provided in accordance with 7.2.1.

58._____ Each portion of trapped dry system piping that is subject to freezing is provided a drain, 7.2.3.

59._____ A waterflow test connection is provided if a waterflow alarm is provided, 7.2.4.

60._____ Type and the diameter of pipe is provided. Pipe shall comply with Tables 5.2.1.1 and 5.2.2.2, 5.2.1.

61._____ At least 1 in. steel pipe or at least ¾ in. for other than steel pipe is used in the sprinkler system, 8.4.3.

62._____ Network systems are allowed ½ in. nonmetallic pipe or copper pipe with listed special fittings when in compliance with 10 conditions specified in 8.4.3.3.

Additional Comments:

Review Date: _____ Approved or Disapproved FD Reviewer: _____

Review Date: _____ Approved or Disapproved FD Reviewer: _____

Review Date: _____ Approved or Disapproved FD Reviewer: _____

Fire Plan Review and Inspection Guidelines

NFPA 13D Sprinkler System Acceptance Inspection
2006 IFC and 2007 NFPA 13D
This worksheet is for jurisdictions that permit the use of the 2007 NFPA 13D in lieu of IFC's referenced 2002 NFPA 13D.
Date of Inspection: _____ Permit Number: _____
Business/Building Name: _____ Address of Project: _____
Contractor: _____ Contractor's Phone: _____

Reference numbers following worksheet statements represent an NFPA code section unless otherwise specified.

Pass | Fail | NA

1. ___ | ___ | ___ Approved drawing and certification documents on site.
2. ___ | ___ | ___ System is leak tested at normal operating pressure when an FDC is not provided, 4.3.1.
3. ___ | ___ | ___ The one system control valve for both the sprinkler and domestic systems is on. If the sprinkler system has its own control valve, the valve is supervised by one of the three approved methods, 7.1.2.
4. ___ | ___ | ___ Signage is adjacent to the main water shutoff valve: Warning, the water system for this house supplies a fire sprinkler system that depends on certain flows and pressures being available to fight a fire...Don't remove this sign., 6.3.

Riser Room

5. ___ | ___ | ___ Operate the drain valve on the system side of the control valve.

Sprinklers

6. ___ | ___ | ___ Spacing between sprinklers does not exceed 12 ft., sprinklers are not greater than 6 ft. from a wall, and sprinklers are not within 8 ft. of each other unless listing allows it and consult the plans 8.1.3.
7. ___ | ___ | ___ Sprinkler heads are not painted or covered or blocked.
8. ___ | ___ | ___ Proper type and temperature sprinklers are used.
9. ___ | ___ | ___ Escutcheon plates are installed and pendent/upright deflectors are within 1 in. to 4 in. from the ceiling, sidewalls are within 4 in. to 6 in. from the ceiling or all are per their listing, 8.2.1 and 8.2.2.
10. ___ | ___ | ___ Pendent and upright sprinklers in closets can be installed within 12 in. of the ceiling, 8.2.1.3.
11. ___ | ___ | ___ Sprinklers are in all areas as shown on the approved set of plans, 8.6.

Pipe and Support

12. ___ | ___ | ___ Piping layout and pipe size are the same as the plans.
13. ___ | ___ | ___ Pipe hangers and supports are per the manufacturer's requirements.
14. ___ | ___ | ___ Pipe laid on open joists is secured to prevent lateral movement and other piping is secured to restrict movement.
15. ___ | ___ | ___ Pipes in attics are adequately insulated, 7.7.
16. ___ | ___ | ___ Antifreeze (AF) system is installed in accordance with the approved set of plans.

Additional Comments:

Inspection Date: _____   Approved or Disapproved   FD Inspector: _____

Inspection Date: _____   Approved or Disapproved   FD Inspector: _____

Fire Plan Review and Inspection Guidelines

NFPA 13D Sprinkler Installation Certification

Permit #: _____   Date: _____

	Property Protected	System Installer	System Supplier
Business Name:	_____	_____	_____
Address:	_____	_____	_____
	_____	_____	_____
Representative:	_____	_____	_____
Telephone:	_____	_____	_____

Location of Plans: _____

Location of Owner's Manual: _____

1. <u>Certification of System Installation</u>: Complete this section after system is installed, but prior to conducting operational acceptance tests.

   This system installation was inspected and found to comply with the installation requirements of:
   _____ NFPA 13D
   _____ IFC and IBC
   _____ Manufacturer's Instructions
   _____ Other (specify; FM, UL, etc.) _____

   Print Name: _____

   Signed: _____ Date: _____

   Organization: _____

2. <u>Certification of System Operation</u>: All operational features and functions of this system were tested and found to be operating properly in accordance with the requirements of:
   _____ NFPA 13D
   _____ IFC and IBC
   _____ Manufacturer's Instructions
   _____ Other (specify) _____

   Print Name: _____

   Signed: _____ Date: _____

   Organization: _____

Fire Plan Review and Inspection Guidelines

Automatic Sprinkler Protection High-Piled Combustible Storage using 2002 NFPA 13, Chapter 12 and IFC Chapter 23

The requirements for the design of automatic sprinkler systems used for the protection of high-piled combustible storage are found in Chapter 12 of NFPA 13 and Chapter 23 of the IFC. The following worksheets are intended to assist design professionals and fire code officials when a sprinklered building will be used for high-piled combustible storage. These worksheets can assist fire code officials when performing plan reviews and inspection of automatic sprinkler systems used for the protection of high-piled combustible storage. Design professionals may use these forms and provide them as supporting documentation with the shop drawings and hydraulic calculations.

Fire Plan Review and Inspection Guidelines

Commodity and Protection Information

Type of Storage: _____ palletized, _____ solid pile, _____ shelf, _____ bin box, _____ rack
If Based as a Mixed Commodity Based on Worst Case Commodity Class: Y or N or as a Mixed Commodity Y or N
Entire Storage Design is the Design Approach as Lower Commodity Class _____ or Commodity Segregation _____ in Accordance with 5.6.1.2

Permit #   **Business**   **Address:**   **Date:**

Rack ID or Area ID from Floor Plan	Storage Height	Ceiling Height and Ceiling Slope	Commodity Class	Rack: Number of Tiers	In-Rack Spklr: Y or N If Y, Provide Reference	Wet or Dry Sys.	Ceiling Spklr: Std, Lrg Drop, ESFR	Spklr Temp 165 F or 286 F	Spklr K-factor:	Spklr Coverage Area	Design Area and Density	Ceiling Design Curve Used: Provide Reference	Design Adjustments: Provide Code Reference and Percentage

Commodity Class and Type of Storage: The commodity class should be determined from the product inventory provided by the architect or owner. The inventory should describe the product, provide the weight or package volume of the amount of plastic or rubber, how the product is packaged, if stacked on wood or the type of plastic pallet, and if encapsulated or not. This information must accompany this code study so the plan reviewer can verify the commodity classification. State type of storage i.e., pallet, shelf, bin box, or rack.

Column Legend: If the entire storage area is going to use the worst case commodity class still provide a row of design information. Also, use **NA** if a column is not applicable.

Rack ID or Area ID: the architectural floor plans should have designated the rack or storage locations and the inventory commodity classification for each rack or storage area. From that information denote in this column the designation of the rack or area to be covered by the sprinkler design specified in the same row of the code study. Provide a copy of the architectural floor plan that shows rack and its commodity class.

Commodity Class: The commodity class should be determined from the product inventory provided by the design professional. The inventory should describe the product, provide the weight or package volume of the amount of plastic or rubber, how the product is packaged, if stacked on wood or the type of plastic pallet, and if encapsulated or not. This information must accompany this code study so the plan reviewer can verify the commodity classification.

Sprinkler (Spklr) Used: Indicate the type of ceiling sprinklers that are installed. If in-rack sprinklers are provided, indicate the type of intermediate level sprinkler used. Provide the equipment data sheet for the sprinklers used to protect the high-piled combustible storage.

Ceiling Design Curve Used: Provide the NFPA code reference, table, and associated figure reference.

Design Adjustments: Provide the NFPA code reference, table, and associated figure reference and the adjustment percentage from which your design adjustments are taken. These design adjustment percentages are related to storage height, dry or wet system, encapsulated or not, extra in-rack sprinklers, and footnotes.

In-Rack Sprinkler If Y, Provide Reference: Provide the NFPA code reference, table and associated figure reference used for the in-rack sprinkler design.

Fire Plan Review and Inspection Guidelines

NFPA 13 Sprinkler System for Storage Up To a Height of 25 Feet
Does not include Foam Systems
Plan Review Worksheet
2006 IFC and 2002 NFPA 13

Date of Review: _____ Permit Number: _____

Business/Building Name: _____ Address of Project: _____

Designer Name: _____ Designer's Phone: _____

Contractor: _____ Contractor's Phone: _____

No. of Sprinklers: _____ Occupancy Classification: _____

Reference numbers following worksheet statements represent an NFPA code section unless otherwise specified.

Worksheet Legend: ✓ or OK = acceptable   N = need to provide   NA = not applicable

1. _____ A minimum of three sets of drawings are provided. The plans declare the design is based on the 2002 edition year of NFPA 13.
2. _____ Equipment is listed for intended use and compatible with the system; specification data sheets are provided.
3. _____ Storage information worksheets are provided that indicate the storage height, ceiling slope, commodity classes, how commodities are packaged, type of packing materials used, type of pallets, whether pallet loads are encapsulated or banded, aisle widths, type of racks, flue spacing, if racks will be equipped with slatted or solid shelves, the sprinkler data sheets, design curve(s) and/or figure(s) used, if design area adjustments were made, and the storage location for each commodity class, IFC 2301.3.

Drawings shall detail the following (14.1.3.1- 14.1.3.44):
General:
4. _____ Type of system is noted; __ wet, __ dry, __preaction, __ deluge.
5. _____ Scale: a common scale shall be used and plan information shall be legible.
6. _____ Plot plan details illustrate the water supply pipe diameters, lengths and fittings to the building.
7. _____ The location of fire partitions and fire walls, and building cross sectional views.
8. _____ Occupancy class and or use of each room or area, 5.1.1.
9. _____ Total area protected by each system for each floor is provided.
10. _____ Dimensions for system piping, sprinkler spacing and branch line spacing, and elevation changes.
11. _____ Equipment symbol legend and the North orientation arrow are provided.
12. _____ Area limitations for hazard classification is 40,000 sq. ft. for storage, 8.2.1.
13. _____ Hydrant flow test determining water supply capacity at 20 PSI residual pressure is provided.
14. _____ Hydraulic calculations are provided with summary, detail worksheets and graph sheet, 14.3.
15. _____ Dry pipe system capacity in gallons is provided _____gal., not to be greater than 750 gal. unless the requirements of 7.2.3.2 or 7.2.3.3 are met, 7.2.3.
16. _____ All water supply valves and flow switches are supervised, IFC 903.4.
17. _____ Exterior flow alarm location is detailed. Note: if electric, it shall be listed for outdoor use, IFC 904.3.2.
18. _____ Backflow prevention device, when required, pressure loss data is provided in the hydraulic calculations.

Sprinklers:
19. _____ Total number of each type of sprinkler and their temperatures are noted, 8.3.2.1.
20. _____ Minimum temperature for ceiling sprinkler for high-piled storage is 150 deg., 8.3.2.7.
21. _____ Ceiling and roof cross sectional views are provided, 14.1.3.
22. _____ For each type of sprinkler the K factor and temperature rating are provided, 14.1.3(12).
23. _____ Each sprinkler coverage area is within its square footage and dimension limitations in accordance with its listing, 8.6.2.2, Table 8.6.2.2.1 (c and d).
24. _____ Control mode specific application, large drop, early suppression fast response sprinklers installation design comply with the standard and listing limitations, 6.1.1 and 8.4.1- 8.4.9.

Large Drop Sprinklers:
25. _____ Large drop sprinkler coverage area and maximum spacing is in accordance with Table 8.11.2.2.1.
26. _____ The minimum area of protection for large drop sprinklers is 80 sq. ft., 8.11.2.3.
27. _____ Maximum perpendicular distance to the walls shall be in accordance with 8.11.3.2.
28. _____ 8 ft. is the minimum spacing between sprinklers, 8.11.3.4.

29. _____ Large drop sprinklers installed in a building of unobstructed construction shall be in accordance with 8.11.4.1.1 and a cross sectional view is provided.
30. _____ Large drop sprinklers installed in a building of obstructed construction shall be in accordance with 8.11.4.1.2. and a cross sectional view is provided.
31. _____ Sprinkler position in relation to the roof or hanging obstructions shall comply with 8.11.5.1.
32. _____ For obstructions 8 in. or less, sprinklers are positioned in accordance with 8.11.5.2.1.3.
33. _____ Depending on the diameter of the branch lines, large drop sprinklers shall be located in accordance with 8.11.5.2.2.
34. _____ Sprinkler position when more than 24 in. directly above the bottom of an obstruction shall be in accordance with 8.11.5.3.4.
35. _____ For obstructions running parallel with a branch line, the sprinkler placement shall be in compliance with 8.11.5.3.5.
36. _____ Sprinklers deflectors shall be located at least 3 ft. above the top of storage, 8.11.6.
37. _____ Sprinkler position when above an obstruction more than 24 in. wide complies with Table 8.11.5.3.2. and Figure 8.11.5.3.2, 8.11.5.3.2.

General Storage Requirements:
38. _____ When a structure has two or more hazard areas or design methods that are nonseparated and are located adjacent to each other, the most hydraulic demanding area shall be extended as required by Section 12.1.5.
39. _____ Dry pipe and preaction system design areas shall be increased 30 percent but shall not exceed the design area specified in 12.1.6.1. This does not apply to preaction system designed in accordance with 12.1.6.3.
40. _____ Ceiling slope is detailed, cross sectional view provided, and does not exceed a 2 in 12 slope, 12.1.7.
41. _____ When adjustments are made to the design area, the designer shall provide an analysis explaining how the adjustments were made, 12.1.8.
42. _____ Design for idle pallets is in accordance with 12.1.9.
43. _____ Miscellaneous and Class I through IV sommodity storage up to 12 ft. and Group A plastic, rolled paper, tires shall be protected accordance with Figure 12.1.10 and Table 12.1.10.1.

General In-Rack Storage Requirements:
44. _____ In-rack sprinklers are ordinary temperature and the K-factor, orientation, temperature rating and type complies with 8.13.2.1 and 8.13.2.1.2.
45. _____ The area limitation for in-rack sprinkler systems is limited to 40,000 sq. ft. of floor area, 8.13.1.
46. _____ When adjacent to a heat source the sprinkler temperature rating is in accordance with 8.13.2.2.
47. _____ When in-rack sprinklers are required for Class I –IV commodities, water shields are provided when required by Section 8.13.3.1.
48. _____ When required for plastics storage, water shields are provided for in-rack sprinklers when required by 8.13.3.2.
49. _____ When in-rack sprinklers are required for miscellaneous storage, the number of sprinklers calculated and the discharge pressure complies with Section 12.1.12.
50. _____ Sprinklers K-factors for storage applications; rack, tire, rollpaper, and baled cotton are in compliance with 12.1.13 based on the required discharge density.

Palletized, Solid Pile, Bin Box, Shelf Storage for Class I-IV Commodities:
51. _____ Hose connections are provided for storage exceeding 12 ft., 12.2.2.1.3.
52. _____ The minimum design area/density for wet and for dry systems complies with 12.2.2.1.4.
53. _____ The design for area/density using ordinary or high temperature sprinklers is based on Figures 12.2.2.1.5.1 and 12.2.2.1.5.2 respectively, 12.2.2.1.5.1 and 12.2.2.1.5.2.
54. _____ If the density is adjusted due to storage height, it is in accordance with Figure 12.2.2.1.5.3 and calculations are provided.

Palletized and Solid Pile for Class I-IV Commodities:
55. _____ Large drop sprinklers protecting palletized or solid-pile Class I –IV commodities is in accordance with Tables 12.2.2.2.1(a) and (b), 12.2.2.2.
56. _____ The number of large drop sprinklers calculated for ordinary hazard and miscellaneous storage for the different types of systems complies with 12.2.2.2.2.1.
57. _____ Large drop and control mode specific application sprinklers protecting open wood joist construction shall be in accordance with 12.2.2.2.2.2 – 12.2.2.2.2.4.
58. _____ Large drop or control mode specific application sprinkler placement and branch line diameter limitations comply with 12.2.2.2.2.6.

59. _____ ESFR sprinklers protecting palletized or solid-pile Class I –IV commodities are in accordance with Table 12.2.2.3.1.
60. _____ The number of ESFR sprinklers calculated for the most hydraulically demanding design area and the number of calculated branch lines complies with 12.2.2.3.3.
61. _____ ESFR sprinklers installed above and below obstructions are hydraulically calculated in accordance with 12.2.2.3.4.

Palletized, Solid Pile, Bin Box, Shelf Storage for Plastic and Rubber Commodities:
62. _____ Sprinkler design protecting plastics stored up to 25 ft. are selected in accordance with Figure 12.2.3.1.1.
63. _____ The designer has provided design information for solid pile storage: closed or open array, clearance between storage, clearance between storage and the roof, pile height, and stable or unstable piles, 12.2.3.1.2. and 12.2.3.1.5.1
64. _____ Class B, free-flowing Group A, and Class C plastics are protected as the commodity specified in 12.2.3.1.3 and .4.
65. _____ Design area and density are based on Table 12.2.3.1.6 and Figure 12.2.3.1.1, 12.2.3.1.6.
66. _____ When determining the design area, the designer provided a calculations illustrating how the design area adjustments were made, 12.2.3.1.7.
67. _____ Large drop or control mode sprinklers K factor, storage height coverage, number of design sprinklers, and hose stream demand are in accordance with Table 12.2.3.2.1(a) or (b), 12.2.3.2.
68. _____ Large drop or control mode sprinklers for open wood joist construction, large drop sprinkler pressure, design area, and hose stream demand are in compliance with 12.2.3.2.2.2.
69. _____ Branch line diameters comply with 12.2.3.2.2.6.
70. _____ ESFR sprinklers protection design is in accordance with Table 12.2.3.3.1.
71. _____ The number of ESFR sprinkler for most hydraulically demanding design area and the number of branch lines calculated complies with 12.2.3.3.3.
72. _____ ESFR sprinklers installed above and below obstructions are hydraulically calculated in accordance with 12.2.2.3.4., 12.2.3.3.4.

Rack Storage of Commodities
General:
73. _____ Structural columns that are within storage racks are protected when required by 12.3.1.7.
74. _____ Solid shelving with an area of 20 sq. ft. up to 64 sq. ft., is protected by automatic sprinklers in accordance with 12.3.1.9.1.
75. _____ Solid shelving with an area exceeding 64 sq. ft. or the storage level exceeds 6 ft. with solid shelves, sprinkler protection is in compliance with 12.3.1.9.2.
76. _____ Where in-rack sprinklers protect a higher commodity hazard and the rack contains mixed commodities the in-rack sprinkler design shall comply with 12.3.1.11.2.
77. _____ When required horizontal barriers within the rack shall comply with Section 12.3.1.12.
78. _____ For storage 25 ft. or less a longitudinal flue is not required in double and multi-row racks so long as solid shelves are not present. Transverse flues required in single, double, and multi-row racks are provided and detailed or noted, 12.3.1.13.

Rack Storage of Class I through IV Commodities Stored up to 25 ft.:
79. _____ The density/area design for miscellaneous storage is in accordance with 12.1.10.
80. _____ The density/area design for storage exceeding the height of miscellaneous storage is in accordance with 12.3.2.
81. _____ The sprinkler design for single or double-row racks is derived from Table 12.3.2.1.2 using Figures 12.3.2.1.1(a)-(g), 12.3.2.1.2. If aisles are less than 3.5 ft. consider the racks as multiple-row racks.
82. _____ The sprinkler design for multiple-row racks 16 ft. or less in depth and with aisles 8 ft. or wider is derived from Table 12.3.2.1.3 using Figures 12.3.2.1.2(a)-(g), 12.3.2.1.3. The sprinkler design for multiple-row racks more than 16 ft in depth and with aisles less than 8 ft. is derived from Table 12.3.2.1.4 using Figures 12.3.2.1.2(a)-(g), 12.3.2.1.4.
83. _____ If the design density curve requirements are adjusted, the designer provided calculations explaining how the design area adjustments were made in accordance with 12.3.2.1.5.
84. _____ Hose stream demands were added to the hydraulic calculations in accordance with Table 12.3.2.1.6.
85. _____ Large drop and specific application control mode sprinklers shall be calculated is in accordance with Table 12.3.2.2.1(a) or (b), 12.3.2.2.1. When in-rack sprinklers are required by Table 12.3.2.2.1(a) or (b) the in-rack sprinkler design criteria is in accordance with 12.3.2.4.
86. _____ Large drop or control mode sprinklers under open wood joist construction shall operate at a minimum pressure specified in 12.3.2.2.3.2(A).

87. _____ Large drop K-11.2 sprinklers installed in open wood joist channel construction shall comply with, 12.3.2.2.3.2(B).
88. _____ The branch line diameter for pre-action automatic sprinklers shall comply with 12.3.2.2.3.6.
89. _____ ESFR sprinkler orientation, K-factor, and minimum pressure design for single, double, and multiple-row racks without solid shelves is in accordance with Table 12.3.2.3.1, 12.3.2.3.1.
90. _____ ESFR sprinklers are not used with solid shelves or combustible open-top cartons, 12.3.2.3.1.1.
91. _____ The number of ESFR sprinkler for most hydraulically demanding design area and the number of calculated branch lines complies with 12.3.2.3.4.
92. _____ ESFR sprinklers installed above and below obstructions are hydraulically calculated in accordance with 12.3.2.3.5.
93. _____ In-rack sprinklers are provided for single and double-row racks without solid shelves when required by Table 12.3.2.1.2, 12.3.4.1.1.
94. _____ In-rack sprinklers are provided for multiple-row rack less than 16 ft. deep and aisles 8 ft. wide or greater when required by Table 12.3.2.1.3, 12.3.2.4.1.2.
95. _____ In-rack sprinklers are provided for multiple-row rack more than 16 ft. deep and aisles up to 8 ft. wide when required by Table 12.3.2.1.4, 12.3.2.4.1.3.
96. _____ In-rack sprinklers for only one level are positioned at the first tier level at or above the halfway height of the storage, 12.3.2.4.1.4.
97. _____ In-rack sprinklers for only two levels are positioned at the first tier storage level at or above 1/3 and 2/3 the height of the storage, 12.3.2.4.1.5.
98. _____ In-rack maximum horizontal sprinkler spacing for single and double-row racks protecting nonencapsulated and encapsulated products are in accordance with 12.3.2.4.2.1.
99. _____ In-rack maximum horizontal sprinkler spacing and area limitations for multiple-row racks protecting nonencapsulated and encapsulated products are in compliance with 12.3.2.4.2.2.
100. _____ At least 6 in. of vertical clearance is detailed between the in-rack sprinkler deflector and the top of the tier of storage, 12.3.2.4.2.5.
101. _____ The in-rack water demand has been calculated based on the quantity of sprinklers specified for its storage conditions in accordance with Section 12.3.2.4.3, and the in-rack sprinkler discharge pressure is in compliance with 12.3.2.4.4.

Rack Storage of Plastics Stored up to 25 ft. and Ceiling Clearances up to 10 ft.:

102. _____ Control mode sprinkler design for single, double, and multiple-row racks for aisles 3.5 ft. wide or greater is in accordance with the decision tree in Figure 12.3.3.1.1.
103. _____ For control mode sprinklers, the ceiling water design density/area demand for Group A plastics packaged in cartons, encapsulated or not in single, double, or multiple-row racks are determined from Figures 12.3.3.1.5(a)-(f) and the designer has provided design rationale and plan details, 12.3.3.1.5.
104. _____ For control mode sprinkler design requirements protecting plastic commodities stored in single, double, or multiple-row racks at varying heights and ceiling clearances, Figures 12.3.3.1.5(a)–(f) have been used as specified in Sections 12.3.3.1.6 through 12.3.3.1.11.
105. _____ Large drop and specific control mode sprinkler design quantity and minimum operating pressure for single, double, and multiple-row racks without solid shelves for unexpanded plastics is as specified in Tables 12.3.3.2.1(a) or (b) and when required in-rack sprinklers are hydraulically designed and positioned in accordance with 12.3.2.4, 12.3.3.2.
106. _____ Large drop K-11.2 sprinklers positioned under open wood joist construction operate at a minimum pressure specified in 12.3.3.2.3.2.
107. _____ ESFR sprinkler orientation, K-factor, and minimum pressure design for single, double, and multiple-row racks without solid shelves for plastics are in accordance with Table 12.3.3.3.1, 12.3.3.3.1.
108. _____ ESFR sprinklers are not used with solid shelves or combustible open-top cartons, 12.3.3.3.1.1.
109. _____ The number of ESFR sprinkler for most hydraulically demanding design area and the number of branch calculated lines complies with 12.3.3.3.3.
110. _____ ESFR sprinklers installed above and below obstructions are hydraulically calculated in accordance 12.3.3.3.4.
111. _____ At least 6 in. of vertical clearance is provided between the in-rack sprinkler deflector and the top of the tier of storage, 12.3.3.4.2.1.
112. _____ In-rack sprinklers are provided for single and double-row racks without solid shelves when required by Table 12.3.2.1.2, 12.3.4.1.1.
113. _____ In-rack horizontal sprinkler spacing is provided and detailed on the plans as specified in Figures 12.3.3.1.5(a) and (f), 12.3.3.4.2.2.

Fire Plan Review and Inspection Guidelines

114. ____ The in-rack water demand calculations are provided and based on the number of sprinklers for each tier 12.3.2.4.3 and the discharge pressure complies with 12.3.3.4.4.

This worksheet should be used in conjunction with the worksheet for NFPA 13 automatic sprinkler systems for the following system components or features:
Other Coverage Areas
Pipe Support and Hangers
Pipe and Valves
Seismic Bracing
Fire Department Connection
Hydraulic Calculations
Wet System
Dry System
Preaction or Deluge
Combined Dry Pipe and Preaction
Valves
Miscellaneous

Additional Comments:

Review Date: _____ Approved or Disapproved   FD Reviewer: _____
Review Date: _____ Approved or Disapproved   FD Reviewer: _____
Review Date: _____ Approved or Disapproved   FD Reviewer: _____

Fire Plan Review and Inspection Guidelines

NFPA 13 Sprinkler System for Storage Up To 25 Feet
Does not include Foam Systems
Plan Review Worksheet
2006 IFC and 2007 NFPA 13

This worksheet is for jurisdictions that permit the use of the 2007 NFPA 13 in lieu of IFC's referenced 2002 NFPA 13.

Date of Review: _____ Permit Number: _____

Business/Building Name: _____ Address of Project: _____

Designer Name: _____ Designer's Phone: _____

Contractor: _____ Contractor's Phone: _____

No. of Sprinklers: _____ Occupancy Classification: _____

Reference numbers following worksheet statements represent an NFPA code section unless otherwise specified.

Worksheet Legend: ✓ or OK = acceptable   N = need to provide   NA = not applicable

1. _____ A minimum of three sets of drawings are provided. The plans declare the design is based on the 2007 edition of NFPA 13.
2. _____ Equipment is listed for intended use and compatible with the system; specification data sheets are provided.
3. _____ Storage information worksheets are provided that indicate the storage height, ceiling slope, commodity classes, how commodities are packaged, type of packing materials used, type of pallets, whether pallet loads are encapsulated or banded, aisle widths, type of racks, flue spacing, if racks will be equipped with slatted or solid shelves, the sprinkler data sheets, design curve(s) and/or figure(s) used, if design area adjustments were made, and the storage location for each commodity class IFC 2301.3.

Drawings shall detail the following (22.1.3.1-22.1.3.46):
General:

4. _____ Type of system is noted; __ hydraulic calc, __ wet, __ dry, __preaction, __ deluge. The plans declare the design 2007 edition year of NFPA 13.
5. _____ Scale: a common scale shall be used and plan information shall be legible.
6. _____ Plot plan details illustrate the water supply pipe diameters, lengths and fittings to the building.
7. _____ The location of fire partitions and fire walls, and building elevation views.
8. _____ Occupancy class and or use of each room or area. Classify a room if the hazard is different than adjacent areas or rooms, 5.1.–5.6.
9. _____ Full height cross sectionals are provided, include ceiling construction as needed for clarification.
10. _____ Total area protected by each system for each floor is provided.
11. _____ Dimensions for system piping, sprinkler spacing and branch line spacing, and elevation changes.
12. _____ Equipment symbol legend and the compass point are provided.
13. _____ Area limitations for hazard classification is 40,000 sq. ft. for high-piled storage, 8.2.1.
14. _____ Hydrant flow test determining water supply capacity at 20 PSI residual pressure is provided.
15. _____ Hydraulic calculations are provided with summary, detail worksheets, and graph sheet, 22.3.
16. _____ Dry pipe system capacity in gallons is provided _____gal., not to be greater than 750 gal. unless the requirements of 7.2.3.2 or 7.2.3.3 are met, 7.2.3.
17. _____ All water supply valves and flow switches are supervised, IFC 903.4 refer to exceptions.
18. _____ Exterior flow alarm location is detailed. Note: if electric, it shall be listed for outdoor use, IFC 904.3.2.
19. _____ Backflow prevention device pressure loss data is provided in the hydraulic calculations.

Sprinklers:
20. _____ Total number of each type of sprinkler and their temperatures are noted, 8.3.2.2 to 8.3.2.5.
21. _____ Minimum temperature for ceiling sprinkler for high-piled storage is 150 deg., 8.3.2.7.
22. _____ Sprinkler locations are correct, ceiling and roof cross sectional views are provided for clarification, 22.1.3(45).
23. _____ For each type of sprinkler the K factor and temperature rating are provided, 22.1.3(12).
24. _____ Each sprinkler coverage area is within its square footage limitations and dimension limitations in accordance with its listing, 8.6.2.2, Table 8.6.2.2.1 (c and d).
25. _____ Control mode specific application, large drop, early suppression fast response sprinklers installation design comply with the standard and listing limitations, 6.1.1 and 8.4.1 – 8.4.10.
26. _____ Standard sprinkler spacing from vertical obstructions complies with Table 8.6.5.1.2 and for floor mounted obstructions, Table 8.6.5.2.2, 8.6.5.1.2 and 8.6.5.2.2.

27. _____ ESFR sprinkler spacing from ceiling obstructions complies with Table 8.12.5.1.1

General Storage Requirements:
28. _____ Ceiling slope is detailed, cross sectional view provided, and does not exceed a 2 in 12 slope, 12.1.2.
29. _____ Storage design requirements are based on the absence of draft curtains and roof vents, 12.1.
30. _____ If a structure has two or more hazard areas or design methods that are non-separated and adjacent to each other, the more demanding area extends coverage in accordance with 12.3.
31. _____ Dry pipe and preaction system design areas are increased 30 percent but not to exceed the area specified in 12.5.1 and 12.5.2.
32. _____ When adjustments are made to the design area, the designer provided calculations explaining and showing how the adjustments were made, 12.7.7.
33. _____ Design for idle pallets is in accordance with 12.12.
34. _____ Densities up to 0.20 GPM/ft^2 are permitted to use sprinklers with a 5.6 K-factor or larger, 12.6.1.
35. _____ Densities 0.21 GPM/ft^2 to 0.34 GPM/ft^2 protecting rack, tire, roll paper and baled cotton storage with upright and pendent standard response sprinklers are to use sprinklers with a 8.0 K-factor or larger, 12.6.2.
36. _____ Densities greater than 0.34 GPM/ft^2 protecting rack, tire, roll paper, and baled cotton storage with upright and pendent standard response sprinklers require sprinklers with a 11.2 K-factor or larger, 12.6.3
37. _____ If the design area is adjacent a combustible concealed space then the minimum design area is in compliance with 12.9.1 unless the concealed space meets the criteria of Section 12.9.2 (1) – (9), 12.9.1.

Large Drop Sprinklers:
38. _____ Large drop sprinkler coverage area and maximum spacing is in accordance with Table 8.11.2.2.1, which depends on the construction situation listed in the table.
39. _____ The minimum coverage area for large drop sprinklers is not less than that specified in 8.11.2.3.
40. _____ Maximum perpendicular distance to the walls is not greater than ½ of allowable distance between sprinklers, 8.11.3.2.and Table 8.11.2.2.1.
41. _____ The minimum spacing between large drop sprinklers is not less that specified in 8.11.3.4.
42. _____ Large drop sprinklers deflectors are between 6 in. and 8 in. below the ceiling of unobstructed construction cross sectional view provided, 8.11.4.1.1.
43. _____ Large drop sprinklers deflectors position complies with Table 8.11.5.1.2, cross sectional view(s) provided, 8.11.4.1.2.
44. _____ Large drop sprinkler position adjacent ceiling or hanging obstructions complies with Table 8.11.5.1.2 unless sprinklers are placed on both sides of the obstructions.
45. _____ For obstructions 8 in. or less, large drop sprinklers are positioned at least three times the maximum obstruction dimension away unless 8.11.5.1.2 or 8.11.5.1.3 are met, 8.11.5.2.1.3.
46. _____ Large drop sprinklers position in relation to specific sized branch lines complies with 8.11.5.2.2.
47. _____ Large drop sprinkler position when more than 24 in. directly above an obstruction complies with Table 8.11.5.3.2. and Figure 8.11.5.3.2, and the design approach is detailed, 8.11.5.3.4.
48. _____ For obstructions running parallel with a branch line, large drop sprinkler placement as detailed in Figure 8.11.5.3.5.
49. _____ At least a 3 ft. clearance between large drop sprinklers deflectors and storage is provided, 8.11.6.
50. _____ Large drop sprinkler position when more than 24 in. directly above an obstruction complies with Table 8.11.5.3.2 and Figure 8.11.5.3.2.

General In-Rack Storage Requirements:
51. _____ In-rack sprinklers are ordinary temperature, K factor, orientation, temperature rating and type complies with 8.13.2.1.
52. _____ One in-rack sprinkler system is limited to 40,000 sq. ft. of floor area including aisle ways, 8.13.1.
53. _____ When adjacent to a heat source the sprinkler temperature rating is in accordance with 8.3.2 and 8.13.2.2.
54. _____ When in-rack sprinklers are required for Class I –IV commodities, water shields are provided when required by Section 8.13.3.1.
55. _____ When required for plastics storage, water shields are provided for in-rack sprinklers when required by 8.13.3.2.
56. _____ When in-rack sprinklers are required for miscellaneous storage, the number of sprinklers calculated and the discharge pressure complies with Section 13.3.
57. _____ Sprinklers K-factors for storage applications; rack, tire, rollpaper, and baled cotton are in compliance with 12.6.1-.3 based on the discharge density required.

Palletized, Solid Pile, Bin Box, Shelf Storage for Class I-IV Commodities:
58. _____ Hose connections are provided for storage exceeding 12 ft., 12.2.2.

Fire Plan Review and Inspection Guidelines

59._____ Storage not exceeding 12 ft. is protected as miscellaneous storage in accordance with Table 13.2.1, 14.2.3.1.
60._____ When the height of the storage is more than 12 feet the minimum design area/density for wet systems is 0.15 GPM/2,000 sq. ft. and for dry systems is 0.15 GPM/2,600 sq. ft., 14.2.4.5.
61._____ The design for area/density using ordinary or high temperature sprinklers is based on Figures 14.2.4.1 and .2 respectively, 14.2.4.1, .2.
62._____ If the density is adjusted due to storage height, it is in accordance with Figure 14.2.4.3 and calculations are provided, 14.2.4.3.

Palletized and Solid Pile for Class I-IV Commodities:
63._____ Large drop sprinklers protecting palletized or solid-pile Class I –IV commodities is in accordance with Tables 14.3.1(a) and (b), 14.3.1.
64._____ The number of large drop sprinklers calculated for ordinary hazard and miscellaneous storage for the different types of systems complies with 14.3.3.
65._____ Large drop or specific application control mode sprinklers for open wood joist construction shall be designed based on a minimum 50 PSI design pressure unless the wood joists are constructed with fire stopping complying with 14.3.4.2.
66._____ ESFR sprinklers protecting palletized or solid-pile Class I –IV commodities are in accordance with Table 14.4.1.
67._____ The number of ESFR sprinkler for most hydraulically demanding design area and the number of branch lines complies with 14.4.3.
68._____ ESFR sprinklers installed above and below obstructions are hydraulically calculated in accordance with, 14.4.4.

Palletized, Solid Pile, Bin Box, Shelf Storage for Plastic and Rubber Commodities:
69._____ Sprinklers design protecting plastics stored up to 25 ft. is selected in accordance with Figure 15.2.1 and justification is provided by the designer.
70._____ The designer has provided design information concerning storage conditions, i.e. closed or open array, clearance between storage, clearance between storage and ceiling/roof, pile height, and stable or unstable piles, 15.1.2 and 15.2.
71._____ Class B, free-flowing Group A, and Class C plastics are protected as the commodity specified 15.2.3, .4.
72._____ When a density/area design method is used the design area and density are based on Table 15.2.5(a) or (b) and Figure 15.2.1, 15.2.5.
73._____ When determining the design area, the designer provided a design sheet explaining and showing how design area adjustments were made, 15.2.7.
74._____ Large drop or specific application control mode sprinklers K-factor, storage height coverage, number of design sprinklers, and hose stream demand are in accordance with Table 15.3.1(a) or (b), 15.3.1.
75._____ Large drop or specific application control mode sprinklers for open wood joist construction, large drop sprinkler pressure, design area, and hose stream demand are in compliance with 15.3.3.
76._____ ESFR sprinklers protection design is in accordance with Table 15.4.1.
77._____ The number of ESFR sprinkler for most hydraulically demanding design area and the number of branch lines complies with 15.4.3.
78._____ ESFR sprinklers installed above and below obstructions are hydraulically calculated in accordance 15.4.4.

Commodities Stored on Racks:
General:
79._____ Structural steel columns that are within storage racks are protected in accordance with 16.1.4.
80._____ Solid shelving exceeding 20 sq. ft. up to 64 sq. ft., sprinkler protection is within 6 ft. vertically of each intermediate level and is provided at the ceiling, 16.1.6.1.
81._____ Solid shelving exceeding 64 sq. ft. or the storage level exceeds 6 ft. with solid shelves, sprinkler protection is in compliance with 16.1.6.2.
82._____ Where in-rack sprinklers protect a higher commodity hazard then the adjacent areas in the same rack, that in-rack sprinkler design complies with 16.1.8.2.
83._____ When required and provided the horizontal barriers within the rack are made of wood or sheet metal, within 2 in. around rack uprights, and the construction details are provided, 16.9.9.
84._____ For storage 25 ft. or less a longitudinal flue is not required in double and multiple-row racks as long as solid shelves are not present. Transverse flues required in single, double, and multiple-row racks are provided and detailed or noted, 16.1.10.
85._____ For storage exceeding 25 ft. the longitudinal flue space in double-row racks and the transverse flue space between loads and rack uprights in single, double, and multiple-row racks are in compliance with 16.1.11.1 and detailed or noted.

86. _____ For storage exceeding 25 ft., vertical clear space is provided between the sprinkler deflector and the top of a tier of storage, the face sprinklers positioned from rack uprights and the aisle face of storage for single, double, and multiple-row racks in accordance with 16.1.11.2.
87. _____ Sprinklers locations in relation to the longitudinal and transverse flue space intersections, the horizontal load beams or horizontal rack components and uprights are in accordance with 16.1.11.2.

Rack Storage of Class I through IV Commodities Stored up to 25 ft.:
88. _____ Rack storage not exceeding 12 ft. is protected as miscellaneous storage in accordance with Table 13.2.1, 16.2.1.2.
89. _____ The density/area design for storage exceeding the height of miscellaneous storage is in accordance with 16.2.1.3.
90. _____ The sprinkler design for single or double-row racks is derived from Table 16.2.1.3.2 using Figures 16.2.1.3.2(a)–(g), 16.2.1.3.2. If aisles are less than 3.5 ft. consider the racks as multiple-row racks.
91. _____ The sprinkler design for multiple-row racks 16 ft. or less in depth and with aisles 8 ft. or wider is derived from Table 16.2.1.3.3.1 using Figures 16.2.1.3.2(a)–(g), 16.2.1.3.3.1.
92. _____ The sprinkler design for multiple-row racks more than 16 ft. in depth and with aisles less than 8 ft. is derived from Table 16.2.1.3.3.2 using Figures 16.2.1.3.2(a)–(g), 16.2.1.3.3.2.
93. _____ If the design density curve requirements are adjusted, the designer provided calculations explaining and showing how design area adjustments were made in accordance with 16.2.1.3.4.
94. _____ Hose stream demands were added to the hydraulic calculations in accordance with Table 16.2.1.3.5.
95. _____ Large drop and control mode sprinkler quantity and minimum pressure design for single, double, and multiple-row racks without solid shelves is in accordance with Table 16.2.2.1(a) or (b), 16.2.2.1. When in-rack sprinklers are required by Table 16.2.2.1(a) or (b) the in-rack sprinkler design criteria is in accordance with 16.2.4.
96. _____ Large drop or specific application control mode sprinklers under open wood joist construction operate at a minimum pressure specified in 16.2.2.4.1.
97. _____ Large drop K-11.2 sprinklers can use the lower pressures offered in Table 16.3.2.1(a) when open wood joist channel construction is completely, to its depth, firestopped for intervals that are not greater than 20 ft., 16.2.2.4.2.
98. _____ ESFR sprinkler orientation, K-factor, and minimum pressure design for single, double, and multiple-row racks without solid shelves is in accordance with Table 16.2.3.1, 16.2.3.1.
99. _____ ESFR sprinklers are not used with solid shelves or combustible open-top cartons, 16.2.3.2.
100. _____ The number of ESFR sprinkler for most hydraulically demanding design area and the number of branch lines complies with 16.2.3.4.
101. _____ ESFR sprinklers installed above and below obstructions are hydraulically calculated in accordance with 16.2.3.5.
102. _____ In-rack sprinklers are provided for single- and double-row racks without solid shelves when required by Table 16.2.1.3.2, 16.2.4.1.1.
103. _____ In-rack sprinklers are provided for multiple-row rack less than 16 ft. deep and aisles 8 ft. wide or greater when required by Table 16.2.1.3.3.1, 16.2.4.1.2.
104. _____ In-rack sprinklers are provided for multiple-row rack more than 16 ft. deep and aisles up to 8 ft. wide when required by Table 16.2.1.3.3.2, 16.2.4.1.3.
105. _____ In-rack sprinklers for only one level are positioned at the first tier level at or above the half way height of the storage 16.2.4.1.4.
106. _____ In-rack sprinklers for only two levels are positioned at the first tier storage level at or above 1/3 and 2/3 the height of the storage, 16.2.4.1.5.
107. _____ In-rack maximum horizontal sprinkler spacing for single and double-row racks protecting nonencapsulated products are in accordance with Table 16.2.4.2.1 and the maximum spacing for encapsulated products is in accordance with 16.2.4.2.1.
108. _____ In-rack maximum horizontal sprinkler spacing and coverage area for multiple-row racks protecting nonencapsulated and encapsulated Class I – IV commodities is in accordance with 16.2.4.2.2.
109. _____ At least 6 in. of vertical clearance is detailed between the in-rack sprinkler deflector and the top of the tier of storage, 16.2.4.2.4, .5.
110. _____ The in-rack water demand has been calculated based on the quantity of sprinklers specified for its storage conditions in accordance with Section 16.2.4.3.1, discharge pressure complies with 16.2.4.4 and calculations are provided.
111. _____ Slatted shelving density/area design, commodity coverage, flue spacing, aisle width, slat dimensions are in accordance with Section 16.2.5.

Rack Storage of Plastics Stored up to 25 ft. and Ceiling Clearances up to 10 ft.:

112. ____ Structural steel columns that are within storage racks are protected in accordance with 17.1.4.
113. ____ Sprinkler design decision for single, double, and multiple-row racks for aisles 3.5 ft. wide or greater is in accordance with Figure 17.1.2.1.
114. ____ Group A plastics not exceeding 5 ft. are protected as miscellaneous storage in accordance with Table 13.2.1, 17.2.1.1.
115. ____ Solid shelving exceeding 20 sq. ft. up to 64 sq. ft., is protected by automatic sprinklers in accordance with 17.1.5.1.
116. ____ Solid shelving exceeding 64 sq. ft. and each storage level that exceeds 6 ft., sprinkler protection is in compliance with 17.1.5.2.
117. ____ For control mode sprinklers, the ceiling water design density/area demand for Group A plastics packaged in cartons (including encapsulated or nonencapsulated commodities), in single, double, or multiple-row racks are determined from Figures 17.2.1.2(a)-(f) and the designer has provided design rationale and plan details, 17.2.1.2.
118. ____ For control mode sprinkler design requirements protecting plastic commodities stored in single, double, or multiple-row racks at varying heights and ceiling clearances, Figures 17.2.1.2(a)-(f) have been used as specified in Sections 17.2.1.3 through 17.2.1.7.
119. ____ Large drop and specific control mode sprinkler design quantity and minimum operating pressure for single, double, and multiple-row racks without solid shelves for unexpanded plastics is as specified in Tables 17.2.2.1(a) or (b) and when required in-rack sprinklers are hydraulically designed and positioned in accordance with 17.2.4, 17.2.2.
120. ____ Large drop K factor 11.2 sprinklers under open wood joist construction operate at a minimum pressure specified in 17.2.2.4.1.
121. ____ ESFR sprinkler orientation, K-factor, and minimum pressure design for single, double, and multiple-row racks without solid shelves for plastics cartoned or not and expanded or not are in accordance with Table 12.3.3.3.1, 12.3.3.3.1.
122. ____ ESFR design is in accordance with Table 17.2.3.1.
123. ____ ESFR sprinklers are not used with solid shelves or combustible open-top cartons, 17.2.3.1.1.
124. ____ The number of ESFR sprinkler for most hydraulically demanding design area and the number of branch lines complies with 17.2.3.3.
125. ____ ESFR sprinklers installed above and below obstructions are hydraulically calculated in accordance with 17.2.3.4.
126. ____ At least 6 in. of vertical clearance is detailed between the in-rack sprinkler deflector and the top of the tier of storage, 17.2.4.2.1.
127. ____ Sprinkler locations and spacing in the racks comply with Figures 17.2.1.2(a)–(f) and the designer has provided which Figure(s) are used, 17.2.4.2.1 and .2.
128. ____ The in-rack water demand calculations are provided and based on the number of sprinklers for each tier 17.2.4.3 and the discharge pressure complies with 17.2.4.4.

The requirements associated with the subjects listed below that pertain to this storage and building design are to be reviewed using the NFPA 13 Sprinkler worksheet:
Other Coverage Areas
Pipe Support and Hangers
Pipe and Valves
Seismic Protection
Fire Department Connection
Hydraulic Calculations
Wet System
Dry System, Grid System is not Permitted, 7.2
Preaction or Deluge
Combined Dry Pipe and Preaction
Valves
Miscellaneous

Fire Plan Review and Inspection Guidelines

Additional Comments:

Review Date: _____ Approved or Disapproved   FD Reviewer: _____
Review Date: _____ Approved or Disapproved   FD Reviewer: _____
Review Date: _____ Approved or Disapproved   FD Reviewer: _____

Fire Plan Review and Inspection Guidelines

NFPA 13 Sprinkler System for Storage Greater Than 25 Feet and Rubber Tires
Does not include Foam Systems
Plan Review Worksheet
2006 IFC and 2002 NFPA 13

Date of Review: _____ Permit Number: _____

Business/Building Name: _____ Address of Project: _____

Designer Name: _____ Designer's Phone: _____

Contractor: _____ Contractor's Phone: _____

No. of Sprinklers: _____ Occupancy Classification: _____

Reference numbers following worksheet statements represent an NFPA code section unless otherwise specified.

Worksheet Legend: ✓ or **OK** = acceptable   **N** = need to provide   **NA** = not applicable

1. _____ A minimum of three sets of drawings are provided. The plans declare the design is based on the 2002 edition of NFPA 13.
2. _____ Equipment is listed for intended use and compatible with the system; specification data sheets are provided.
3. _____ Storage information worksheets are provided that indicate the storage height, ceiling slope, commodity classes, how commodities are packaged, type of packing materials used, type of pallets, whether pallet loads are encapsulated or banded, aisle widths, type of racks, flue spacing, if racks will be equipped with slatted or solid shelves, the sprinkler data sheets, design curve(s) and/or figure(s) used, if design area adjustments were made, and the storage location for each commodity class, IFC 2301.3.

Drawings shall detail the following (14.1.3.1- 14.1.3.44):
General:

4. _____ Type of system is noted: __ wet, __ dry, __preaction, __ deluge.
5. _____ Scale: a common scale shall be used and plan information shall be legible.
6. _____ Plot plan details illustrate the water supply pipe diameters, lengths and fittings to the building.
7. _____ The location of partitions and fire walls, and building cross sectional views.
8. _____ Occupancy class and or use of each room or area, 5.1.1.
9. _____ Total area protected by each system per floor is provided.
10. _____ Dimensions for system piping, sprinkler spacing and branch line spacing and elevation changes.
11. _____ Equipment symbol legend and the compass point are provided.
12. _____ Area limitations for hazard classification is 40,000 sq. ft. for storage, 8.2.1.
13. _____ Hydrant flow test determining water supply capacity at 20 PSI residual pressure is provided.
14. _____ Hydraulic calculations are provided with summary, detail worksheets and graph sheet, 14.3.
15. _____ Dry pipe system capacity in gallons is provided _____gal., not to be greater than 750 gal. unless the requirements of 7.2.3.2 or 7.2.3.3 are met, 7.2.3.
16. _____ All water supply valves and flow switches are supervised, IFC 903.4.
17. _____ Exterior flow alarm location is detailed. Note: if electric, it shall be listed for outdoor use, IFC 904.3.2.
18. _____ When required, backflow prevention device pressure loss data is provided in the hydraulic calculations.

Sprinklers:

19. _____ Total number of each type of sprinkler and their temperatures are noted, 8.3.2.1.
20. _____ Minimum temperature for ceiling sprinkler for high-piled storage is 150°, 8.3.2.7.
21. _____ Sprinkler locations are correct, ceiling and roof cross sectional views are provided for clarification, 14.1.3.
22. _____ Each sprinkler coverage area is within its square footage and dimension limitations in accordance with its listing, 8.6.2.2, Table 8.6.2.2.1(c and d).
23. _____ Control mode specific application, large drop, early suppression fast response sprinklers installation design comply with the standard and listing limitations, 6.1.1 and 8.4.1- 8.4.9.

Large Drop Sprinklers:

24. _____ Large drop sprinkler coverage area and maximum spacing is in accordance with Table 8.11.2.2.1.
25. _____ The minimum area of protection for large drop sprinklers is 80 sq. ft., 8.11.2.3.
26. _____ Maximum perpendicular distance to the walls shall be in accordance with 8.11.3.2.
27. _____ 8 ft. is the minimum spacing between sprinklers, 8.11.3.4.

28. _____ Large drop sprinklers installed in a building of unobstructed construction shall be in accordance with 8.11.4.1.1 and a cross sectional view is provided.
29. _____ Large drop sprinklers installed in a building of obstructed construction shall be in accordance with 8.11.4.1.2. and a cross sectional view is provided.
30. _____ Sprinkler position in relation to the roof or hanging obstructions shall comply with 8.11.5.1.
31. _____ For obstructions 8 in. or less, sprinklers are positioned in accordance with 8.11.5.2.1.3. Depending on the size of the branch lines, large drop sprinklers are permitted are located in accordance with 8.11.5.2.2.
32. _____ Sprinkler position when more than 24 in. directly above the bottom of an obstruction is in accordance with 8.11.5.3.4.
33. _____ For obstructions running parallel with a branch line, the sprinkler placement is in compliance with 8.11.5.3.5.
34. _____ Sprinklers deflectors have a minimum clearance of 3 ft. above the top of storage, 8.11.6.
35. _____ Sprinkler position when above an obstruction more than 24 in. wide complies with Table 8.11.5.3.2. and Figure 8.11.5.3.2, 8.11.5.3.2.

General Storage Requirements:
36. _____ When a structure has two or more hazard areas or design methods that are nonseparated and are located adjacent to each other, the most hydraulic demanding area shall be extended as required by Section 12.1.5.
37. _____ Dry pipe and preaction system design areas are increased 30 percent but not exceed 6,000 sq. ft., 12.1.6.1. This does not apply to preaction system designed in accordance with 12.1.6.3.
38. _____ Ceiling slope is detailed, cross sectional view provided, and does not exceed a 2 in 12 slope, 12.1.7.
39. _____ When adjustments are made to the design area, the designer provided calculations explaining the basis of the calculations, 12.1.8.
40. _____ Design for idle pallets is in accordance with 12.1.9.

General In-Rack Storage Requirements:
41. _____ In-rack sprinklers are ordinary temperature and the K-factor, orientation, temperature rating and type complies with 8.13.2.1 and 8.13.2.2.
42. _____ The area of protection for an in-rack sprinkler system is limited to 40,000 sq. ft., 8.13.1.
43. _____ When adjacent to a heat source the sprinkler temperature rating is in accordance with 8.13.2.2.
44. _____ When in-rack sprinklers are required for Class I –IV commodities, water shields are provided when required by Section 8.13.3.1.
45. _____ When required for plastics storage, water shields are provided for in-rack sprinklers when required by 8.13.3.2.
46. _____ Sprinklers K-factors for storage applications; rack, tire, rolled paper, and baled cotton are in compliance with 12.1.13 based on the discharge density required.

Palletized and Solid Pile for Class I-IV Commodities:
47. _____ Hose connections are provided for storage exceeding 12 ft., 12.2.2.1.3.
48. _____ ESFR sprinklers protecting palletized or solid-pile Class I –IV commodities are in accordance with Table 12.2.2.3.1.
49. _____ The number of ESFR sprinkler for most hydraulically demanding design area and the number of branch lines calculated complies with 12.2.2.3.3.
50. _____ ESFR sprinklers installed above and below obstructions are hydraulically calculated in accordance with 12.2.2.3.4.

Palletized, Solid Pile for Plastic and Rubber Commodities:
51. _____ ESFR sprinklers protection design is in accordance with Table 12.2.3.3.1.
52. _____ The number of ESFR sprinkler for most hydraulically demanding design area and the number of branch lines calculated complies with 12.2.3.3.3.
53. _____ ESFR sprinklers installed above and below obstructions are hydraulically calculated in accordance with 12.2.2.3.4., 12.2.3.3.4.

Rack Storage of Commodities
General:
54. _____ Hose connections are provided for Class I-IV commodities exceeding 12 ft., 12.3.1.3.
55. _____ When required the building steel columns that are within storage racks are protected in accordance with 12.3.1.7.
56. _____ Solid shelving with an area exceeding 20 sq. ft. up to 64 sq. ft., is protected by automatic sprinklers in accordance with 12.3.1.9.1.
57. _____ Solid shelving exceeding 64 sq. ft. or the storage level exceeds 6 ft. with solid shelves, sprinkler protection is in compliance with 12.3.1.9.2.

58._____ Where in-rack sprinklers protect a higher commodity hazard and the rack contains mixed commodities the in-rack sprinkler design shall comply with 12.3.1.11.2.
59._____ When required horizontal barriers are required they shall comply with Section 12.3.1.12.
60._____ For storage exceeding 25 ft., the longitudinal flue space in double-row racks and the transverse flue space between loads and rack uprights in single, double, and multiple-row racks are provided and detailed or noted in accordance with 12.3.1.14.
61._____ For storage exceeding 25 ft., vertical clear space is provided between the sprinkler deflector and the top of a tier of storage, the face sprinklers positioned from rack uprights and the aisle face of storage for single, double, and multiple-row racks in accordance with 12.3.1.14.2.
62._____ Sprinklers locations in relation to the longitudinal and transverse flue space intersections, the horizontal load beams or horizontal rack components and uprights are in accordance with 12.3.1.14.2.

Rack Storage of Class I through IV Commodities for Greater Than 25 ft.:
63._____ The design requirements for control mode sprinklers protecting single, double-row racks with nonencapsulated products is in compliance with 12.3.4.1.1 in regards to the type of shelving used, aisles width, sprinkler clearance from the top of storage, sprinkler temperature, design area, and discharge density for specific commodities.
64._____ Increase density design 25 percent for encapsulated commodities, 12.3.4.1.2.
65._____ The design requirements for control mode sprinklers protection for multiple-row racks with nonencapsulated products is in compliance with 12.3.4.1.3 in regards to the type of shelving used, aisles width, sprinkler clearance from the top of storage, sprinkler temperature, design area, and discharge density for specific commodities.
66._____ For control mode water design density/area demand the hose demand from Table 12.3.4.1.5 has been added, 12.3.4.1.5.
67._____ Large drop and specific control mode sprinkler design quantity and minimum operating pressure and in-rack criteria for single, double, and multiple-row racks without solid shelves for Class I though IV commodities are as specified in Table 12.3.4.2.1 and when required, in-rack sprinklers are hydraulically designed and positioned in accordance with 12.3.4.4.
68._____ Large drop K-factor-11.2 sprinklers positioned under open wood joist construction shall operate at a pressure specified 12.3.4.2.3.2.
69._____ ESFR protection pertaining to sprinkler location, K-factor, minimum pressure design, storage and ceiling height, and in-rack criteria for single, double, and multiple-row racks without solid shelves for specific commodities are in compliance with 12.3.4.3.1 and 12.3.4.3.2, and when required, in-rack sprinklers are hydraulically designed and positioned in accordance with 12.3.4.3.4.
70._____ ESFR sprinklers shall not be used with solid shelves or combustible open-top cartons, 12.3.4.3.1.1.
71._____ The number of ESFR sprinkler for most hydraulically demanding design area and the number of branch lines calculated shall comply with 12.3.4.3.3.
72._____ ESFR sprinklers installed above and below obstructions are hydraulically calculated in accordance 12.3.4.3.5.
73._____ In-rack sprinklers protection involving double-row racks without solid shelves, with ceiling to the top of storage clearance up to 10 ft. complies with Table 12.3.4.1.1 and are designed in accordance with Figures 12.3.4.4.1.1(a)-(j). The same criteria are used when single-row racks area mixed in with double-row racks.
74._____ The location of in-rack sprinklers comply with 12.3.4.4.1.1.
75._____ In-rack sprinklers protection involving single-row racks without solid shelves, with ceiling to the top of storage clearance up to 10 ft. complies with Figures 12.3.4.4.1.2(a)-(e), 12.3.4.4.1.2.
76._____ In-rack sprinklers protection involving multiple-row racks with ceiling clearance to the top of storage up to 10 ft. complies with Table 12.3.4.1.3 and sprinkler locations are in accordance with Figures 12.3.4.4.1.2(a)-(c), and the top level of in-rack sprinklers locations are in compliance with 12.3.4.4.1.3 for specific commodity classes.
77._____ In-rack sprinklers are spaced and staggered vertically and horizontally as specified in 12.3.4.4.2.1.
78._____ In-rack sprinklers for double-row racks are horizontally spaced in the horizontal space at or near vertical locations in accordance with 12.3.4.4.2.2.
79._____ In-rack sprinklers for multiple-row racks are horizontally spaced in accordance with 12.3.4.4.2.3.
80._____ The in-rack water demand calculations are provided and based on one of four design approaches specified in 12.3.4.4.3 and the discharge quantity is in compliance with 12.3.4.4.4.

Rack Storage of Plastics Stored Greater Than 25 ft. for Single, Double, and Multiple-row Racks:
81. _____ Control mode ceiling sprinkler density/area design for Group A plastics are in accordance with Table 12.3.5.1.1 and if single-row racks are intermixed with double-racks the in-rack sprinkler locations are based on Figures 12.3.5.1.2(a) and (b), 12.3.5.1.1 and 12.3.5.1.2.
82. _____ Control mode sprinklers in single-row racks are located in accordance with Figures 12.3.5.1.2.1(a)–(c), 12.3.5.1.2.1.
83. _____ Hose stream demands were added to the hydraulic calculations in accordance with Table 12.3.5.1.3.
84. _____ ESFR sprinkler orientation, K-factor, and minimum pressure design for single, double, and multiple-row racks for plastics cartoned or not and expanded or not are in accordance with Table 12.3.5.3.1, 12.3.5.3.1.
85. _____ ESFR sprinklers are not used with solid shelves or combustible open-top cartons, 12.3.5.3.1.1.
86. _____ The number of ESFR sprinkler for most hydraulically demanding design area and the number of branch lines calculated complies with 12.3.5.3.3.
87. _____ ESFR sprinkler coverage is supplemented with in-rack sprinklers when required by Table 12.3.5.3.1, the hydraulic design, pressure the horizontal sprinkler spacing, and sprinkler locations are in compliance with 12.3.5.3.4.
88. _____ ESFR sprinklers installed above and below obstructions are hydraulically calculated in accordance with 12.3.5.3.5.
89. _____ In-rack sprinklers are provided for double-row racks without solid shelves with a ceiling clearance to the top of storage of up to 10 ft. and are located in accordance with Figures 12.3.5.1.2(a) or (b) and the top level of in-rack sprinklers are in compliance with 12.3.5.4.1.1.
90. _____ In-rack sprinklers are provided for single-row racks without solid shelves with a ceiling clearance to the top of storage of up to 10 ft. and are located in accordance with Figures 12.3.5.1.2.1(a), (b), or (c), 12.3.5.4.1.2.
91. _____ In-rack sprinklers are provided for multiple-row racks without solid shelves with a roof/ceiling clearance to the top of storage of up to 10 ft. and are located in accordance with Figures 12.3.4.4.1.3(a)-(f), 12.3.5.4.1.3.
92. _____ In-rack sprinklers are provided in accordance with Figure 12.3.5.4.1.4 for single and double-row racks without solid shelves and the aisle width and the design density/area are in compliance with 12.3.5.4.1.4.
93. _____ Double-row rack horizontal and vertical sprinkler spacing are provided and detailed on the plans as specified in Figures 12.3.5.1.2(a) or (b), 12.3.5.4.2.1.
94. _____ The vertical clearance specified in 12.3.5.4.2.2 between the in-rack sprinkler deflector and the top of the tier of storage is detailed.
95. _____ The in-rack water demand calculations are in compliance with 12.3.5.4.3 and the discharge flow for each sprinkler complies with 12.3.5.4.4.

Rubber Tire Storage:
96. _____ Ceiling does not exceed a 2 in 12 slope, 12.4.1.
97. _____ When required, unprotected steel building columns are protected by sprinklers in accordance with Section 12.4.1.1.1 unless tires stored in fixed racks have in-rack sprinklers, 12.4.1.2.
98. _____ Water demand for sprinkler protection and hose stream is based on one of three approaches in Section 12.4.1.4.
99. _____ Miscellaneous tire storage shall be protected in accordance with the requirements in Section 12.1.10.
100. _____ The design density/area and sprinkler temperature for ceiling systems using standard, large drop, or ESFR sprinklers are in accordance with 12.4.2.
101. _____ When provided, in-rack sprinklers spacing, water flow demand, and discharge pressure comply with Section 12.3 unless modified according to 12.4.3.2–12.4.3.4, 12.4.3.

The requirements associated with the subjects listed below that pertain to this storage and building design are to be reviewed using the NFPA 13 Sprinkler worksheet:
Other Coverage Areas
Pipe Support and Hangers
Pipe and Valves
Seismic Protection
Fire Department Connection
Hydraulic Calculations
Wet System
Dry System, Grid System is not Permitted, 7.2

Preaction or Deluge
Combined Dry Pipe and Preaction
Valves
Miscellaneous

Additional Comments:

Review Date: _____   Approved or Disapproved   FD Reviewer: _____
Review Date: _____   Approved or Disapproved   FD Reviewer: _____
Review Date: _____   Approved or Disapproved   FD Reviewer: _____

Fire Plan Review and Inspection Guidelines

NFPA 13 Sprinkler System for Storage Greater Than 25 Feet and Rubber Tires
Does not include Foam Systems
Plan Review Worksheet
2006 IFC and 2007 NFPA 13

This worksheet is for jurisdictions that permit the use of the 2007 NFPA 13 in lieu of IFC's referenced 2002 NFPA 13.

Date of Review: _____ Permit Number: _____

Business/Building Name: _____ Address of Project: _____

Designer Name: _____ Designer's Phone: _____

Contractor: _____ Contractor's Phone: _____

No. of Sprinklers: _____ Occupancy Classification: _____

Reference numbers following worksheet statements represent an NFPA code section unless otherwise specified.

Worksheet Legend: ✓ or OK = acceptable   N = need to provide   NA = not applicable

1. _____ A minimum of three sets of drawings are provided. The plans declare the design is based on the 2007 edition of NFPA 13.
2. _____ Equipment is listed for intended use and compatible with the system; specification data sheets are provided.
3. _____ Storage information worksheets are provided that indicate the storage height, ceiling slope, commodity classes, how commodities are packaged, type of packing materials used, type of pallets, whether pallet loads are encapsulated or banded, aisle widths, type of racks, flue spacing, if racks will be equipped with slatted or solid shelves, the sprinkler data sheets, design curve(s) and/or figure(s) used, if design area adjustments were made, and the storage location for each commodity class IFC 2301.3.

Drawings shall detail the following (22.1.3.1- 22.1.3.1.46):
General:
4. _____ Type of system is noted; __ wet, __ dry, __preaction.
5. _____ Scale: a common scale shall be used and plan information shall be legible.
6. _____ Plot plan details illustrate the water supply pipe diameters, lengths, and fittings to the building.
7. _____ The location of fire partitions, fire walls and building elevation views.
8. _____ Occupancy class and or use of each room or area 5.1.1.
9. _____ Full height cross sectionals are provided, include ceiling construction as needed for clarification.
10 _____ Total area protected by each system for each floor is provided.
11. _____ Dimensions for system piping, sprinkler spacing and branch line spacing, and elevation changes.
12. _____ Equipment symbol legend and the compass point are provided.
13. _____ Area limitations for hazard classification is 40,000 sq. ft. for storage, 8.2.1.
14. _____ Hydrant flow test determining water supply capacity at 20 PSI residual pressure is provided.
15. _____ Hydraulic calculations are provided with summary, detail worksheets, and graph sheet, 22.3.
16. _____ Dry pipe system capacity in gallons is provided _____gal., not to be greater than 750 gal. unless the requirements of 7.2.3.2 or 7.2.3.3 are met, 7.2.3.
17. _____ All water supply valves and flow switches are supervised, IFC 903.4 refer to exceptions.
18. _____ Exterior flow alarm location is detailed. Note: if electric, it shall be listed for outdoor use, IFC 904.3.2.
19. _____ When required, backflow prevention device pressure loss data is provided in the hydraulic calculations.

Sprinklers:
20. _____ Total number of each type of sprinkler is noted, see other permitted temperature ratings from 8.3.2.2 to 8.3.2.5.
21. _____ Minimum temperature for ceiling sprinkler for general rack, tire, and rolled paper storage is 150 deg., 8.3.2.7.
22. _____ Sprinkler locations are correct, ceiling and roof cross sectional views are provided for clarification, 22.1.3(45).
23. _____ Each sprinkler coverage area is within its square footage limitations and dimension limitations in accordance with its listing, 8.6.2.2, Table 8.6.2.2.1 (c and d).
24. _____ Control mode specific application, large drop, early suppression fast response sprinklers installation design comply with the standard and listing limitations, 6.1.1 and 8.4.1 – 8.4.10.

25. _____ Standard sprinkler spacing from vertical obstructions complies with Table 8.6.5.1.2 and for floor mounted obstructions, Table 8.6.5.2.2, 8.6.5.1.2 and 8.6.5.2.2.
26. _____ ESFR sprinkler spacing from ceiling obstructions complies with Table 8.12.5.1.1.

General Storage Requirements:
27. ____ Ceiling slope is detailed, cross sectional view provided, and does not exceed a 2 in 12 slope, 12.1.2.
28. ____ Storage design requirements are based on the absence of draft curtains and roof vents, 12.1.
29. ____ If a structure has two or more hazard areas or design methods that are non-separated and adjacent to each other, the more demanding area extends coverage in accordance with 12.3.
30. ____ Dry pipe and preaction system design areas are increased 30 percent but not to exceed the area specified in 12.5.1 and 12.5.2.
31. ____ When adjustments are made to the design area, the designer provided calculations explaining and showing how the adjustments were made, 12.7.7.
32. ____ Design for idle pallets is in accordance with 12.12.
33. ____ Densities up to 0.20 GPM/ft^2 are permitted to use sprinklers with K-factors specified in 12.6.1.
34. ____ Densities 0.21 GPM/ft^2 to 0.34 GPM/ft^2 protecting rack, tire, roll paper, and baled cotton storage are protected with upright and pendent sprinklers with K-factors as specified in 12.6.2.
35. ____ Densities greater than 0.34 GPM/ft^2 protecting rack, tire, roll paper, and baled cotton storage are protected with upright and pendent sprinklers with K-factors as specified in 12.6.3.
36. ____ If the design area is adjacent a combustible concealed space then the minimum design area is in compliance with 12.9.1 unless the concealed space meets the criteria of Section 12.9.2 (1) – (9), 12.9.1.

Large Drop Sprinklers:
37. _____ Large drop sprinkler coverage area and maximum spacing is in accordance with Table 8.11.2.2.1.
38. _____ The minimum coverage area for large drop sprinklers is not less than that specified in 8.11.2.3.
39. _____ Maximum perpendicular distance to the walls is not greater than 1/2 of allowable distance between sprinklers, 8.11.3.2.and Table 8.11.2.2.1.
40. _____ The minimum spacing between large drop sprinklers is not less that specified in 8.11.3.4.
41. _____ Large drop sprinklers deflectors are between 6 in. and 8 in. below the ceiling of unobstructed construction cross sectional view provided, 8.11.4.1.1.
42. _____ Large drop sprinklers deflectors position complies with Table 8.11.5.1.2, cross sectional view(s) provided, 8.11.4.1.2.
43. _____ Large drop sprinkler position adjacent ceiling or hanging obstructions complies with Table 8.11.5.1.2 unless sprinklers are placed on both sides of the obstructions.
44. _____ For obstructions 8 in. or less, large drop sprinklers are positioned at least three times the maximum obstruction dimension away unless 8.11.5.1.2 or 8.11.5.1.3 are met, 8.11.5.2.1.3.
45. _____ Large drop sprinklers position in relation to specific sized branch lines complies with 8.11.5.2.2.
46. _____ Large drop sprinkler position when more than 24 in. directly above an obstruction complies with Table 8.11.5.3.2. and Figure 8.11.5.3.2, and the design approach is detailed, 8.11.5.3.4.
47. _____ For obstructions running parallel with a branch line, large drop sprinkler placement as detailed in Figure 8.11.5.3.5.
48. _____ At least a 3 ft. clearance between large drop sprinklers deflectors and storage is provided, 8.11.6.
49. ____ Large drop sprinkler position when more than 24 in. directly above an obstruction complies with Table 8.11.5.3.2 and Figure 8.11.5.3.2.

General In-Rack Storage Requirements:
50. _____ In-rack sprinklers are ordinary temperature, K-factor, orientation, temperature rating and type complies with 8.13.2.1.
51. _____ One in-rack sprinkler system is limited to 40,000 sq. ft. of floor area including aisle ways, 8.13.1.
52. _____ When adjacent to a heat source the sprinkler temperature rating is in accordance with 8.3.2 and 8.13.2.2.
53. _____ When in-rack sprinklers are required for Class I –IV commodities, water shields are provided when required by Section 8.13.3.1.
54. _____ When required for plastics storage, water shields are provided for in-rack sprinklers when required by 8.13.3.2.
55. _____ When in-rack sprinklers are required for miscellaneous storage, the number of sprinklers calculated and the discharge pressure complies with Section 13.3.
56. _____ Sprinklers K-factors for storage applications; rack, tire, rolled paper, and baled cotton storage are in compliance with 12.6.1-.3 based on the discharge density required.

Palletized, Solid Pile, Bin Box, Shelf Storage for Class I-IV Commodities:
57. _____ Hose connections are provided for storage exceeding 12 ft., 12.2.2.

58. _____ Storage not exceeding 12 ft. is protected as miscellaneous storage in accordance with Table 13.2.1, 14.2.3.1.
59. _____ When the height of the storage is more than 12 feet the minimum design area/density for wet systems is 0.15 GPM/2,000 sq. ft. and for dry systems is 0.15 GPM/2,600 sq. ft., 14.2.4.5.
60. _____ The design for area/density using ordinary or high temperature sprinklers is based on Figures 14.2.4.1 and .2 respectively, 14.2.4.1, .2.
61. _____ If the density is adjusted due to storage height, it is in accordance with Figure 14.2.4.3 and calculations are provided, 14.2.4.3.

Palletized and Solid-Pile for Class I-IV Commodities:
62. _____ Large drop sprinklers protecting palletized or solid pile Class I –IV commodities is in accordance with Tables 14.3.1(a) and (b), 14.3.1.
63. _____ Large drop and specific application control mode sprinklers protecting ordinary hazard and miscellaneous storage shall be calculated based on a minimum 15 sprinklers for wet systems and 25 sprinklers for preaction or dry systems, 14.3.3.
64. _____ Large drop or specific application control mode sprinklers for open wood joist construction shall be designed based on a minimum 50 PSI design pressure unless the wood joists are constructed with fire stopping complying with 14.3.4.2.
65. _____ ESFR sprinklers protecting palletized or solid pile Class I –IV commodities are in accordance with Table 14.4.1.
66. _____ ESFR sprinkler most hydraulically demanding design area is a minimum of 12 sprinklers, 4 sprinklers on 3 branch lines, 14.4.3.
67. _____ ESFR sprinkler placement above and below obstructions includes in the hydraulic calculations up to 2 sprinklers from one level along with the sprinklers in the other level, 14.4.4.

Palletized, Solid Piled Storage of Plastic and Rubber Commodities:
68. _____ Sprinklers design protecting plastics stored up to 25 ft. is selected in accordance with Figure 15.2.1.
69. _____ Design information addressing the storage conditions, i.e. closed or open array, clearance between storage, clearance between storage and ceiling/roof, pile height, and stable or unstable piles, shall be provided 15.1.2 and 15.2.
70. _____ Class B, free-flowing Group A, and Class C plastics are protected in accordance with Section 15.2.3, and 15.2.4.
71. _____ When a density/area design method is used, the design area and density are based on Table 15.2.5(a) or (b) and Figure 15.2.1, 15.2.5.
72. _____ When determining the design area, the designer shall provide documentation explaining any design area adjustments, 15.2.7.
73. _____ Large drop or specific application control mode sprinklers K-factor, storage height coverage, number of design sprinklers, and hose stream demand are in accordance with Table 15.3.1(a) or (b), 15.3.1.
74. _____ Large drop or specific application control mode sprinklers for open wood joist construction, large drop sprinkler pressure, design area, and hose stream demand are in compliance with 15.3.3.
75. _____ ESFR sprinklers protection design is in accordance with Table 15.4.1.
76. _____ ESFR sprinkler most hydraulically demanding design area is a minimum of 12 sprinklers, 4 sprinklers on 3 branch lines, 15.4.3.
77. _____ ESFR sprinkler placement above and below obstructions includes in the hydraulic calculations up to 2 sprinklers from one level along with the sprinklers in the other level, 15.4.4.

Rack Storage of Commodities
General:
78. _____ Structural steel columns that are within storage racks are protected in accordance with 16.1.4.
79. _____ Solid shelving exceeding 20 sq. ft. up to 64 sq. ft., sprinkler protection is within 6 ft. vertically of each intermediate level and is provided at the ceiling, 16.1.6.1.
80. _____ Solid shelving exceeding 64 sq. ft. or the storage level exceeds 6 ft. with solid shelves, sprinkler protection is in compliance with 16.1.6.2.
81. _____ Where in-rack sprinklers protect a higher commodity hazard then the adjacent areas in the same rack, that in-rack sprinkler design complies with 16.1.8.2.
82. _____ When required and provided the horizontal barriers within the rack are made of wood or sheet metal, within 2 in. around rack uprights, and the construction details are provided, 16.9.9.
83. _____ For storage 25 ft. or less a longitudinal flue is not required in double and multiple-row racks as long as solid shelves are not present. Transverse flues required in single, double, and multiple-row racks are provided and detailed or noted, 16.1.10.

84. _____ For storage exceeding 25 ft. the longitudinal flue space in double-row racks and the transverse flue space between loads and rack uprights in single, double, and multiple-row racks are in compliance with 16.1.11.1 and detailed or noted.
85. _____ For storage exceeding 25 ft., vertical clear space is provided between the sprinkler deflector and the top of a tier of storage, the face sprinklers positioned from rack uprights and the aisle face of storage for single, double, and multiple-row racks in accordance with 16.1.11.2.
86. _____ Sprinklers locations in relation to the longitudinal and transverse flue space intersections, the horizontal load beams or horizontal rack components and uprights are in accordance with 16.1.11.2.

Rack Storage for Class I through IV Commodities for Greater Than 25 ft.:

87. _____ For control mode sprinkler density protecting single, double, and multiple-row racks with nonencapsulated products without solid shelves with aisles 4 ft. or greater and ceiling clearances up to 10 ft., the ceiling water design density/area demand using ordinary temperature sprinklers shall have a minimum discharge density as required by Section 16.3.1.1 for single and double row racks or Section 16.3.1.2 for double row racks. Double row racks shall also comply with Table 16.3.1.1 and Annex C.23 and for multiple-row racks refer to Table 16.3.1.2 and Section 16.3.1.2. Increase density 25 percent for encapsulated commodities, 16.3.1.1.1 and 16.3.1.2.1.
88. _____ For control mode water design density/area demand the hose demand from Table 16.3.1.3 has been added, 16.3.1.3.
89. _____ The design requirements for control mode sprinklers protection for single, double-row racks with nonencapsulated products is in compliance with 16.3.1.1 in regards to the type of shelving used, aisles width, sprinkler clearance from the top of storage, sprinkler temperature, design area, and discharge density for specific commodities. Other design criteria for single and double-row racks refer to Table 16.3.1.1 and Annex C.23 and for multiple-row racks refer to Table 16.3.1.2 and Section 16.3.1.2. Increase density 25 percent for encapsulated commodities, 16.3.1.1.1 and 16.3.1.2.1.
90. _____ Large drop and specific control mode sprinkler design quantity and minimum operating pressure and in-rack criteria for single, double, and multiple-row racks without solid shelves for Class I though IV commodities are as specified in Table 16.3.2.1(a) and (b) when required, in-rack sprinklers are hydraulically designed and positioned in accordance with 16.3.4.
91. _____ Large drop K-11.2 sprinklers positioned under open wood joist construction operate at a pressure specified 16.3.2.4.1.
92. _____ Large drop K-11.2 sprinklers can use the lower pressures offered in Table 16.3.2.1(a) when open wood joist channel construction is completely, to its depth, fire-stopped for intervals that are not greater than 20 ft., 16.3.2.4.2.
93. _____ ESFR sprinkler orientation, K-factor, and minimum pressure design and in-rack criteria for single, double, and multiple-row racks without solid shelves for Class I though IV commodities are as specified in Table 16.3.3.1, 16.3.3 and when required, in-rack sprinklers are hydraulically designed and positioned in accordance with 16.3.3.3-.5.
94. _____ ESFR sprinklers are not used with solid shelves or combustible open-top cartons, 16.3.3.2.
95. _____ The number of ESFR sprinkler for most hydraulically demanding design area and the number of branch lines complies with 16.3.3.4.
96. _____ ESFR sprinklers installed above and below obstructions are hydraulically calculated in accordance with 16.3.3.6.
97. _____ In-rack sprinklers are provided for double-row racks without solid shelves with ceiling clearance to the top of storage up to 10 ft. when required by Table 16.3.1.1 and are designed in accordance with Figures 16.3.4.1.1(a)-(j), and the top level of in-rack sprinklers are within the distance specified for specific commodities in accordance with 16.3.4.1.1.1. The same criteria are used when single-row racks area mixed in with double-row racks.
98. _____ In-rack sprinklers are provided for single-row racks without solid shelves with ceiling clearance to the top of storage up to 10 ft. are designed in accordance with Figures 16.3.4.1.2(a)-(e), 16.3.4.1.2.
99. _____ In-rack sprinklers are provided for multiple-row racks with ceiling clearance to the top of storage up to 10 ft. when required by Table 16.3.1.2 and are designed in accordance with Figures 16.3.4.1.3(a)-(c), and the top level of in-rack sprinklers are within the distance specified for specific commodities in accordance with 16.3.4.1.3.
100. ____ In-rack sprinklers are spaced and staggered vertically and horizontally as specified in 16.3.4.2.1.
101. ____ In-rack sprinklers for double-row racks are horizontally spaced in the horizontal space at or near vertical locations in accordance with 16.3.4.2.2.
102. ____ In-rack sprinklers for multiple-row racks are horizontally spaced in accordance with 12.3.4.4.2.3.

Fire Plan Review and Inspection Guidelines

103. ____ The in-rack water demand calculations are provided and based on one of four design approaches specified in 16.3.4.3 and the discharge quantity is in compliance with 16.3.4.3.1.

Rack Storage of Plastics Stored Greater Than 25 ft. for Single, Double and Multiple-row Racks:

104. ____ Structural steel columns that are within storage racks are protected in accordance with 17.1.4.
105. ____ Sprinkler design decision for single, double, and multiple-row racks for aisles 3.5 ft. wide or greater is in accordance with Figure 17.1.2.1.
106. ____ Solid shelving exceeding 20 sq. ft. up to 64 sq. ft., is protected by automatic sprinklers in accordance with 17.1.5.1.
107. ____ Solid shelving exceeding 64 sq. ft. or the storage level exceeds 6 ft. with solid shelves, sprinkler protection is in compliance with 17.1.5.2.
108. ____ Control mode ceiling sprinkler density/area design for Group A plastics (including encapsulated or nonencapsulated commodities) are in accordance with Table 17.3.1.1 and if single-row racks are intermixed with double-racks the in-rack sprinkler locations are based on Figures 17.3.1.2(a) or (b), 17.3.1.1 and .2.
109. ____ Control mode sprinklers in single-row racks are located in accordance with Figures 17.3.1.2.1(a)-(c), 17.3.1.2.1.
110. ____ Hose stream demands were added to the hydraulic calculations in accordance with Table 17.3.1.3.
111. ____ ESFR sprinkler orientation, K-factor, and minimum pressure design for single, double, and multiple-row racks for plastics cartoned or not and expanded or not are in accordance with Table 17.3.3.1, 17.3.3.1.
112. ____ ESFR sprinklers are not used with solid shelves or combustible open-top cartons, 17.3.3.1.1.
113. ____ The number of ESFR sprinkler for most hydraulically demanding design area and the number of branch lines calculated complies with 17.3.3.3.
114. ____ ESFR sprinkler coverage is supplemented with in-rack sprinklers when required by Table 17.3.3.1, the hydraulic design, pressure the horizontal sprinkler spacing, and sprinkler locations are in compliance with 17.3.3.4.
115. ____ ESFR sprinklers installed above and below obstructions are hydraulically calculated in accordance with 17.3.3.5.
116. ____ In-rack sprinklers are provided for double-row racks without solid shelves with a ceiling clearance to the top of storage of up to 10 ft. and are designed in accordance with Figures 17.3.1.2(a) or (b) and the top level of in-rack sprinklers are within 10 ft. of the top of the storage, 17.3.4.1.1.
117. ____ In-rack sprinklers are provided for single-row racks without solid shelves with a ceiling clearance to the top of storage of up to 10 ft. and are designed in accordance with Figures 17.3.1.2.1(a), (b), or (c), 17.3.4.1.2.
118. ____ In-rack sprinklers are provided for multiple-row racks without solid shelves with a roof/ceiling clearance to the top of storage of up to 10 ft. and are designed in accordance with Figures 17.3.4.1.3(a)-(f), 17.3.4.1.3.
119. ____ In-rack sprinklers are provided in accordance with Figure 17.3.4.1.4 for single and double-row racks without solid shelves and the aisle width and the design density/area are in compliance with 17.3.4.1.4.
120. ____ Double-row rack horizontal and vertical sprinkler spacing are provided and detailed on the plans as specified in Figures 17.3.1.2(a) or (b), 17.3.4.2.1.
121. ____ The vertical clearance specified in 17.3.4.2.2 between the in-rack sprinkler deflector and the top of the tier of storage is detailed.
122. ____ The in-rack water demand calculations when only one level of sprinklers are provided and when sprinklers are provided on two or more levels are in compliance with 17.3.4.3 and the discharge quantity for each sprinkler complies with 17.3.4.4.

Rubber Tire Storage:

123. ____ Structural steel building columns are protected by sprinklers in accordance with Section 18.2.1 unless tires stored in fixed racks have in-rack sprinklers, 18.2.3.
124. ____ Water demand for sprinkler protection and hose stream is based on one of three approaches in Section 18.3.
125. ____ Miscellaneous tire storage design is based on Section 13.2.1.
126. ____ The density/area and sprinkler temperature for ceiling system is in accordance with Table 18.4(a) for standard sprinklers, Table 18.4(c) for large drop sprinklers, and Table 18.4(d) for ESFR sprinklers, 18.4.
127. ____ Where provided, in-rack sprinklers are designed in accordance with Chapter 17 unless modified according to 18.5.2-.4, 18.5.1.
128. ____ In-rack sprinkler spacing is a maximum of 8 ft., water demand is based on the 12 most remote sprinklers when only one level of sprinklers is provided and with a minimum operating pressure of 30 PSI, 18.5.

Fire Plan Review and Inspection Guidelines

The requirements associated with the subjects listed below that pertain to this storage and building design are to be reviewed using the NFPA 13 Sprinkler worksheet:

Other Coverage Areas
Pipe Support and Hangers
Pipe and Valves
Seismic Protection
Fire Department Connection
Hydraulic Calculations
Wet System
Dry System, Grid System is not Permitted, 7.2
Preaction or Deluge
Combined Dry Pipe and Preaction
Valves
Miscellaneous

Additional Comments:

Review Date: _____   Approved or Disapproved   FD Reviewer: _____
Review Date: _____   Approved or Disapproved   FD Reviewer: _____
Review Date: _____   Approved or Disapproved   FD Reviewer: _____

Prefinal and Certificate of Occupancy Inspection Requirements For Contractors
Contractors Worksheet

Sprinkler System Test Requirements

1. All certification forms and documents are required to be on the site for review:
   - ___ Plans
   - ___ Permit
   - ___ A system hydrostatic test is required before calling for an inspection as well as the completion of with the items on this pretest form. Use the Acceptance Inspection worksheet for the pretest.
   - ___ Installation certification is completed, use the form contained in this book.
2. ___ A person familiar with installation must be present to perform the test.
3. ___ Owner's representative approval is needed for the time and date of testing.
4. ___ All areas are accessible.
5. ___ Hydrostatic testing and the flow test should be done during the same inspection.
6. ___ If Items 1-5 are incomplete, the inspection will be cancelled and another inspection request is required. A reinspection fee may be assessed.

Prior to the next approval test:

7. ___ When there are device additions, contractor must provide:
   - ___ As-builts and new calculations shall be submitted for review and approval.
   - Note: New plan review will be submitted as "supplemental information" and proof of the additional review fee payment is required.
8. ___ A reinspection fee may be assessed if the system and paperwork are not ready.

Fire Plan Review and Inspection Guidelines

Water Mist System
Contents

1. Plan Review Worksheet. 163

2. Voltage Drop Calculation Form 167

3. Acceptance Test Checklist. 168

4. Installer Certification Form for Final Permit Inspection. 170

5. Contractor Prefinal and Certificate of Occupancy Inspection Requirements. 171

Fire Plan Review and Inspection Guidelines

Water Mist System
Plan Review Worksheet
2006 IFC and 2003 NFPA 750

Date of Review: _____ Permit Number: _____

Business/Building Name: _____ Address of Project: _____

Designer Name: _____ Designer's Phone: _____

Contractor: _____ Contractor's Phone: _____

System Application: Local ___   Total Compartment ___   Zoned ___   **Design:** Prengineered ___ Engineered ___
System Type: Low Pressure___   Intermediate Pressure___   High Pressure___
Nozzle Types: Auto ___   Nonauto ___   Hybrid ___

Numbers following worksheet comments represent a NFPA code section unless otherwise specified.

Worksheet Legend: ✓ or OK = acceptable   N = need to provide   NA = not applicable
 1._____ Three sets of drawings are provided.
 2._____ Equipment is listed for intended use and compatible with the system and equipment data sheets are provided.

Plan Set Shall Provide and Detail the Following (11.1):
General:
 3._____ Scale: a common scale shall be used and plan information is legible.
 4._____ Description of the water and gas storage containers including internal volume, design pressure at standard temperature and pressure, 11.1.5 (9).
 5._____ Building dimensions, location of fire partitions and fire walls.
 6._____ Description of the hazards or occupancies being protected and if these areas are occupied, 11.1.5 (7)
 7._____ Full height cross sections, which include ceiling construction.
 8._____ System application, nozzle type, operation method, and media type.
 9._____ Device and nozzle location, provide sectional view detailing detectors position.
 10._____ Type of devices and detail proper device wiring for detectors, horns, etc.
 11._____ Equipment symbol legend and compass point.
 12._____ Water mist control panel location is detailed and connected to the building fire alarm system, if the building is equipped with such a system.
 13._____ Sequence of operation for operation of the water mist system, 11.1.5 (22).

Detection System Riser, 5.10.1.1 and 11.1.5:
 14._____ Riser diagram shows the number and type of devices, audible, visual, release, shutdown, and discharge controls, per circuit, zone ID, a dedicated 120 AC power supply, batteries, panel, etc.

Point to Point System Wiring Diagram, 11.1.5(23):
 15._____ Interconnection and wire routing to identified devices and controls per circuit.
 16._____ Indicate the number of conductors and wire gauge for each circuit.
 17._____ Identify separate zones, circuits, and end of line resistor locations.

Alarm Indicating Circuit Load Consumption of Circuits, 11.1.5:
 18._____ Quantity of signaling devices, current consumption, and end-of-line voltage for each circuit.
 19._____ Based on the approximate length of each circuit and the conductor amperage, determine the resistance for each 1,000 feet of wire using National Electrical Code ampacity values or those specified by the manufacturer of the conductors.
 20._____ Show the formula and acceptable circuit limits on the drawing or on an attached sheet including:
 21._____ A. Standby power consumption of all current drawing devices multiplied by the hours required by NFPA (24 hours) including power consumption of the control panel modules.
 22._____ B. Power consumption of all devices on standby power; including door holders, relays, smoke detectors, etc.
 23._____ C. Alarm power consumption of all current drawing devices multiplied by the minutes required by NFPA (5 minutes).
 24._____ D. Formula format for battery calculations.

System Devices:
25. _____ Preengineered Water Mist system layout meets the manufacturer's listing requirements and a specification/design manual is provided, 5.10.2.1. Alarm initiating and signaling devices are installed in accordance with NFPA 72.
26. _____ Equipment and detectors are listed for use and the listing data sheets are provided, 5.1.1.
27. _____ Two sources of electrical power are provided (24 hr minimum standby power), 5.10.2.2.
28. _____ Emergency release device is provided and detailed, unless each nozzle is thermally activated, 5.10.3.5.
29. _____ Normal manual control(s) for activation is detailed to be accessible, labeled, and mounted 4 ft. or less above the floor level, 5.10.3.6.
30. _____ Pneumatic control lines are protected against damage and supervised, refer to the exceptions, 5.10.4.3.
31. _____ When automatic activation is provided, the method is designed in compliance with 5.10.1.2.

Battery Calculation Sheet Includes, 5.10.1.1 and 5.10.2.2.:
32. _____ Standby power consumption of all current drawing devices multiplied by the hours required by NFPA (24 hours), including power consumption of the control panel modules.
33. _____ Power consumption: Transfer to secondary power (UPS or generator), batteries provide no power loss for 15 minutes.
34. _____ Primary batteries shall be sized to at least 100 percent of maximum normal load.

Water Mist Information:
35. _____ Type of system, system application, type of nozzles, operation method, and media type.
36. _____ Design objective and hazard classifications are provided, 8.3.1 and 8.4.2.
37. _____ Components subject to corrosion are protected, 5.1.3.
38. _____ If required a FDC is detailed on the discharge side of pressure source and prior to the filter/strainer, 10.5.5.

Calculations:
39. _____ System hydraulic and atomizing medium calculations are provided in accordance with Chapter 9.
40. _____ Hydraulic calculation nodes match plan nodes.
41. _____ Hydraulic junction points balance within the pressure specified in 9.3.5 and equivalent pipe lengths are in accordance with 9.3.6.
42. _____ Nozzle pressures are within limitations specified by the manufacturer, 9.4.1.2.
43. _____ The results of the hydraulic and pneumatic calculations at the supply point and at the nozzle are provided, 9.4.6.
44. _____ The water supply is designed for the largest single hazard or group of hazards, 10.2.
45. _____ A volume and pressure of the propellant gas is in accordance with Section 9.4.1.1.

Atomizing Media:
46. _____ For twin fluid systems the atomizing media source shall be in accordance with 10.6.1.
47. _____ Pump capacity is in accordance with 10.5.2.2.
48. _____ A test connection is detailed for testing the pump in accordance with 10.5.2.3
49. _____ When used, an air compressor is listed for fire service use, 10.6.8.1.
50. _____ When used as the dedicated air supply, the compressor is connected to a backup power supply, 10.6.8.2.

Containers and Piping:
51. _____ Pressurized water and atomizing media containers shall meet the construction requirements of the American Society of Mechanical Engineers Boiler and Pressure Vessel Code, Section VIII, *Unfired Pressure Vessels* or in accordance with U.S. Department of Transportation requirements., 5.2.2.2.
52. _____ Gas and water containers are sized for required quantities, 5.2.1, and are not located where environmental or mechanical damage will occur, 7.5.4.
53. _____ When required in a seismic design category, documentation explaining seismic bracing for atomizing media containers shall be provided, 5.2.2.1.
54. _____ Containers that are pressurized shall be equipped with a pressure relief device, 5.2.2.5.
55. _____ Manifolded containers shall be interchangeable and have the some volume and discharge pressure, 5.2.3.
56. _____ Low pressure storage cylinder detail shows the liquid level and pressure gauges, and high/low pressure supervisory alarm, 7.5.6.4.

57._____ Pressure gauges are detailed on all pressurized cylinders, both sides of pressure regulator valve, pressurized side of the supply connections and system control valves, and air supplies for dry systems, 7.8.5.
58._____ Pipe or tube: type of material, sizes, pressure rating, if used in low, intermediate, or high PSI system, and pipe specifications are provided, 5.3.2.1.
59._____ Bending criteria for Type K and L copper pipe is noted on plans which is in accordance with 5.3.6.
60._____ Fittings are either listed or meet the referenced ANSI or ASTM standard for the given application. Specifications or equipment data sheets are provided, 5.4.
61._____ Screwed unions are limited to pipe diameters of 2 inches or less, 5.4.2.3.
62._____ One-piece reducers are used and noted on plans, 5.4.2.4.
63._____ When required, an FDC is detailed and interfaces on the pressure side of the system, refer to exceptions 10.5.5.

Hangers:
64._____ Hangers are listed for their intended use or in accordance with 5.5.
65._____ Types of hangers and hanger locations on structural elements are detailed on plans, 7.3.7.1. Low pressure water pipe is hung in accordance with NFPA 13.
66._____ Armovers to nozzles are detailed and the supports shown for steel pipe and tube length greater than what is specified in 7.3.7.2.

Seismic Bracing (Based on the requirements in NFPA 13):
67._____ Seismic bracing is designed, detailed, and seismic calculations are provided, NFPA13.
68._____ Lateral sway brace spacing complies with 9.3.5.3.
69._____ A seismic separation assembly for piping is provided at building seismic joints, 9.3.3.
70._____ Longitudinal sway brace spacing complies with 9.3.5.4.
71._____ A 4-way sway brace is provided at the top of the riser, 9.3.5.5.
72._____ Longitudinal and lateral bracing is provided for each run of pipe between the change of direction in accordance with 9.3.5.11.
73._____ Branch lines and end sprinklers are restrained against vertical and lateral movement, 9.3.6.3.
74._____ Calculations for seismic bracing is provided, 9.3.5.6.–9.3.5.11.

Nozzles:
75._____ Nozzles: All design and installation listing data for each nozzle is provided. The information shall include: specific hazard objectives, flow rate, space height; protection distance, spacing, coverage area, and pressures; delivery time, spacing from walls, compartment volume, and thermal classification, etc., 5.6.1 and 7.2.
76._____ Thermal nozzles: nozzle temperature rating and the maximum ambient temperatures are provided and comply with Table 5.6.7.1.
77._____ Number, type, and the placement of spare nozzles are noted on plans, 5.6.7.
78._____ Nozzles with waterway dimensions less than 51 microns use the type of water specified in 10.5.1.7.

Valves:
79._____ Valves are listed for the intended use, equipment data sheets are provided and valve signage is provided.
80._____ A monitored or locked indicating valve is provided for each source of water supply, 7.8.1.8.
81._____ Water pressure regulating valve (PRV) is provided for any portion of the system with the potential to exceed the maximum system pressure rating and it opens at the percentage of system-rated pressure specified in 7.8.3.1.
82._____ Water pressure relief valve size and location is detailed and in compliance with 7.8.3.1.3.
83._____ Indicating valve location is detailed and in compliance with 7.8.3.1.4.
84._____ A water flow test valve is detailed and designed to meet the equivalent flow of PRV, 7.8.3.1.6.
85._____ Compressed gas PRV is detailed when the supply pressure is higher than the operating pressure, 7.8.3.2.2.
86._____ Check valve is detailed between the system and the potable water connection, 7.8.4.3.
87._____ Pressure gauges are detailed on the pressurized side of control valves and supply connections, 7.8.5.

Strainers:
88._____ Strainers and filters are listed for their use and the listing data sheets are provided, 5.8.1.
89._____ Pipeline strainer and filter designs have a flush-out connection, 5.8.5.
90._____ Number, type, and placement of spare strainers and filters are noted on the plans, 5.8.8
91._____ Strainers and filters are detailed at each water supply connection or system riser, 7.7.
92._____ Strainer and filter ratings or mesh openings are of a percentage of the nozzle waterway dimension as specified in 10.5.1.6.

Fire Plan Review and Inspection Guidelines

Pumps and Controllers:
93. _____ An automatic pump is provided and detailed, 10.5.2.1.
94. _____ Pump capacity is in accordance with 10.5.2.2.
95. _____ A test connection is detailed for testing the pump in accordance with 10.5.2.3
96. _____ Pumps: design information and details include pump capacity, over pressure relief, method of automatic start and shutoff and water supply method, 5.9.
97. _____ Pumps are sized to provide the water flow rate and system demand, 10.5.2.2.
98. _____ Pump operation and functions are supervised at a constantly attended location, the method and what is supervised on the electrical and diesel pumps are noted on plans, 10.5.2.3.
99. _____ Power supply for pump driver complies with NFPA 20 except for being fed with an independent service feed, 5.9.2.
100. ____ Pump controller is a listed fire pump controller, 5.9.3.1.

Test Connector:
101. ____ It is detailed and it is sized not less than the largest nozzle, located at the most hydraulically remote point of the system, 7.10

Additional Comments:

Review Date: _____ Approved or Disapproved FD Reviewer: _____

Review Date: _____ Approved or Disapproved FD Reviewer: _____

Review Date: _____ Approved or Disapproved FD Reviewer: _____

Fire Plan Review and Inspection Guidelines

Voltage Drop Calculations for Notification Appliance Circuit (NAC): _____

Each NAC shall have its voltage drop determined. This sheet shall be used for one NAC but every NAC should have a sheet completed and submitted with each permit application.

STEP 1: complete the following to provide data for determining the resistance of the conductor in Step 2

Wire length is from fire alarm control
panel to the end of the fire alarm circuit = _____ ft. X 2 = _____ ft.
Wire Size = # _____ AWG (American Wire Gauge)
Resistance (R) = _____ OHMS for a given 1,000 ft. of the conductor specified

Step 2: complete the following to determine the total resistance (OHMS) for a NAC

(R) = Total Wire Resistance

From Step 1 divide the OHMS by 1,000, which will convert the conductor resistance to OHMS in each linear foot of wire

Determine OHMS per foot = _____ ft. = _____ OHMS/ft.
                                 1,000

Take the total feet of wire from Step 1 and OHMS/ft. from the line above and put both in the equation below

Circuit resistance = _____ ft. X _____ OHMS/ per ft. = _____ (R) Total OHMS

Step 3: complete the following to determine the total alarm notification device amperage and devices may be rated in milliamps

(I) = Alarm Appliance Amperage

A. No. of Alarm Appliances = _____   B. Current amperage each _____ = A x B _____ (I)
A. No. of Alarm Appliances = _____   B. Current amperage each _____ = A x B _____ (I)
A. No. of Alarm Appliances = _____   B. Current amperage each _____ = A x B _____ (I)
A. No. of Alarm Appliances = _____   B. Current amperage each _____ = A x B _____ (I)
                                                                     Total _____ (I)

Step 4: complete the following to determine the total voltage drop for the branch circuit

Voltage (E) = (I) X (R) from totals in Steps 2 and 3 above
(E) = _____ (I) X _____ (R)
     = _____ (E) (shall not exceed 4.4)

Step 5: complete the following to determine if enough voltage is available to operate fire alarm notification devices

Maximum allowable voltage drop: notification devices cannot drop below their Nameplate Operating Voltage (NOV) range. As of 5/1/2004 UL required indicating devices to operate within their NOV. The UL NOV standard is 16VDC to 33VDC, consult the 2002 NFPA 72 Handbook 7.3 for more information. Fire Alarm Control Units (FACU) are tested to UL 864 and are required to operate at the end of useful battery life, 20.4 V.

Allowable voltage drop is 20.4 V (FACU) - 16 VDC (NOV) = 4.4 V

If (E) from Step 4 exceeds 4.4 V then the NAC is not compliant with NFPA 72

Take (E) from Step 4 and put in the equation below

Voltage Drop = 20.4 V - _____ (E) = _____ V (shall not be less than 16V)

Fire Plan Review and Inspection Guidelines

Water Mist System Acceptance Inspection
2006 IFC and 2003 NFPA 750

Date of Inspection: _____ Permit Number: _____

Business/Building Name: _____ Address of Project: _____

Contractor: _____ Contractor's Phone: _____

<u>Pass | Fail | NA</u>

1. ____ | ____ | ____ If water supply is from a city or private system, the system is flushed in accordance with NFPA 750 12.2.1.
2. ____ | ____ | ____ Approved drawing is on site.
3. ____ | ____ | ____ Control panel and component type match approved plans.
4. ____ | ____ | ____ Piping, nozzles, and component locations are the same as on the approved plan.
5. ____ | ____ | ____ Detection zones are verified and comply with those identified on the plans.
6. ____ | ____ | ____ Verify dedicated 120 AC branch circuit and labeling, NFPA 750 5.10.1.1(1).
7. ____ | ____ | ____ Device locations comply with the approved plans.
8. ____ | ____ | ____ Devices are properly wired, NFPA 750 12.2.4.2.5.
9. ____ | ____ | ____ Type and gauge of wires or cables comply with the approved plans.
10. ____ | ____ | ____ If a monitoring service is provided, the agency received signals, trouble, alarm, and supervisory, IFC 904.4.3, NFPA 750 12.2.5.1.
11. ____ | ____ | ____ Two monitoring circuits are provided and both are tested for sending signals to monitoring company, NFPA 750 5.10.1.1(1).
12. ____ | ____ | ____ Verify proper operation of auxiliary operations and/or ventilation shutdown, NFPS 750 12.2.5.4.
13. ____ | ____ | ____ Battery stress test: system switched to battery operation 24 earlier then activate audible circuit for 5 minutes or 8 minutes if the battery was not switched 24 hours earlier.
14. ____ | ____ | ____ Emergency and manual release devices are single action, tested to operate the system, and mounted at proper height and location, NFPA 750 5.10.3.5 and 12.2.4.2.7.
15. ____ | ____ | ____ Under primary and secondary power operational tests are performed including: NFPA 750 7.9.2 and 12.2.4.2.
16. ____ | ____ | ____   A. power light on and in normal condition.
17. ____ | ____ | ____   B. <u>supervisory signals</u>: water flow, pressure switches, and valves.
18. ____ | ____ | ____   C. silence switches.
19. ____ | ____ | ____   D. trouble signals and panel light operate for each circuit tested when disconnecting wires from devices.
20. ____ | ____ | ____   E. trouble and alarm reset switches operate.
21. ____ | ____ | ____   F. if provided, a 2nd initiating zone overrides silence switch.
22. ____ | ____ | ____   G. audibles and visuals operate.
23. ____ | ____ | ____   H. panel lamp test switch operates, if available.
24. ____ | ____ | ____   I. field zone signals correspond with panel zones.
25. ____ | ____ | ____   J. detection devices operate properly, NFPA 750 12.2.5.3.
26. ____ | ____ | ____ The water mist system activates the building fire alarm (if provided).
27. ____ | ____ | ____ As-built drawings are required when system installation is not the same as the plans.
28. ____ | ____ | ____ Piping and nozzles are secured so no unacceptable movement occurs, check prior to the pressure test.
29. ____ | ____ | ____ Pipe and tube is marked as required by Section 5.3.5.1 and shall not be painted or concealed until it is approved by the fire code official, 5.3.5.3.
30. ____ | ____ | ____ Low pressure piping and attached appurtenances are hydrostatically tested for at least 2 hours at 200 PSI and without the loss of any pressure, NFPA 750 12.2.2.2.

31. ____|____|____ Intermediate and high pressure piping and attached appurtenances are hydrostatically tested for 2 hours at 150 percent of working pressures and without the loss of any pressure, NFPA 750 12.2.2.2.2.
32. ____|____|____ In addition to a hydrostatic pressure test, dry and preaction systems shall be pneumatically pressure tested. The minimum test duration is 24 hours at a minimum pressure of 40 PSIG. The pressure drop shall be in accordance with the limits specified in NFPA 750 12.2.3.
33. ____|____|____ Nozzle locations and orientation are in accordance with the approved plans, NFPA 750 12.2.4
34. ____|____|____ Release circuit is tested at the storage container, NFPA 750 12.2.5.2.
35. ____|____|____ If provided, abort switch is tested, NFPA 750 12.2.4.2.8.
36. ____|____|____ A copy of the water mist system installation certification is provided by the installer.
37. ____|____|____ Confirm that the water mist system piping is not used as an electrical ground, 12.2.4.2.13.
38. ____|____|____ System design signage is permanently posted at the protected area's control valve, NFPA 750 11.4.4.
39. ____|____|____ If required, a FDC is provided in compliance with the approved plans, NFPA 750 10.5.5
40. ____|____|____ If provided, system pressure pumps are performance tested and supervisory conditions are tested, NFPA 750 10.5.2

Additional Comments:

Inspection Date: _____   Approved or Disapproved   FD Inspector: _____

Inspection Date: _____   Approved or Disapproved   FD Inspector: _____

Inspection Date: _____   Approved or Disapproved   FD Inspector: _____

Fire Plan Review and Inspection Guidelines

Water Mist Installation Certification

Permit #: _____   Date: _____

	Property Protected	System Installer	System Supplier
Business Name:	_____	_____	_____
Address:	_____	_____	_____
	_____	_____	_____
Representative:	_____	_____	_____
Telephone:	_____	_____	_____

Location of Plans: _____

Location of Owner's Manual: _____

1. <u>Certification of System Installation:</u> Complete this section after system is installed, but prior to conducting operational acceptance tests.

    This system installation was inspected and found to comply with the installation requirements of:
    - _____ NFPA 750
    - _____ IFC
    - _____ Manufacturer's Instructions or Listing
    - _____ Other (specify; FM, UL, etc.) _____

    Print Name: _____

    Signed: _____ Date: _____

    Organization: _____

2. <u>Certification of System Operation:</u> All operational features and functions of this system were tested and found to be operating properly in accordance with the requirements of:
    - _____ NFPA 750
    - _____ IFC
    - _____ Manufacturer's Instructions or Listing
    - _____ Other (specify) _____

    Print Name: _____

    Signed: _____ Date: _____

    Organization: _____

Prefinal and Certificate of Occupancy Inspection Requirements For Contractors
Contractors Worksheet

Water Mist Test Requirements

1. All certification forms and documents are required to be on the site for review:
   - ___ Plans
   - ___ Permit
   - ___ A system hydrostatic test is required before calling for an inspection as well as the completion of with the items on this pretest form. Use the Acceptance Inspection worksheet for the pretest.
   - ___ Installation certification is completed, use the form contained in this book.
2. ___ A person familiar with installation must be present to perform the test.
3. ___ Owner's representative approval is needed for the time and date of testing.
4. ___ All areas are accessible.
5. ___ Hydrostatic testing and the flow test should be done during the same inspection.
6. ___ If Items 1-5 are incomplete, the inspection will be cancelled and another inspection request is required. A reinspection fee may be assessed.

Prior to the next approval test:

7. ___ When there are device additions, contractor must provide:
   - ___ As-builts and new calculations shall be submitted for review and approval.

   Note: New plan review will be submitted as "supplemental information" and proof of the additional review fee payment is required.
8. ___ A reinspection fee may be assessed if the system and paperwork are not ready.

Carbon Dioxide
Contents

1. Plan Review Worksheet. 175

2. Voltage Drop Calculation Form 177

3. Acceptance Test Checklist. 178

4. Installer Certification Form. 180

5. Contractor Prefinal and Certificate of Occupancy Inspection Requirements 181

Fire Plan Review and Inspection Guidelines

Carbon Dioxide Alternative Fire-Extinguishing System
Plan Review Worksheet
2006 IFC and 2000 NFPA 12

Date of Review: _____ Permit Number: _____

Business/Building Name: _____ Address of Project: _____

Designer Name: _____ Designer's Phone: _____

Contractor: _____ Contractor's Phone: _____

System Manufacturer: _____ Model: _____

Reference numbers following worksheet statements represent an NFPA code section unless otherwise specified.

Worksheet Legend: ✓ or **OK** = acceptable **N** = need to provide **NA** = not applicable

1. _____ Three sets of drawings.
2. _____ Equipment is listed for intended use and specifications listing sheets are required.

Floor Plan Showing (1.7.2.2) :

3. _____ Scale: a common scale is used and the plan information is legible.
4. _____ Equipment/symbol legend is provided.
5. _____ Sectional view of the room with floor and ceiling assemblies is provided.
6. _____ Total number of all electrical devices are noted on the plans.
7. _____ Detector and nozzle location: provide sectional view detailing detector and nozzle positions.
8. _____ Type of devices and detail devices, detector, horn, wiring, etc.
9. _____ Control panel is at a constantly attended location and is connected to the building fire alarm system, if such a system is required.
10. _____ Type and gauge of wire or cable are provided.
11. _____ Riser diagram: number and type of devices for circuit, zone ID, dedicated branch circuit, batteries, panel, etc.

Point to Point System Wiring Diagram:

12. _____ Interconnection of identified devices and controls are shown.
13. _____ Indicate the number of conductors in each circuit run.
14. _____ Identify zones, circuits, and end-of-line resistor locations.

Show Alarm Indicating Circuit Load Consumption (Voltage Drop) of Circuits on Drawing:

15. _____ Quantity of signaling devices, current consumption, end-of-line voltage for each circuit, and the lowest nameplate voltage range for audible and visual notification devices.
16. _____ Approximate length of each circuit and resistance of wire, use National Electrical Code ampacity values or provide manufacturer data sheet for the selectec conductors.
17. _____ Provide calculations indicating the voltage drop for each circuit is within acceptable limits, 1.8.6.1.
18. _____ Audible and visual predischarge alarms shall be provided within the protected area, 1.8.5.
19. _____ Show means of activation of magnetic door-releasing hardware and/or ventilation shutdown.
20. _____ Show the primary and secondary sources of power, 1-8.6.

Battery Calculation Sheet Showing:

21. _____ Standby power consumption of all current drawing devices times the hours required by NFPA, 24 hours, including power consumption of the control panel modules.
22. _____ Power consumption of all devices on standby power; including door holders, relays, smoke detectors, etc.

CO_2 System Design Information:

23. _____ Type of system: flooding or local application; preengineered or engineered; high or low pressure.
24. _____ Preengineered CO_2 system layout meets manufacturer's listing requirements and a design specification manual is provided.
25. _____ Type of hazard protected and agent quantity, 1-5.2, 1-9.
26. _____ Volume of area protected: _____ cu. ft., if area above the ceiling is not included then drop ceilings shall be secured in place.
27. _____ Volume of CO_2 _____ percent, which includes a 20 percent safety factor, is adequate for the hazard, and the concentration level for surface fires is not less then 34 percent, 2–3.2.
28. _____ Concentration level calculations for surface fires, enclosed or unenclosed areas are provided, 2–2.1.1.

Fire Plan Review and Inspection Guidelines

29. _____ Concentration level calculations for deep-seat fires, enclosed or unenclosed areas are provided, 2–4.4.
30. _____ Calculations: provide nozzle flow rate and orifice size and coverage area, pipe lengths/type and equivalent pipe lengths, gas storage temperature, reference points, gas quantity, volume of the area protected, 3-4, 3-5.
31. _____ High pressure system design quantity shall be increased by 40 percent, 3-3.1.1.
32. _____ Discharge time for local application is 30 seconds, 3-3.3.
33. _____ Discharge time within 60 seconds for surface fires, 2-5.2.1.
34. _____ Discharge time for high pressure system is calculated as required by 3-2.3.
35. _____ Discharge time for deep seated fires shall be within 7 minutes and 30 percent concentration within 2 minutes, 2-5.2.3.
36. _____ Pressure relief venting calculations are provided unless venting is not necessary, 2-6.2.
37. _____ Local Application: Rate by area calculations are provided for flat surfaces or low-level hazards, 3-4.
38. _____ Local Application: Rate by volume calculations are provided for 3-D irregular objects, 3-5.1.

Operating Devices, Control Devices and Alarms (1-7, 1-8):
39. _____ Devices are listed for their use and specification sheets are provided.
40. _____ Detail location of emergency release, manual pull device, 1-7.3.5
41. _____ Automatic detection and actuation shall be used, 1-8.1.1.
42. _____ Manual pull device is distinct in appearance and not more than 4 ft. above the floor, note or detail, 1-8.3.4.
43. _____ Automatic fuel or power shutoff is provided, see exception, 1-8.3.8.
44. _____ Predischarge alarms or indicators for system operation are provided, 1-8.5.
45. _____ Audible and visual pre-discharge alarms are provided, 1-8.5.
46. _____ Abort switches are not permitted, 1-8.3.10.
47. _____ A "lockout valve" for the system is provided, 1-6.1.7.
48. _____ Alarms indicating failure of supervised devices or equipment are provided and detailed, 1-8.5.3.
49. _____ Pre-discharge alarm is provided when a hazard to personnel exists, 1-8.5 and sound pressure levels shall be 15 dBA above ambient or 5 dBA above maximum noise level, 90 dBA is a minimum.

Piping, Fittings, Valves and Nozzles:
50. _____ Sizes, type of material, pipe schedule, ASTM specifications, fittings, valves, and nozzle listing data and working pressure ratings are provided, 1-10.
51. _____ Sections of closed piping are equipped with pressure relief devices, 1-10. 2.2.
52. _____ Method of securing pipe is detailed, 1-10.2.

Miscellaneous:
53. _____ Detail location of warning and instruction signage at entrances and inside the protected area, 1-6.1.2.
54. _____ Agent containers are close to or within hazard area and are approved for that use, 1-9.4.
55. _____ Location of agent storage containers is accessible, 1-9.4.
56. _____ Storage container securing system is detailed for manifolded systems, 1-9.4.
57. _____ Nozzle specification sheet for proper application and installation is provided, 1-10.4.
58. _____ Protected area or room is properly sealed and so noted on plans, 2-2.2
59. _____ Forced-air ventilation is designed to be shutoff if its operation may affect the performance of the system, 2-2.2.2.
60. _____ Nozzle coverage areas are in accordance with the criteria established for the specific hazard, 3-4.3, 3-4.4.

Additional Comments:

Review Date: _____   Approved or Disapproved   FD Reviewer:_____

Review Date: _____   Approved or Disapproved   FD Reviewer:_____

Voltage Drop Calculations for Notification Appliance Circuit (NAC): _____

Each NAC shall have its voltage drop determined. This sheet shall be used for one NAC but every NAC should have a sheet completed and submitted with each permit application.

STEP 1: complete the following to provide data for determining the resistance of the conductor in Step 2

    Wire length is from fire alarm control
    panel to the end of the fire alarm circuit = _____ ft. X 2 = _____ ft.
    Wire Size = #_____ AWG (American Wire Gauge)
    Resistance (R) = _____ OHMS for a given 1,000 ft. of the conductor specified

Step 2: complete the following to determine the total resistance (OHMS) for a NAC

    (R) = Total Wire Resistance

From Step 1 divide the OHMS by 1,000, which will convert the conductor resistance to OHMS in each linear foot of wire

    Determine OHMS per foot = _____ ft. ÷ 1,000 = _____ OHMS/ft.

Take the total feet of wire from Step 1 and OHMS/ft. from the line above and put both in the equation below

    Circuit resistance = _____ ft. X _____ OHMS/ per ft. = _____ (R) Total OHMS

Step 3: complete the following to determine the total alarm notification device amperage and devices may be rated in milliamps

    (I) = Alarm Appliance Amperage

A. No. of Alarm Appliances = _____	B. Current amperage each _____	= A x B _____ (I)
A. No. of Alarm Appliances = _____	B. Current amperage each _____	= A x B _____ (I)
A. No. of Alarm Appliances = _____	B. Current amperage each _____	= A x B _____ (I)
A. No. of Alarm Appliances = _____	B. Current amperage each _____	= A x B _____ (I)
		Total _____ (I)

Step 4: complete the following to determine the total voltage drop for the branch circuit

    Voltage (E) = (I) X (R) from totals in Steps 2 and 3 above
    (E) = _____ (I) X _____ (R)
    = _____ (E) (shall not exceed 4.4)

Step 5: complete the following to determine if enough voltage is available to operate fire alarm notification devices

Maximum allowable voltage drop: notification devices cannot drop below their Nameplate Operating Voltage (NOV) range. As of 5/1/2004 UL required indicating devices to operate within their NOV. The UL NOV standard is 16VDC to 33VDC, consult the 2002 NFPA 72 Handbook 7.3 for more information. Fire Alarm Control Unit (FACU) are tested to UL 864 and are required to operate at the end of useful battery life, 20.4 V.

    Allowable voltage drop is 20.4 V (FACU) - 16 VDC (NOV) = 4.4 V

    If (E) from Step 4 exceeds 4.4 V then the NAC is not compliant with NFPA 72

    Take (E) from Step 4 and put in the equation below

    Voltage Drop = 20.4 V - _____ (E) = _____ V (shall not be less than 16V)

Fire Plan Review and Inspection Guidelines

Carbon Dioxide Alternative Fire-Extinguishing System Acceptance Inspection
2006 IFC and 2000 NFPA 12

Date of Inspection: _____ Permit Number: _____

Business/Building Name: _____ Address of Project: _____

Contractor: _____ Contractor's Phone: _____

Reference numbers following worksheet statements represent an NFPA code section unless otherwise specified.

Pass | Fail | NA

1. ___ | ___ | ___ Received CO_2 system installation certification from installer.
2. ___ | ___ | ___ Control panel and components match approved plans.
3. ___ | ___ | ___ Approved drawing on site, as-builts are required when the installation is substantially different when compared to the approved plans.
4. ___ | ___ | ___ Control panel, piping, nozzles, and other component locations are the same as shown on the plans.
5. ___ | ___ | ___ Zones are properly identified on panel(s).
6. ___ | ___ | ___ Verify the automatic detection and actuation components are connected to a power supply complying with Section 1.8.6.
7. ___ | ___ | ___ Device location same as plans.
8. ___ | ___ | ___ Devices are located in all areas required by the code.
9. ___ | ___ | ___ Devices are properly wired.
10. ___ | ___ | ___ Type and gauge of wire or cable match plans.
11. ___ | ___ | ___ 24 hour monitoring service agency received signals.
12. ___ | ___ | ___ Verify proper operation of magnetic door-releasing hardware and/or ventilation shutdown.
13. ___ | ___ | ___ Battery stress test: system switched to battery operation 24 hours earlier, then activate audible circuit for 5 minutes.
14. ___ | ___ | ___ Pull stations are distinct appearance, identified as to function, and mounted at proper height and location which is no more that 4 ft. above the floor.
15. ___ | ___ | ___ Under primary and secondary power operational tests are performed including:
    ___ | ___ | ___ A. power light on and in normal condition.
    ___ | ___ | ___ B. supervisory signals: water flow and pressure switches, valves, etc.
    ___ | ___ | ___ C. silence switches.
    ___ | ___ | ___ D. trouble signals and panel light operate for each circuit tested, disconnect wires from devices.
    ___ | ___ | ___ E. trouble and alarm reset switches operate.
    ___ | ___ | ___ F. a second initiating zone overrides silence switch.
    ___ | ___ | ___ G. second audible and visual operation.
    ___ | ___ | ___ H. panel lamp test switch operates.
    ___ | ___ | ___ I. field zone signals correspond with panel zones.
    ___ | ___ | ___ J. detection devices operate.
16. ___ | ___ | ___ Activation of the CO_2 system activates the building fire alarm, if a fire alarm system is required.
17. ___ | ___ | ___ Piping and nozzles are restrained so no unacceptable movement occurs, prior to the pressure test.
18. ___ | ___ | ___ Nozzles shall not be installed at a location were injury can occur.
19. ___ | ___ | ___ Release circuit is tested at the storage container.
20. ___ | ___ | ___ Pull stations activate system.
21. ___ | ___ | ___ Protected area is properly sealed.
22. ___ | ___ | ___ Warning and instruction signage is properly posted.
23. ___ | ___ | ___ Conduct a flow test with nitrogen or inert gas to verify unobstructed pipes and nozzles.
24. ___ | ___ | ___ An enclosure or room integrity (room leakage) test may be required by the AHJ.

Additional Comments:

Inspection Date: _____   Approved or Disapproved   FD Inspector: _____

Inspection Date: _____   Approved or Disapproved   FD Inspector: _____

Inspection Date: _____   Approved or Disapproved   FD Inspector: _____

Fire Plan Review and Inspection Guidelines

Carbon Dioxide Alternative Fire-Extinguishing Installation Certification

---

Permit #: _____   Date: _____

	Property Protected	System Installer	System Supplier
Business Name:	_____	_____	_____
Address:	_____	_____	_____
	_____	_____	_____
Representative:	_____	_____	_____
Telephone:	_____	_____	_____

Type of System: _____

Location of Plans: _____

Location of Owner's Manual: _____

1. **Certification of System Installation:** Complete this section after system is installed, but prior to conducting operational acceptance tests. Check wiring for opens, ground faults, and improper branching.

    This system installation was inspected and was found to comply with the installation requirements of:
    - _____ NFPA 12
    - _____ Article 760 of National Electrical Code
    - _____ NFPA 72
    - _____ Manufacturer's Instructions
    - _____ Other (specify: FM, UL, etc.) _____

    Print Name: _____

    Signed: _____ Date: _____

    Organization: _____

2. **Certification of System Operation:** All operational features and functions of this system were tested and found to be operating properly in accordance with the requirements of:
    - _____ NFPA 12
    - _____ Design Specifications
    - _____ NFPA 72
    - _____ Manufacturer's Instructions
    - _____ Other (specify) _____

    Print Name: _____

    Signed: _____ Date: _____

    Organization: _____

Fire Plan Review and Inspection Guidelines

Prefinal and Certification of Occupancy Inspection Requirements For Contractors
Contractors Worksheet

Carbon Dioxide System Test Requirements

1. All certification forms and documents are required to be on the site for review:
   - ___ Plans
   - ___ Permit
   - ___ Prefinal detector test and a device location inspection are required as well as the completion of the items on this pretest form. Use the Acceptance Inspection worksheet for the pretest.
   - ___ Installation certification is completed, use the form contained in this book.
2. ___ Person familiar with installation must be present to perform the test.
3. ___ Owner's representative approval needed for time and date of testing.
4. ___ All rooms or areas are unlocked and accessible.
5. ___ Device placement is in accordance with the plan, verification to have been done by the contractor.
6. ___ If Items 1-5 are incomplete, the inspection will be cancelled and another inspection request will be required. A reinspection fee may be assessed.

Prior to the next approval test:

7. ___ When additional devices, e.g., nozzles, initiating or indicating devices, etc., are installed, the contractor must provide:
   - ___ As-builts and new calculations for review and approval.
   - Note: A new plan review will be submitted as "supplemental information" and proof of the additional review fee payment is required.
8. ___ The reinspection fee may be assessed if the system and required documentation are not ready.

Fire Plan Review and Inspection Guidelines

Halon
Contents

1. Plan Review Worksheet. 185

2. Voltage Drop Calculation Form 187

3. Acceptance Test Checklist. 188

4. Installer Certification Form...................................... 189

5. Contractor and Certification of Occupancy Inspection Requirements 190

6. Semiannual Service Test Form. 191

Halon 1301 Systems
Plan Review Worksheet
2006 IFC, 2002 NFPA 72, and 2004 NFPA 12A

Date of Review: _____ Permit Number: _____

Business/Building Name: _____ Address of Project: _____

Designer Name: _____ Designer's Phone: _____

Contractor: _____ Contractor's Phone: _____

System Manufacturer: _____ Model: _____

Reference numbers following worksheet statements represent an NFPA code section unless otherwise specified.

Worksheet Legend: ✓ or OK = acceptable   N = need to provide   NA = not applicable

1. _____ Three sets of drawings.
2. _____ Equipment is listed for its intended use and listing data sheets are provided.
3. _____ Design manual for pre-engineered system is provided.

Floor Plan Details Are Provided for the Following Items, NFPA 12A Chapter 5:

4. _____ Scale: use a common scale and provide room dimensions.
5. _____ Equipment symbol legend.
6. _____ Cross sectional view of the room with floor and ceiling assemblies.
7. _____ Floor plan and cross sectional views detail alarm initiating devices, audible signaling devices, and the location and position of agent nozzles.
8. _____ Type of devices are specified and wiring circuits are detailed.
9. _____ Control panel location is at a constantly attended location and tied into the building fire alarm system if provided.
10. _____ Type and gauge of wire or cable is noted for each circuit.
11. _____ Riser diagram details the number and type of devices for each circuit, zone IDs, a dedicated 120 volt AC branch circuit, secondary power supply, panel, etc.

Point to Point System Wiring Diagram Details the Following Items:

12. _____ Interconnections of identified devices and controls are shown.
13. _____ Indicate number of conductors in each circuit.
14. _____ Identify each circuit zone and end of line resistor locations.

Provide Alarm Notification Circuit Load Consumption (Voltage Drop) for Each Circuit on the Plan:

15. _____ Quantity of signaling devices, current consumption, end of line voltage for each circuit, and the lowest nameplate operating voltage range for audible and visual notification devices.
16. _____ Approximate length of each circuit and resistance of wire, use National Electrical Code resistance specification or provide manufacturer specification sheet.
17. _____ Voltage drop calculations for each notification circuit and acceptable voltage drop limits are provided on the plans or on an attachment.
18. _____ Audible and visual discharge alarm devices are provided within the protected area, 4.3.5.2.
19. _____ When provided, the means of activation of magnetic door-releasing hardware and for ventilation shutdown, IFC 904.3.3.
20. _____ A primary and at least a 24 hour secondary power supply are provided and detailed, 4.3.2.2.
21. _____ The main power supply for the system is on a dedicated branch circuit, 6.7.2.3.3 and 2002 NFPA 72 4.4.1.4.2.
22. _____ Preengineered Halon Gas System piping and nozzle arrangement meets the manufacturer's listing requirements and the manufacturer's installation manual.

Battery Load Calculations:

23. _____ Standby power consumption of all current drawing devices multiplied by the number of hours required by NFPA 72 (24 hours) including power consumption of the control panel modules.
24. _____ Power consumption of all devices on standby power; including door holders, relays, smoke detectors, and etc.

Fire Plan Review and Inspection Guidelines

25. _____ Alarm power consumption of all current drawing devices multiplied by the number of minutes required by NFPA 72, 5 minutes.
26. _____ Battery load calculations are provided.

System Design Information:
27. _____ The type of system design, inerting or flame extinguishment is specified.
28. _____ The type of hazard protected and agent quantity is specified.
29. _____ Inerting is required when the quantity of fuel would develop up to or exceed 50 percent of the LEL throughout the enclosure.
30. _____ The design concentration is specified as: inert plus a 10 percent safety factor or flame extinguishment plus a 20 percent safety factor, 5.4.1.1.1. Consult Table 5.4.1.1.1. for inerting concentrations and Table 5.4.1.2 for design requirements for systems designed to perform flame extinguishment.
31. _____ For fires in solid materials consult Annex I and for solid surface fires use a minimum of 5 percent concentration and consult Annex J, 5.4.1.
32. _____ For total flooding show the formula from 5.5.1 and the results, adjust for altitude when greater then 3,000 ft., 5.6.

Distribution:
33. _____ Calculations show the discharge time to be in compliance with Section 5.7.1.2.
34. _____ Calculations specify the nozzle flow rate and orifice size, pipe lengths, type, and equivalent lengths, gas temperature, reference points, gas quantity, and cubic footage of area protected, 5.2.2.
35. _____ The location of storage container(s) are easily accessible and not permitted above the ceiling, 4.1.3.
36. _____ The storage container securing system is detailed in accordance with the manufacturer's listing manual, 4.1.3.4.
37. _____ Agent containers are close to or within the hazard area, 4.1.3.2.
38. _____ When manifolded halocarbon agent containers are the same size and charge, 4.1.4.5, .6.
39. _____ Pipe joints are either screwed or flanged and any other pipe joint shall be listed or approved for the application, 4.2.2.
40. _____ Piping material types are noted on the plans and are compatible with environment and the agent. Cast iron, steel pipe or fittings conforming to ASTM A120 or nonmetallic pipe shall not be used, 4.2.1.2.
41. _____ The type of fittings and pressure ratings are noted on the plans and comply with Section 4.2.3.
42. _____ Listing data sheets are provided that specifying type, size area coverage, height limits, and minimum PSI, etc. in conformance with 4.2.5 and 5.8.

Operating Devices, Control Devices and Alarms:
43. _____ Device and circuit design is in accordance with NFPA 72, NFPA 12A 4.3.1.
44. _____ Devices are listed for their use and listing data sheets are provided, 4.3.2.1.
45. _____ Detailed is the location of the normal manual release pull device, 4.3.3.5.
46. _____ Automatic detection and actuation shall be used, 4.3.2.
47. _____ The emergency single action manual pull device is detailed, distinct in appearance and the mounting height is not less than 4 ft. and it does not exceed 5 ft. above the floor, 4.3.3.5 – 4.3.3.7 and IFC 904.3.2.
48. _____ Manual operating devices are signed for the hazard they protect and the sign location is noted on the plans, 4.3.3.10.
49. _____ Detail the location of warning and instruction signs at entrances and inside protected area, 4.3.5.5.
50. _____ Alarms or indicators for system operation are provided in accordance with Section 4.3.5, IFC 904.3.4.
51. _____ Audible and highly visible alarms are provided for warning of discharge 4.3.5.2, and audible levels are in accordance with NFPA 72.
52. _____ If abort switches are used, they located in protected area near the exit, be a constant manual pressure type, tied into visual and audible devices, and are detailed on plans, 4.3.5.3.
53. _____ Alarms, distinctive from other alarms, indicating failure of supervised devices or equipment are provided and detailed, 4.3.5.4.
54. _____ If discharge time delays are proposed, they are only permitted when personnel must evacuate first, 4.3.5.6.

Other:
55. _____ The protected area or room is properly sealed and noted on plans, 5.3.1.2.
56. _____ Confirm that the force-air ventilation connected to the halon system to shut down or close automatically if it would adversely affect the performance of the system, 5.3.1.3.

Review Date: _____ Approved or Disapproved FD Reviewer: _____

Review Date: _____ Approved or Disapproved FD Reviewer: _____

Fire Plan Review and Inspection Guidelines

Line Voltage Drop Calculations for Notification Appliance Circuit (NAC): _____

Each NAC shall have its voltage drop determined. This sheet shall be used for one NAC but every NAC should have a sheet completed and submitted with each permit application.

STEP 1: complete the following to provide data for determining the resistance of the conductor in Step 2

Wire length is from fire alarm control
panel to the end of the fire alarm circuit = _____ ft. X 2 = _____ ft.

Wire Size = #_____ AWG (American Wire Gauge)

Resistance (R) = _____ OHMS for a given 1,000 ft. of the conductor specified

Step 2: complete the following to determine the total resistance (OHMS) for a NAC

(R) = Total Wire Resistance

From Step 1 divide the OHMS by 1,000, which will convert the conductor resistance to OHMS in each linear foot of wire

Determine OHMS per foot = _____ ft. / 1,000 = _____ OHMS/ft.

Take the total feet of wire from Step 1 and OHMS/ft. from the line above and put both in the equation below

Circuit resistance = _____ ft. X _____ OHMS/ per ft. = _____ (R) Total OHMS

Step 3: complete the following to determine the total alarm notification device amperage and devices may be rated in milliamps

(I) = Alarm Appliance Amperage

A. No. of Alarm Appliances = _____   B. Current amperage each _____ = A x B _____ (I)
A. No. of Alarm Appliances = _____   B. Current amperage each _____ = A x B _____ (I)
A. No. of Alarm Appliances = _____   B. Current amperage each _____ = A x B _____ (I)
A. No. of Alarm Appliances = _____   B. Current amperage each _____ = A x B _____ (I)
                                                                       Total _____ (I)

Step 4: complete the following to determine the total voltage drop for the branch circuit

Voltage (E) = (I) X (R) from totals in Steps 2 and 3 above

(E) = _____ (I) X _____ (R)

 = _____ (E) (shall not exceed 4.4)

Step 5: complete the following to determine if enough voltage is available to operate fire alarm notification devices

Maximum allowable voltage drop: notification devices cannot drop below their Nameplate Operating Voltage (NOV) range. As of 5/1/2004 UL required indicating devices to operate within their NOV. The UL NOV standard is 16VDC to 33VDC, consult the 2002 NFPA 72 Handbook 7.3 for more information. Fire Alarm Control Units (FACU) are tested to UL 864 and are required to operate at the end of useful battery life, 20.4 V.

Allowable voltage drop is 20.4 V (FACU) - 16 VDC (NOV) = 4.4 V

If (E) from Step 4 exceeds 4.4 V then the NAC is not compliant with NFPA 72

Take (E) from Step 4 and put in the equation below

Voltage Drop = 20.4 V - _____ (E) = _____ V (shall not be less than 16V)

Fire Plan Review and Inspection Guidelines

Halon System Acceptance Inspection
2006 IFC, 2002 NFPA 72, and 2004 NFPA 12A

Date of Inspection: _____ Permit Number: _____

Business/Building Name: _____ Address of Project: _____

Contractor: _____ Contractor's Phone: _____

Reference numbers following worksheet statements represent an NFPA code section unless otherwise specified.

Pass | Fail | NA

1. ___ | ___ | ___ Received halon system certification from installer.
2. ___ | ___ | ___ Control panel and components match approved plans.
3. ___ | ___ | ___ Approved drawing is on site. As-builts are required if installation does not match plans.
4. ___ | ___ | ___ Control panel, devices, and other component locations are same as shown on plans.
5. ___ | ___ | ___ Zones are properly identified on control panel(s).
6. ___ | ___ | ___ Verify the system is on a dedicated 120 AC branch circuit.
7. ___ | ___ | ___ The location of warning and instruction signage matches the approved plans.
8. ___ | ___ | ___ Devices are properly wired.
9. ___ | ___ | ___ Type and gauge of wire match the approved plans.
10. ___ | ___ | ___ If provided, a 24 hour monitoring agency received signals and conveyed the type of signals received.
11. ___ | ___ | ___ Battery load test is performed by placing the system on battery operation 24 hours earlier, then activate audible circuit for 5 minutes, 6.7.2.4.4.
12. ___ | ___ | ___ Verify proper operation of magnetic door-release hardware and/or ventilation shutdown, when provided.
13. ___ | ___ | ___ Under primary and then secondary power, operational tests are performed, including:
    ___ | ___ | ___ A. power light on and in normal condition.
    ___ | ___ | ___ B. each initiating and detection device is tested per the manufacturer's instructions.
    ___ | ___ | ___ C. trouble signals and panel lights operate for each circuit tested; disconnect wires from devices.
    ___ | ___ | ___ D. trouble and alarm reset switches operate.
    ___ | ___ | ___ F. field zone signals correspond with panel zones.
    ___ | ___ | ___ G. door release devices and other system's emergency shutdowns operate.
14. ___ | ___ | ___ The system activates the building fire alarm, if provided.
15. ___ | ___ | ___ An enclosure or room integrity leakage test is witnessed, 6.7.2.2.
16. ___ | ___ | ___ Wiring is secure and protected from possible damage.
17. ___ | ___ | ___ The normal and emergency manual pull stations are duel action, distinct in appearance, properly identified and mounted from 4 ft. to 5 ft above the floor. Both manual pull stations override any system Abort Switch.
18. ___ | ___ | ___ Verify systems with abort switches are deadman type requiring constant pressure.
19. ___ | ___ | ___ Piping and nozzles are restrained so no unacceptable movement occurs, check prior to the pressure test and verify the pipe and nozzles match the approved plans, 6.7.2.1.
20. ___ | ___ | ___ Piping is pneumatically tested for 10 minutes at 150 PSIG and verify the pressure drop does not exceed 20 percent of the test pressure, 6.7.2.1.12.
21. ___ | ___ | ___ The release circuit is tested at the storage container and end of line resistors are checked, 6.7.2.4.
22. ___ | ___ | ___ The manual pull station activates the systems (discharge of system is not required for this).
23. ___ | ___ | ___ Conduct a puff test with nitrogen or inert gas to verify pipe and nozzles are unobstructed, 6.7.2.1.13.
24. ___ | ___ | ___ Verify the agent quantity and fill pressure to the cylinder nameplate specifics, 4.1.4.2 and 6.1.

Inspection Date: _____ Approved or Disapproved FD Inspector: _____

Inspection Date: _____ Approved or Disapproved FD Inspector: _____

Fire Plan Review and Inspection Guidelines

Halon System Installation Certification

Permit #: _____ Date: _____

	Property Protected	System Installer	System Supplier
Business Name:	_____	_____	_____
Address:	_____	_____	_____
	_____	_____	_____
Representative:	_____	_____	_____
Telephone:	_____	_____	_____

Location of Plans: _____

Location of Owner's Manual: _____

1. **Certification of System Installation:** Complete this section after system is installed, but prior to conducting operational acceptance tests.

   This system installation was inspected and found to comply with the installation requirements of:
   - _____ NFPA 72
   - _____ NFPA 12A
   - _____ Article 760 of the National Electrical Code
   - _____ Manufacturer's Instructions
   - _____ Other (specify; FM, UL, etc.) _____

   Print Name: _____

   Signed: _____ Date: _____

   Organization: _____

2. **Certification of System Operation:** All operational features and functions of this system were tested and found to be operating properly in accordance with the requirements of:
   - _____ NFPA 72
   - _____ NFPA 12A
   - _____ Design Specifications
   - _____ Manufacturer's Instructions
   - _____ Other (specify) _____

   Print Name: _____

   Signed: _____ Date: _____

   Organization: _____

Prefinal and Certification of Occupancy Inspection Requirements For Contractors
Contractors Worksheet

Halon Test Requirements

1. All certification forms and documents are required to be on the site for review:
   - ___ Plans
   - ___ Permit
   - ___ Prefinal sensor/detector test and a device location inspection are required as well as the completion of the items on the pretest form. Use the Acceptance Inspection worksheet for the pretest.
   - ___ Installation certification is completed, use the form contained in this book.
2. ___ Person familiar with installation must be present to perform the test.
3. ___ Owner's representative approval needed for time and date of testing.
4. ___ All rooms are unlocked and accessible.
5. Device placement is in accordance wih the plan, audible verification to have been done by contractor.
6. ___ If Items 1-5 are incomplete, the inspection will be cancelled and another inspection request is required. A reinspection fee may be assessed.

Prior to the next approval test:

7. ___ When there are device additions, the contractor must provide:
   - ___ As-builts and new calculations shall be submitted for review and approval.
   - Note: New plan review will be submitted as "supplemental information" and proof of the additional review fee payment is required.
8. ___ A reinspection fee may be assessed if the system and paperwork are not ready.

Semiannual Service Test Report
Halon System

Inspector: _____ Date: _____
Customer: _____
Customer Address: _____

SYSTEM DETAILS:
( ) New system   ( ) Reinspection
( ) Halon 1301   ( ) Halon 1211   ( ) Other _____

Installation date or most recent modification date: _____

Installed and/or supervised by: (check one)
( ) Mfr. or representative   ( ) Other contractor
( ) Automatic sprinkler contractor   ( ) In-house
( ) Unknown

Check as many as necessary:
( ) Engineered   ( ) Precalculated   ( ) Not sure
( ) Total flooding   ( ) Local application ( ) Combination of both

Actuation provided:
( ) Manual only   ( ) Automatic only   ( ) Automatic and manual

If automatic:
( ) Fixed temperature heat   ( ) Flame   ( ) Ionization
( ) Photo smoke   ( ) Rate of rise, heat det. ( ) Combination
( ) Single zone   ( ) Cross zoned   ( ) Other

If other or combination, describe: _____
_____

Interlocks provided:
( ) Air handling equipment   ( ) Door releases   ( ) None
( ) Duct dampers   ( ) Process shutdown   ( ) Other

Describe other or need for additional interlocks: _____
_____

Maintenance program:
Tag date of last system maintenance checkout: _____

HAZARD DETAILS:
Area, process or equipment protected: _____
Are drawings and calculation sheets of installation and modification available for use:
( ) Yes   ( ) No   ( ) Unknown
Has protected hazard changed since system installation or most recent system modification:
( ) Yes   ( ) No   ( ) Unknown

Fire Plan Review and Inspection Guidelines

TESTING AND INSPECTION OF SYSTEMS:

Control Panel:
Were all control panel operating functions exercised	( ) Yes ( ) No
Did they all function properly	( ) Yes ( ) No
Describe deficiencies_____	
Exercise all supervisory functions	( ) Yes ( ) No
Did they all function properly	( ) Yes ( ) No
Describe deficiencies_____	

Power Supply: ( ) No power supply
Fuses and disconnects provided	( ) Yes ( ) No
Fuses and disconnects lockable	( ) Yes ( ) No
Describe deficiencies_____	

Emergency Power: ( ) No emergency power
Battery condition	( ) Good ( ) Fair ( ) Poor
Comments_____	
Charger operable	( ) Yes ( ) No
Automatic changeover to emergency power operable	( ) Yes ( ) No
If battery is not provided and power source from generator is provided, was it exercised	( ) Yes ( ) No
Did it operate properly	( ) Yes ( ) No
Generator provided	( ) Yes ( ) No

Detectors: ( ) No detectors
Test all (if restorable type)
Did all detectors function properly	( ) Yes ( ) No
Electric	( ) Yes ( ) No
Were all smoke detectors cleaned, adjusted, and sensitivity checked	( ) Yes ( ) No

Pneumatic: ( ) No pneumatic
Were all pneumatic rate of rise detectors tested for operation using manometer	( ) Yes ( ) No
Were they checked using heat	( ) Yes ( ) No
Were mercury check valves tested using manometer	( ) Yes ( ) No

Link Type: ( ) No link type
Terminal link removed/cut, etc. for operation check	( ) Yes ( ) No
Link changed	( ) Yes ( ) No

Time Delay: ( ) No time delay
Was it exercised	( ) Yes ( ) No
How_____	
Was the time limit checked	( ) Yes ( ) No
How many seconds_____	
If time is electric, will it complete its cycle if wiring between control panel and detector is interrupted	( ) Yes ( ) No

Alarms: ( ) No alarms
Did all audible alarms operate	( ) Yes ( ) No
Did all visual alarm signals operate	( ) Yes ( ) No

Is there a predischarge alarm ( ) Yes ( ) No
Are suitable warning signs provided for alarms ( ) Yes ( ) No

Selector (Directional) Valves: ( ) No selector valves
Were all valves tested ( ) Yes ( ) No
With gas ( ) Yes ( ) No
Manually only ( ) Yes ( ) No
Did they reset properly ( ) Yes ( ) No

Release Devices:
Dampers: ( ) No dampers
  Did they close completely ( ) Yes ( ) No
Doors: ( ) No release dampers
  Were any blocked ( ) Yes ( ) No
  Did they all operate ( ) Yes ( ) No

Equipment Shutdown:
Were all equipment shutdown features exercised ( ) Yes ( ) No
  If not, why not_____
  Name of representative who refused_____
Were all necessary or desirable equipment shutdowns provided ( ) Yes ( ) No
  Describe deficiencies_____

Manual Releases:
Mechanical: ( ) No mechanical releases
  Was pull force and length of pull acceptable ( ) Yes ( ) No
  Were all mechanical pulls operated ( ) Yes ( ) No
  Describe deficiencies_____
Electric: ( ) No electric releases
  Were all electric releases tested ( ) Yes ( ) No
  Did they all function correctly ( ) Yes ( ) No
Were all releases properly located so as to be readily accessible in case of fire ( ) Yes ( ) No
Will a single manual release provide full operation of system including selector valves if any ( ) Yes ( ) No
Are all manual releases clearly marked and identified ( ) Yes ( ) No

Piping:
Is piping securely and adequately supported per Manufacturer's recommendations ( ) Yes ( ) No

Nozzles:
Were nozzle orifices clean ( ) Yes ( ) No
Were nozzles properly secured ( ) Yes ( ) No

Agent Containers:
Physical condition ( ) OK ( ) Not OK
Signs of corrosion ( ) Yes ( ) No
Were contents checked by weight or other acceptable method ( ) Yes ( ) No
Were all containers full to design level ( ) Yes ( ) No

Fire Plan Review and Inspection Guidelines

Were containers securely bracketed	( ) Yes ( ) No
Was there a net weight change since last inspection	( ) Yes ( ) No
Date of manufacture or last hydro test_____	
Were cylinder connectors secure and in good condition	( ) Yes ( ) No
On mechanical release system, were weights and cables operable	( ) Yes ( ) No
With explosive release devices, were they within replacement date	( ) Yes ( ) No
Did they appear in good condition	( ) Yes ( ) No
Were they checked for electrical continuity	( ) Yes ( ) No

Tests:

Was a partial discharge test performed	( ) Yes ( ) No
Was a full discharge test performed	( ) Yes ( ) No
Was system returned to full service and all disarmed features restored to normal operating service	( ) Yes ( ) No
Was certificate of inspection provided to owner	( ) Yes ( ) No

Describe how system was tested showing that system would have discharged if it had not been disabled; e.g.

1. Explosive squibs replaced with test bulbs and all "fired" as designed.
2. Solenoid removed from cylinder head and it "fired" as designed.

Inspection Comments:

Clean Agent
Contents

1. Plan Review Worksheet 197

2. Voltage Drop Calculation Form 199

3. Acceptance Test Checklist. 200

4. Installer Certification Form...202

5. Contractor Final and Certificate of Occupancy Inspection Requirements 203

Clean Agent System
Plan Review Worksheet
2004 NFPA 2001 and 2002 NFPA 72

Date of Review: _____ Permit Number: _____

Business/Building Name: _____ Address of Project: _____

Designer Name: _____ Designer's Phone: _____

Contractor: _____ Contractor's Phone: _____

System Manufacturer: _____ Model: _____ Occupancy Classification: _____

Reference numbers following worksheet statements represent an NFPA code section unless otherwise specified.

Worksheet Legend: ✓ or **OK** = acceptable   **N** = need to provide   **NA** = not applicable

1. _____ Three sets of drawings.
2. _____ Equipment is listed for intended use and specification listing sheets are required.
3. _____ Type of agent is provided and design manual for preengineered systems.

Floor Plan Detailing the Following Items, Section 5.1.2:

4. _____ Scale: use a common scale, also provide room description and dimensions.
5. _____ An equipment symbol legend is provided.
6. _____ Sectional view of the room with floor and ceiling assemblies and isometric view of the system are provided, 5.1.2.2 (6).
7. _____ A floor plan and a isomeric detail of the agent distribution system including calculation reference points, pipe diameters and lengths, and equivalent length of fittings, 5.1.2.2 (18). detailing device and nozzle locations and positions.
8. _____ Methods for installation of detectors, fire detection devices, alarm signaling devices and selected wiring practices including the selected conductors for these devices, 5.1.2.2 (14), (17) and (23)
9. _____ The system control panel is installed at a constantly attended location and connected to the building fire alarm system, if such a system is required.
10. _____ A sequence of operation for the clean agent system complying with Section 5.1.2.2 (22).
11. _____ Provide calculations for clean agent serving the enclosed volume, the backup power duration and the voltage drop for the number of installed alarm initiating and alarm signaling devices, 5.1.2.2 (25)

Analysis of Hazards to Personnel:

12. _____ If a halocarbon agent is used, means shall be provided to limit personnel to the time exposure and the "No Observed Adverse Exposure Limit" values in Table 1.5.1.2.1(a), 1.5.1.2.1
13. _____ If a halocarbon agent is used and the concentration will exceed the No Observed Adverse Exposure Limit values in Table 1.5.1.2.1, an exposure and means of egress of analysis shall be performed in accordance with Section 1.5.1.2.1(2).
14. _____ If a halocarbon agent is used in an area that is not normally occupied and the design requires a volume of agent that exceeds the "Lowest Observed Adverse Exposure Limit" values in Table 1.5.1.2.1(a) an approved means of limiting exposure of personnel to the discharge shall be provided.
15. _____ If design information is not sufficient to ensure that the Lowest Observed Adverse Limit is not exceeded, additional analysis and engineering controls as prescribed in Section 1.5.1.2.1 (4) shall be provided and approved by the fire code official.
16. _____ If a inert clean agent is used, a predischarge alarm shall be provided and personnel exposure limits shall not exceed the values in Section 1.5.1.3 at the indicate concentrations.
17. _____ Safeguards shall be provided to ensure that personnel are evacuated and exposures are limited prior to the discharge and after the discharge of a clean agent. The safeguards shall provide a means of affecting rescue of individuals trapped within an enclosure protected by a clean agent (Section 1.5.1.4.1).
18. _____ Minimum clearances between clean agent systems components and electrical equipment shall be provided in accordance with the requirements of Sections 1.5.2.1 through 1.5.2.5.
19. _____ Mixing of agents within the same storage container shall be in accordance with the requirement in Section 1.8.

Fire Plan Review and Inspection Guidelines

Electrical Components:
20. _____ A primary and standby source of power shall be to the control united in accordance with Section 6.7.2.4.3 and 6.7.2.4.
21. _____ Demonstrate that the main power supply for the system is on a dedicated branch circuit and properly labeled, 6.7.2.5.4 (1).

Clean Agent Information:
22. _____ Type of system: flooding or local application.
23. _____ Type of hazard protected and agent quantity.
24. _____ Indicate the type of agent: Halocarbon or Inert Gas, 1.5.1.2 and .3, 5.1.2.2 (7).
25. _____ The design concentration is in compliance with Section 5.4.2 for flame extinguishment or Section 5.4.3 for inerting.
26. _____ Volume of area protected is _____ , if the volume above the ceiling is not included then drop ceilings must be secured. An additional amount of clean agent is required if the ventilation system cannot be shutdown prior to discharge, 5.3.5.3.
27. _____ Calculations: provide nozzle flow rate and orifice size, pipe lengths/type and equivalent lengths, gas temperature, reference points, gas quantity, and the volume of the area protected, 5.1.2.2, 5.3.5.2.
28. _____ Discharge time within 10 seconds for halocarbon agents, inert agents up to 1 minute, based on 20% safety factor, 5.7.1.2.
29. _____ Retrofitted clean agents into existing systems shall result in listed or approved design, 1.7.
30. _____ Location of storage container is easily accessible, 4.1.3
31. _____ Storage container securing system is detailed per the manufacturer's listing manual, 4.1.3.4.
32. _____ Agent containers are located near or inside of the hazard area, 4.1.3.2.
33. _____ Manifolded halocarbon agent containers are the same size and charge, inert agent containers may be different sizes, 4.1.4.5.
34. _____ Piping materials types are noted on plans and are compatible with environment and clean agent. Cast iron, steel ASTM A 120, or nonmetallic pipe shall not be used, 4.2.1.
35. _____ Fittings and associated pressure ratings are noted on the plans or provide specification sheet, 4.2.3.
36. _____ Nozzles: provided is listing information for area coverage, height limits, and minimum design pressures, 4.2.5.1.

Operating Devices, Control Devices and Alarms:
37. _____ Devices are listed for their use and equipment data sheets are provided, 4.3.3.2, 4.3.2.1, and 4.3.4.1.
38. _____ Detail location of emergency release manual pull device.
39. _____ Automatic detection and actuation shall be used, 4.3.1.2; manual activation must be approved, 4.3.1.2.1.
40. _____ Manual pull device is distinct in appearance and not more than 4 ft. above the floor, note or detail, 4.3.3.7.
41. _____ Detail location of warning and instruction signage at entrances and inside protected area, 4.3.5.5.
42. _____ Alarms or indicators for system operation are provided, 4.3.5.1.
43. _____ Audible and visual predischarge alarms are provided 4.3.5.2, and audible levels are per NFPA 72.
44. _____ If abort switches are used, they shall be located in protected area near the exit, use a constant manual pressure design, are connected to the alarm signaling devices, and are detailed on the shop drawings, 4.3.5.3.
45. _____ Alarms indicating failure of supervised devices or equipment are provided and detailed, 4.3.5.4.

Other:
46. _____ Protected area or room is properly sealed and the method is noted on plans, 5.3.
47. _____ Forced-air ventilation shuts down if it can affect the performance of the system, 5.3.5.

Additional Comments:

Review Date: _____   Approved or Disapproved   FD Reviewer: _____

Review Date: _____   Approved or Disapproved   FD Reviewer: _____

Review Date: _____   Approved or Disapproved   FD Reviewer: _____

Voltage Drop Calculations for Notification Appliance Circuit (NAC): _____

Each NAC shall have its voltage drop determined. This sheet shall be used for one NAC but every NAC should have a sheet completed and submitted with each permit application.

STEP 1: complete the following to provide data for determining the resistance of the conductor in Step 2

Wire length is from fire alarm control
panel to the end of the fire alarm circuit = _____ ft. X 2 = _____ ft.

Wire Size = # _____ AWG (American Wire Gauge)

Resistance (R) = _____ OHMS for a given 1,000 ft. of the conductor specified

---

Step 2: complete the following to determine the total resistance (OHMS) for a NAC

(R) = Total Wire Resistance

From Step 1 divide the OHMS by 1,000, which will convert the conductor resistance to OHMS in each linear foot of wire

Determine OHMS per foot = _____ ft. / 1,000 = _____ OHMS/ft.

Take the total feet of wire from Step 1 and OHMS/ft. from the line above and put both in the equation below

Circuit resistance = _____ ft. X _____ OHMS/ per ft. = _____ (R) Total OHMS

Step 3: complete the following to determine the total alarm notification device amperage and devices may be rated in milliamps

(I) = Alarm Appliance Amperage

A. No. of Alarm Appliances = _____ B. Current amperage each _____ = A x B _____ (I)
A. No. of Alarm Appliances = _____ B. Current amperage each _____ = A x B _____ (I)
A. No. of Alarm Appliances = _____ B. Current amperage each _____ = A x B _____ (I)
A. No. of Alarm Appliances = _____ B. Current amperage each _____ = A x B _____ (I)

Total _____ (I)

Step 4: complete the following to determine the total voltage drop for the branch circuit

Voltage (E) = (I) X (R) from totals in Steps 2 and 3 above

(E) = _____ (I) X _____ (R)

= _____ (E) (shall not exceed 4.4)

Step 5: complete the following to determine if enough voltage is available to operate fire alarm notification devices

Maximum allowable voltage drop: notification devices cannot drop below their Nameplate Operating Voltage (NOV) range. As of 5/1/2004 UL required indicating devices to operate within their NOV. The UL NOV standard is 16VDC to 33VDC, consult the 2002 NFPA 72 Handbook 7.3 for more information. Fire Alarm Control Unit (FACU) are tested to UL 864 and are required to operate at the end of useful battery life, 20.4 V.

Allowable voltage drop is 20.4 V (FACU) - 16 VDC (NOV) = 4.4 V

If (E) from Step 4 exceeds 4.4 V then the NAC is not compliant with NFPA 72

Take (E) from Step 4 and put in the equation below

Voltage Drop = 20.4 V - _____ (E) = _____ V (shall not be less than 16V)

Fire Plan Review and Inspection Guidelines

Clean Agent System Acceptance Inspection
IFC 904.10, 2004 NFPA 2001, and 2002 NFPA 72

Date of Inspection: _____ Permit Number: _____

Business/Building Name: _____ Address of Project: _____

Contractor: _____ Contractor's Phone: _____

Reference numbers following worksheet statements represent an NFPA code section unless otherwise specified.

Pass | Fail | NA

1. ___ | ___ | ___ Received clean agent system certification from installer.
2. ___ | ___ | ___ Control panel and components match approved plans.
3. ___ | ___ | ___ Approved drawing on site, as-builts required when installation is not the same as the plans.
4. ___ | ___ | ___ Control panel, piping, nozzles, and other components location are the same as shown on the plans.
5. ___ | ___ | ___ Zones are properly identified on panel(s).
6. ___ | ___ | ___ Verify dedicated 120 AC branch circuit and labeling, 6.7.2.4.3.
7. ___ | ___ | ___ Device location same as plans.
8. ___ | ___ | ___ Devices are located in all areas required by the code.
9. ___ | ___ | ___ Devices are properly wired and in raceways, 4.3.1.3.
10. ___ | ___ | ___ Type and gauge of wire or cable match plans.
11. ___ | ___ | ___ 24 hour monitoring service agency received signals and conveys the type signals received, 6.7.2.5.3
12. ___ | ___ | ___ Verify proper operation of magnetic door-releasing hardware and/or ventilation shutdown, 6.7.2.3.
13. ___ | ___ | ___ Battery load test: system switched to battery operation 24 hours earlier, then activate audible circuit for 5 minutes, 6.7.2.5.4(2).
14. ___ | ___ | ___ Pull stations comply with NFPA 72, distinct appearance, and mounted at proper height and location which is no more that 4 ft. above the floor, 4.3.3.7., 6.7.2.4.11
15. ___ | ___ | ___ Under primary and secondary power operational tests are performed including. 6.7.2.5:
    ___ | ___ | ___ A. power light on and in normal condition.
    ___ | ___ | ___ B. supervisory signals: pressure switches, valves, etc..
    ___ | ___ | ___ C. silence switches.
    ___ | ___ | ___ D. trouble signals and panel light operate for each circuit tested, disconnect wires. from devices and end-of-line resistors.
    ___ | ___ | ___ E. trouble and alarm reset switches operate.
    ___ | ___ | ___ F. a second initiating zone overrides silence switch.
    ___ | ___ | ___ G. audible and visual operation.
    ___ | ___ | ___ H. panel lamp test switch operates.
    ___ | ___ | ___ I. field zone signals correspond with panel zones.
    ___ | ___ | ___ J. detection devices and manual pull stations operate.
    ___ | ___ | ___ K. abort switch is in protected area; requires manual pressure and initiates visual/audible devices, 4.3.5.3, 6.7.2.5.
16. ___ | ___ | ___ The clean agent system activates the building fire alarm, if provided, 2002 NFPA 72 5.11.
17. ___ | ___ | ___ Piping and nozzles are restrained so no unacceptable movement occurs, prior to the pressure test.
18. ___ | ___ | ___ Piping pneumatically tested for 10 minutes at 40 PSIG and the pressure drop shall not exceed 20 percent of the test pressure, 6.7.2.2.12.
19. ___ | ___ | ___ Nozzles shall not be installed at a location were injury can occur.
20. ___ | ___ | ___ Release circuit is tested at the storage container.
21. ___ | ___ | ___ Pull stations activate system and override abort switch, 6.7.2.4.13.
22. ___ | ___ | ___ Test connection to a monitoring company or a location receiving signal, 6.7.2.5.1.
23. ___ | ___ | ___ Protected area is properly sealed and tested using a fan and smoke pencil, 6.7.2.3.
24. ___ | ___ | ___ An enclosure or room integrity room leakage test (fan test) may be required, 6.7.2.3.
25. ___ | ___ | ___ Warning and instruction signage is properly posted.

Fire Plan Review and Inspection Guidelines

26. ____ |_____| ____ Perform a flow test with compressed air or inert gas to verify unobstructed pipes and nozzles, 6.7.2.2.13.

Additional Comments:

Inspection Date: _____   Approved or Disapproved   FD Inspector: _____

Inspection Date: _____   Approved or Disapproved   FD Inspector: _____

Inspection Date: _____   Approved or Disapproved   FD Inspector: _____

Clean Agent Installation Certification

Permit #: _____   Date: _____

	Property Protected	System Installer	System Supplier
Business Name:	_____	_____	_____
Address:	_____	_____	_____
	_____	_____	_____
Representative:	_____	_____	_____
Telephone:	_____	_____	_____

Type of System: _____

Location of Plans: _____

Location of Owner's Manual: _____

1. <u>Certification of System Installation:</u> Complete this section after system is installed, but prior to conducting operational acceptance tests. Check wiring for opens, ground faults, and improper branching.

    This system installation was inspected and was found to comply with the installation requirements of:
    - _____ NFPA 2001
    - _____ Article 760 of National Electrical Code
    - _____ NFPA 72
    - _____ Manufacturer's Instructions
    - _____ Other (specify: FM, UL, etc.) _____

    Print Name: _____

    Signed: _____ Date: _____

    Organization: _____

2. <u>Certification of System Operation:</u> All operational features and functions of this system were tested and found to be operating properly in accordance with the requirements of:
    - _____ NFPA 2001
    - _____ Design Specifications
    - _____ NFPA 72
    - _____ Manufacture's Instructions
    - _____ Other (specify) _____

    Print Name: _____

    Signed: _____ Date: _____

    Organization: _____

Prefinal and Certificate of Occupancy
Inspection Requirements For Contractors
Contractors Worksheet

Clean Agent System Test Requirements

1. All certification forms and documents are required to be on the site for review:
   - ___ Plans
   - ___ Permit
   - ___ Prefinal detector test and a device location inspection are required as well as the completion of the items on this pretest form. Use the Acceptance Inspection worksheet for the pretest.
   - ___ Installation certification is completed, use the form contained in this book.
2. ___ Person familiar with installation must be present to perform the test.
3. ___ Owner's representative approval needed for time and date of testing.
4. ___ All rooms or areas are unlocked and accessible.
5. ___ Device placement is in accordance with the plan, verification to have been done by contractor.
6. ___ If Items 1-5 are incomplete, the inspection will be cancelled and another inspection request will be required. A reinspection fee may be assessed.

Prior to the next approval test:

7. ___ When additional devices, e.g., nozzles, initiating or indicating devices, etc., are installed, the contractor must provide:
   - ___ As-builts and new calculations for review and approval.
   - Note: A new plan review will be submitted as "supplemental information" and proof of the additional review fee payment is required.
8. ___ A reinspection fee may be assessed if the system and required documentation are not ready.

Kitchen Hood Suppression Systems
Contents

1. Sample Plan Review Illustration. 207

2. Plan Review Worksheet. 208

3. Acceptance Test Checklist. 210

4. Installer Certification Form. 212

5. Contractor Prefinal and Certificate of Occupancy Inspection Requirements. 213

6. Semiannual Service Inspection Guide. 214

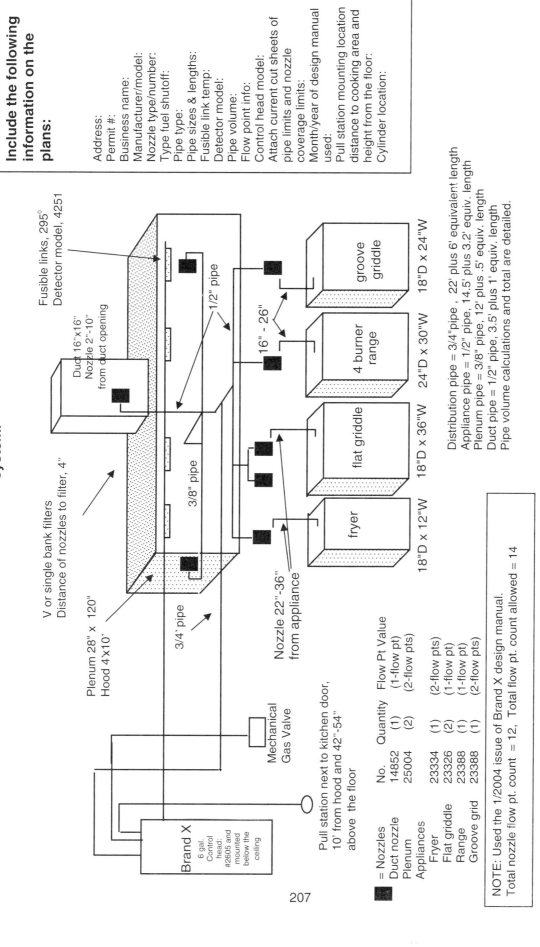

Fire Plan Review and Inspection Guidelines

Kitchen Hood Suppression System
Plan Review Worksheet
2006 IFC, 2006 IMC, 2004 NFPA 17 and 17A, and 2002 NFPA 13

Date of Review: _____   Permit Number: _____

Business/Building Name: _____   Address of Project: _____

Designer Name: _____   Designer's Phone: _____

Contractor: _____   Contractor's Phone: _____

System Manufacturer: _____   Model: _____

Reference numbers following worksheet statements represent an NFPA code section unless otherwise specified.

Worksheet Legend: ✓ or OK = acceptable   N = need to provide   NA = not applicable
1. _____ Three sets of drawings.
2. _____ The fire extinguishing system is listed in accordance with UL 300.

Floor Plan Showing:
3. _____ Scale: a common scale shall be used and plan information is legible.
4. _____ An equipment symbol legend is provided.
5. _____ Cross sectional view of the room and equipment are provided.

Preengineered Wet Chemical and Water Spray Systems:
6. _____ Total number of nozzles provided is _____ and aggregate flow rate is _____.
7. _____ System model is provided and the plan indicates the permissible number of flow points.
8. _____ Description and measurements of the appliances to be protected is provided, 5.1.4.
9. _____ Measurements of hood, plenum, and duct are provided, 5.1.4.
10. _____ Pipe size and length for supply, branches, etc. are provided, and if applicable, the equivalent pipe length of fittings, 6.3.3.
11. _____ Pipe volumes are provided with calculations when required as part of the listing, 6.3.3.
12. _____ The pipe configuration complies with the listed manufacturer's design manual, 6.3.3.
13. _____ Piping and nozzles are adequately braced, 6.3.2.
14. _____ Type of fuel or power shutoff device is described and detailed.
15. _____ Fuel or power shutdown device shall be arranged that it requires manual resetting, IFC 904.11.2.
16. _____ All equipment under the hood shall shutdown when the fire-extinguishing system activates, IFC 904.11.2.
17. _____ Nozzle types are identified and are correct for the appliance hazard, type of use, and coverage area, 6.3.3.
18. _____ Nozzle placement complies with the manufacturer's data sheet, distances from each nozzle to the protected hazard surface are detailed and distance from appliances to filters and duct opening are detailed.
19. _____ Plenum and duct areas are protected in accordance with the manufacturer's design manual.
20. _____ If provided, the fire-extinguishing system is connected to the building fire alarm system, 5.2.1.9.
21. _____ At least one accessible manual pull station is provided in path of egress, 10 ft. to 20 ft. (2006 IMC 509.3) from the hood and 42 in. to 48 in. above the floor level, IFC 904.11.1.
22. _____ The control head model number is identified and the wet chemical container installation location is detailed and complies with Section 5.4.1.
23. _____ Heat detectors or fusible links are located in accordance with the manufacturer's design manual and the detector part number is provided, 6.3.4 (1).
24. _____ Fusible link temperature is in accordance with fire extinguishing systems' listing requirements, 5.6.1.6.
25. _____ Simultaneous activation of systems occurs when protecting common hoods, plenums, and ducts, 5.1.4.

NFPA 13:7.9 Sprinkler Protection:
26. _____ Duct, hood, and appliance configuration(s) are detailed and measurements provided.
27. _____ Sprinkler protection is provided for cooking equipment, plenum area, and the duct(s).
28. _____ Location of duct sprinklers complies with Section 7.9.3.1.
29. _____ Sprinkler spacing in ducts and sprinkler temperature ratings comply with Section 7.9.3.3.
30. _____ Sprinklers are installed above duct collars and the temperature ratings comply with Section 7.9.4.1.
31. _____ The location of sprinklers required in the plenum chamber complies with Section 7.9.5.

32. _____ Sprinklers used to protect deep fat fryers will be listed for that use, IFC 904.11.4.1.
33. _____ The operation of a sprinkler automatically shuts off all sources of fuel and heat to all equipment under the hood.
34. _____ A listed indicating control valve for the water supply is provided, 7.9.9.
35. _____ A listed strainer for the water supply is provided when required by Section 7.9.10.
36. _____ Adequate water pressure and flow is available to operate the system and meet the listing requirements of the sprinklers, pressure and flow information are provided, 7.9.1.
37. _____ A supervised water supply valve is provided, 7.9.1.
38. _____ Sprinklers in ducts are accessible for maintenance, 7.9.7.
39. _____ Sprinklers are a minimum 6 ft. apart unless baffled in accordance with NFPA 13.
40. _____ Sprinklers exposed to temperatures of 300°F or less will be 325-375°F but if temperatures exceed 300°F then a higher temperature sprinkler will be used, 7.9.6.
41. _____ The K-factor for sprinklers installed in ducts, above the duct collar, and in plenum areas are in accordance with Section 5.6.
42. _____ A test connection to verify equipment shutdown is detailed, 7.9.11.

Fire Extinguishers:
43. _____ Solid fuel appliance with firebox volume of 5 cu. ft. or less shall be equipped with at least one 2.5 gallon or two 1.5 gallon K extinguishers. The extinguishers shall be located within 30 ft., IFC 904.11.5.1.
44. _____ Class K extinguisher is within 30 ft. of the appliance. Provide one 1.5 gallon extinguisher for up to four deep fat fryers with a maximum cooking medium capacity of 80 pounds and one additional extinguisher for every additional group of four fryers. For fryers exceeding 6 sq. ft. provide an extinguisher in accordance with the manufacturer recommendations, IFC 904.11.5.

Additional Comments:

Review Date: _____   Approved or Disapproved   FD Reviewer: _____

Review Date: _____   Approved or Disapproved   FD Reviewer: _____

Review Date: _____   Approved or Disapproved   FD Reviewer: _____

Fire Plan Review and Inspection Guidelines

Kitchen Hood Suppression System Acceptance Inspection
2006 IFC, 2006 IMC, 2002 edition of NFPA 13, and 2004 edition of NFPA 17A

Date of Inspection: _____ Permit Number: _____

Business/Building Name: _____ Address of Project: _____

Contractor: _____ Contractor's Phone: _____

Pass | Fail | NA

1. ____|____|____ Approved drawing is on site.
2. ____|____|____ Received system certification from installer.
3. ____|____|____ Manual pull station easily accessible in path of egress.
4. ____|____|____ Manual pull station is 42 in. to 48 in. above floor level.
5. ____|____|____ Manual pull system activates system.
6. ____|____|____ Building fire alarm, if system is available, sounds upon system activation.
7. ____|____|____ Fuel or power shutdown device operates on system activation and all equipment under the hood shall shutdown when the system activates.
8. ____|____|____ Fuel or power shutdown device must be manually reset.
9. ____|____|____ Operation of detection device activates system, perform a nitrogen blow off test.
10. ____|____|____ Pipe size and configuration complies with the approved plans.
11. ____|____|____ Piping and nozzles are secured.
12. ____|____|____ Nozzle types match appliance hazard and type of use as shown on approved plans.
13. ____|____|____ Nozzle placement complies with the approved plans.
14. ____|____|____ Nozzle blow-off caps, when provided, are in place.
15. ____|____|____ Plenum and duct areas are protected in accordance with the approved plans.
16. ____|____|____ Chemical container is accessible and installed in accordance with NFPA 17A, Section 5.4.1.
17. ____|____|____ Pressure gauges are in the operable range.
18. ____|____|____ Maintenance tag is in place.

NFPA 13:7-9 Sprinkler Protection

19. ____|____|____ Duct, hood, and appliances configuration(s) are same as approved plans.
20. ____|____|____ Sprinkler protection provided for cooking equipment, plenum area and the ducts are the same as the approved plans.
21. ____|____|____ The sprinklers listed for protecting the deep fat fryers are provided in accordance with the approved plans.
22. ____|____|____ The operation of a sprinkler will automatically shutoff all sources of fuel and heat to all equipment under the hood, manual reset is required.
23. ____|____|____ A listed indicating control valve for the water supply is provided.
24. ____|____|____ A listed strainer for the water supply is provided when required by NFPA 13, Section 7.9.10.
25. ____|____|____ A system test connection is provided at the end of the system.
26. ____|____|____ Sprinklers in ducts are accessible for maintenance.

All Systems

27. ____|____|____ Listed grease filters are in place and are stamped "Listed Grease Filter" on the side.
28. ____|____|____ All penetrations of the hood are properly sealed.

Miscellaneous

29. ____|____|____ Solid fuel appliance with firebox volume of 5 cu. ft. or less shall be equipped with at least one 2.5 gallon or two 1.5 gallon K-extinguishers. The extinguishers shall be located within 30 ft., IFC 904.11.5.1.

30. ____|_____|____ Class K-extinguisher is within 30 ft. of the appliance. Provide one 1.5 gallon extinguisher for up to four deep fat fryers with a maximum cooking medium capacity of 80 pounds and one additional extinguisher for every additional group of four fryers. For fryers exceeding 6 sq. ft. provide an extinguisher in accordance with the manufacturer recommendations, IFC 904.11.5.

Additional Comments:

Inspection Date: _____ Approved or Disapproved FD Inspector: _____

Inspection Date: _____ Approved or Disapproved FD Inspector: _____

Inspection Date: _____ Approved or Disapproved FD Inspector: _____

Fire Plan Review and Inspection Guidelines

Kitchen Hood Suppression System Installation Certification

Permit #: _____ Date: _____

	Property Protected	System Installer	System Supplier
Business Name:	_____	_____	_____
Address:	_____	_____	_____
	_____	_____	_____
Representative:	_____	_____	_____
Telephone:	_____	_____	_____

Type of System: _____

Location of Plans: _____

Location of Owner's Manual: _____

1. <u>Certification of System Installation:</u> Complete this section after system is installed, but prior to conducting operational acceptance tests.

   This system installation was inspected and found to comply with the installation requirements of:
   _____ NFPA 13 or NFPA 17A
   _____ IFC and IMC
   _____ Manufacturer's Instructions
   _____ Other (specify: FM, UL, etc.) _____

   Print Name: _____
   Signed: _____ Date: _____
   Organization: _____

2. <u>Certification of System Operation:</u> All operational features and functions of this system were tested and found to be operating properly in accordance with the requirements of:
   _____ NFPA 13 or NFPA 17A
   _____ Design Specifications
   _____ IFC and IMC
   _____ Manufacturer's Instructions
   _____ Other (specify) _____

   Print Name: _____
   Signed: _____ Date: _____
   Organization: _____

Prefinal and Certificate of Occupancy Inspection Requirements For Contractors
Contractors Worksheet

Kitchen Hood Suppression System Test Requirements
1. All certification forms and documents are required to be on the site for review:
   ____ Plans

   ____ Permit Prefinal system test and a device location inspection are required as well as the completion of the items on this pretest form. Use the Acceptance Inspection worksheet for the pretest.

   ____ Installation certification is completed, use the form contained in this book.

2. ____ Person familiar with installation must be present to perform the test.
3. ____ Owner's representative approval needed for time and date of testing.
4. ____ The kitchen areas are accessible.
5. ____ Device placement is in accordance with the plan, verification to have been done by contractor.
6. ____ If Items 1-5 are incomplete, the inspection will be cancelled and another inspection request is required. A reinspection fee may be assessed.

Prior to the next approval test:
7. ____ When there are device additions, contractor must provide:

   ____ As-builts and new calculations shall be submitted for review and approval.

   Note: New plan review will be submitted as "supplemental information" and proof of the additional review fee payment is required.

8. ____ A reinspection fee may be assessed if the system and paper work are not ready.

Guidelines for Conducting Semiannual Service Inspection: Kitchen Hood Suppression System

1. A trained person shall conduct the maintenance on the system.

2. Verify the hazard or appliances have not changed.

3. Examine all detectors, expellant gas containers, agent containers, releasing devices, piping, hose assemblies, nozzles, alarms and all auxiliary equipment.

4. Verify that the agent distribution piping is not obstructed.

5. Examine the dry or wet chemical cylinder every 6 years if it is a stored pressure system.

6. Test the detection system, alarms and releasing devices.

7. Replace fixed temperature fusible links at least annually or more frequently if conditions warrant it or if the manufacturer requires it.

8. Hydrostatically test stored pressure system containers every 12 years.

9. Replace or repair defective or damaged parts.

10. Verify a current maintenance tag is attached to the system.

Fire Plan Review and Inspection Guidelines

Standpipe
Contents

1. Plan Review Worksheet 217

2. Acceptance Test Checklist 220

3. Installer Certification Form 222

4. Contractor Prefinal and Certificate of Occupancy Inspection Requirements 223

5. AFSA Inspection Testing, and Maintenance Forms 224

Fire Plan Review and Inspection Guidelines

Standpipe System
Plan Review Worksheet
2006 IFC and 2003 NFPA 14

Date of Review: _____ Permit Number: _____

Business/Building Name: _____ Address of Project: _____

Designer Name: _____ Designer's Phone: _____

Contractor: _____ Contractor's Phone: _____

Occupancy Classification: _____ Class: I   II   III   Type: Dry   Wet   Combination

---

Numbers following worksheet comments represent a **NFPA code section** unless otherwise specified.

<u>Worksheet Legend:</u>   ✓ or OK = acceptable   N = need to provide   NA = not applicable

1. _____ Three sets of drawings are provided.
2. _____ Equipment is listed for intended use, compatible with the system, and equipment data sheets are provided.

Minimum Information:

3. _____ Class of standpipe system and the type of standpipe in accordance with 5.2 and 5.3.
4. _____ Scale: a common scale is used and the plan information shall be clear and legible, 8.1.
5. _____ Plot plan showing supply piping and pipe size from the water source to the building.
6. _____ Equipment symbol legend and compass point.
7. _____ The correct standpipe class is provided for the occupancy and is in accordance with IFC 905.
8. _____ Building dimensions, height, and the location of the fire department connection.

Pipe:

9. _____ The material standard and pipe wall thickness (schedule) for steel pipe assembled using welded or rolled groove method shall comply with the requirements in Section 4.2.3. Steel pipe assembled using threading shall comply with the material standard and pipe wall thickness requirements in Section 4.2.4.
10. _____ Piping shall be supported and anchored in accordance with NFPA 13, Standard for the Installation of Sprinkler Systems, 6.4.

Valves:

11. _____ Valve locations are detailed and data sheets are provided, 4.5.1.
12. _____ The type of indicating or nonindicating shall comply with the design and operational requirements in Section 4.5.1.
13. _____ Connection to the water supply is equipped with the appropriate valve(s) as specified in 6.2.1.
14. _____ Gate valves are provided to permit the isolation of standpipes without interrupting the supply to other standpipes from the same water source, 6.2.2.
15. _____ Combined systems: connection from standpipe to sprinkler system has an individual control valve and check valve detailed, 6.2.5.1.
16. _____ Electric supervision of the valves for water supply, isolation control, and valves in feed mains is provided in accordance with IFC 905.9.
17. _____ All valves are marked or otherwise identified in accordance with the requirements of Section 6.2.8 to indicate the portion of the system they control.

Hose and Cabinets:

18. _____ Specification sheets for hose cabinets, racks and hose are provided, 4.6.
19. _____ Hose for Class II and III systems is listed and complies with the diameter and length requirements of Section 4.6.2.
20. _____ Nozzles for Class II service are listed, 4.6.4.
21. _____ Hose valves and connections comply with the requirements in Section 4.7.
22. _____ Hose cabinets are provided with signage and operating instructions, 4.6.1.1.2.
23. _____ Cabinets with glass shall have a glass breaking device secured to the cabinet, which is detailed or noted on the plans, 4.6.1.2.

24. _____ When a hose cabinet penetrates a fire-resistive assembly, the assembly shall be protected in accordance with IBC 712.3.2 requirements for membrane penetrations, 4.6.1.3.

Hose Connections:
25. _____ Approved pressure regulating device is provided when the residual pressure exceeds the pressure specified in 7.2.1, detail and specification sheets are provided.
26. _____ Hose connections and hose stations are unobstructed and shall be located above the floor in compliance with 7.3.1.
27. _____ The Class I standpipe is detailed showing outlets locations in compliance with 7.3.2 and IFC 905.4.
28. _____ Each Class I standpipe has a roof outlet or an outlet at the highest landing of stairway that has roof access for a roof with less than a 4/12 slope, IFC 905.4.(5).
29. _____ The Class II standpipe is detailed showing outlets locations in compliance with 7.3.3.
30. _____ Class III standpipe outlets are located the same as Class I and II outlets, including the roof outlets, 7.3.4, IFC 905.6.
31. _____ When required, Class II and III standpipe systems are auto- or semiautomatic-wet systems as specified in 5.4.3 and IFC 905.
32. _____ An extra outlet is detailed and provided for the most hydraulically remote standpipe for testing purposes when the roof has less than a 4/12 slope, IFC 905.4.(5).

Fire Department Connection (FDC):
33. _____ Each FDC has swivel fittings that comply with 4.8.2.
34. _____ Each fire department hose connection is provided signage in accordance with 6.3.5.2.
35. _____ If the FDC also supplies the sprinkler system then a sign indicating the system pressure and demand are detailed, 6.3.5.2.2.
36. _____ When a portion of a building is served by an FDC, a sign is detailed to specify which part of the building is being served, 6.3.5.3.
37. _____ Each FDC is provided with a listed check valve, 6.3.2.
38. _____ FDC connections to a specific type of system are located and detailed relative to the control valves in accordance with the criteria listed in 6.3.3.
39. _____ For freezing environments, an automatic drip valve is detailed between the check valve and the FDC, 6.3.4.
40. _____ FDC location is detailed on the street or response side of the building and signage detail complies with 6.3.5.1.
41. _____ The FDC height above finish grade is detailed and complies with 6.3.6.
42. _____ The number of FDCs required for Class I or III standpipe system shall comply with 7.13.1.
43. _____ Each high-rise building zone is provided the number of remotely located FDCs in accordance with 7.13.2.

Protection:
44. _____ Class I and III standpipes and lateral piping supplied from the standpipes are located in stairways or are protected in accordance with 6.1.2.2, IFC 905.4.1, and 905.6.
45. _____ Class I and III lateral piping to hose connections need not be protected in sprinklered buildings, 6.1.2.2.1, IFC 905.4.1 and 905.6.
46. _____ Class II standpipes and risers need not be protected, IFC 905.5.2.
47. _____ Piping exposed to corrosive conditions is corrosion-resistant pipe or provided a protective coating, coating information is provided, 6.1.2.4.
48. _____ Dry standpipes are not concealed unless monitored in accordance with 6.1.1.
49. _____ If piping is subject to freezing, it is detailed how water filled piping will be protected to maintain the water temperature in accordance with 6.1.2.3.
50. _____ If pipe must be installed under the building, details are provided to show method of protecting the pipe in accordance with 6.1.2.6.1.
51. _____ Earthquake bracing is provided and detailed in accordance with NFPA13, 6.1.2.5.

Interconnection:
52. _____ Interconnection between two or more standpipes in the same building is detailed, 7.5.1.
53. _____ Interconnection at the top of the building is detailed when water supply tanks are at the top of the building and check valves are located and detailed in accordance with 7.5.2.

Design Criteria:
54. _____ Each FDC for Class I and II standpipes are designed to provide the system demand, calculations are provided, 7.7.1.
55. _____ When automatic or semiautomatic water supply is required by 5.4 and IFC 905 the standpipe system demand shall comply with 7.7.2 and calculations are provided.

56. ____ Combination automatic sprinkler and standpipe systems shall be calculated in accordance with the requirements in Section 7.10.1.3.

Hydraulic Demand:

57. ____ Class I and III standpipes: Calculations based on the criteria in Section 7.10.1.2 shall be hydraulically calculated to verify the minimum flow rates specified in Sections 7.10.1.1.1, 7.10.1.1.2, 7.10.1.1.3, or 7.10.1.1.4.1 are satisfied.
58. ____ Class II standpipes: Calculations demonstrate the hydraulically most remote hose connection is supplied with the minimum water flow and pressure specified in 7.10.2.2.
59. ____ Maximum flow rate for each hose connection is in accordance with 7.10.3.
60. ____ Pipe schedule standpipe system complies with Table 7.8.2.1. Pipe schedule designs are limited to buildings not classified as a high-rise and equipped with wet standpipe systems, 7.8.2.
61. ____ A drain risers are detailed is provided in accordance with 7.12.1 for a standpipe equipped with pressure-regulating devices 7.12.1.
62. ____ The drain riser detail illustrates a tee as required in Section 7.12.1.1.
63. ____ A drain valve and piping are detailed in accordance with 7.12.2.1.
64. ____ At least a 30 minute water supply is available for any class system, 9.2 and 9.3.
65. ____ Standpipe zoning is designed, detailed, and complies with 7.9.

Additional Comments:

Review Date: _____ Approved or Disapproved FD Reviewer: _____

Review Date: _____ Approved or Disapproved FD Reviewer: _____

Review Date: _____ Approved or Disapproved FD Reviewer: _____

Fire Plan Review and Inspection Guidelines

Standpipe System Acceptance Inspection
2006 IFC and 2003 NFPA 14

Date of Inspection: _____ Permit Number: _____

Business/Bldg Name: _____ Address of Project: _____

Contractor: _____ Contractor's Phone: _____

Numbers following worksheet comments represent a NFPA code section unless otherwise specified.

<u>Pass</u>| <u>Fail</u> | <u>NA</u>

1. ____|____|____ Received standpipe certification from installer.
2. ____|____|____ Plans are on site.
3. ____|____|____ Location and size of standpipes and FDCs comply with the plans.
4. ____|____|____ Outlet valves are functional and hose threads in good condition.
5. ____|____|____ Roof outlets comply with the plans.
6. ____|____|____ Hydraulic calculation information sign is mounted at the system control valve.
7. ____|____|____ Underground pipe from FDC to check valve in inlet pipe is flushed before completing systems, 11.2.
8. ____|____|____ <u>Standpipe test including yard and FDC piping</u>: per NFPA 14 11.4 and 11.5
9. ____|____|____   A. hydrostatic test at 200 PSI for 2 hours, PSI is measured at the lowest point or B.
10. ____|____|____   B. hydrostatic test not less than 50 PSI in excess of maximum pressure; where the maximum pressure is in excess of 150 PSI.
11. ____|____|____   C. where cold weather prevents testing, an air test at 40 PSI for 24 hours shall occur with a pressure loss of only up to 1.5 PSI permitted.
12. ____|____|____   D. flow test: the hydraulically most remote standpipe will verify system design pumping through the FDC, 11.5.
13. ____|____|____   E. a flow test at each roof outlet to verify the required pressure and flow is available, 11.5.
14. ____|____|____   F. maximum flow from a 2½ in. hose connection is 250 GPM, for a 1½ in. connection it is 100 GPM, 11.5.5.
15. ____|____|____ Pressure regulating devices are flow tested to verify proper operation, 11.5.6.
16. ____|____|____ Main drain valve, if provided, is opened until system pressure stabilizes, 11.5.7.
17. ____|____|____ No shutoff valve is in the FDC.
18. ____|____|____ Check valve is near the FDC connection to the system.
19. ____|____|____ The pipe between the FDC and the check valve has an automatic drip.
20. ____|____|____ The FDC is 18 in. to 48 in. above finish grade and signed.
21. ____|____|____ A standpipe drain is provided at the lowest point and it drains to the exterior.
22. ____|____|____ Standpipes are located in noncombustible stair enclosures or equivalent construction.
23. ____|____|____ Hose connections are readily accessible, 3 ft. to 5 ft. above the floor, and caps are tight.
24. ____|____|____ Standpipes having listed pressure-regulating devices will be flow tested to verify the PSI setting.
25. ____|____|____ Water filled pipe exposed to freezing conditions is protected from freezing.
26. ____|____|____ All manual valves shall be fully opened, fully closed, and supervised or secured.
27. ____|____|____ Riser supports are provided at the lowest level, alternate levels and at the top.
28. ____|____|____ Connection to the water supply has a listed indicating-type valve and a check valve located close to the supply.
29. ____|____|____ Lateral runs from standpipe to the hose valve over 18 in. are provided with hangers.
30. ____|____|____ Horizontal standpipe hangers do not exceed a 15 ft. spacing.
31. ____|____|____ Attached 1½ in. hose is free from mildew, cuts, abrasions, and couplings, gaskets, and nozzles are undamaged and without obstructions.
32. ____|____|____ Multiple Class I and III standpipes are interconnected at the bottom.
33. ____|____|____ Automatic and semiautomatic-dry systems are tested by initiating a flow from the hydraulically most remote hose connection and water is delivered in 3 minutes and each remote control device is tested per the manufacturer's instructions, 11.5.8.1.

Additional Comments:

Inspection Date: _____   Approved or Disapproved   FD Inspector: _____
Inspection Date: _____   Approved or Disapproved   FD Inspector: _____
Inspection Date: _____   Approved or Disapproved   FD Inspector: _____

Fire Plan Review and Inspection Guidelines

Standpipe Installation Certification

Permit #: _____   Date: _____

	Property Protected	System Installer	System Supplier
Business Name:	_____	_____	_____
Address:	_____	_____	_____
	_____	_____	_____
Representative:	_____	_____	_____
Telephone:	_____	_____	_____

Location of Plans: _____

Location of Owner's Manual: _____

1. <u>Certification of System Installation:</u> Complete this section after system is installed, but prior to conducting operational acceptance tests.

   This system installation was inspected and found to comply with the installation requirements of:
   - _____ NFPA 14
   - _____ IFC and IBC
   - _____ Manufacturer's Instructions
   - _____ Other (specify; FM, UL, etc.) _____

   Print Name: _____
   Signed: _____ Date: _____
   Organization: _____

2. <u>Certification of System Operation:</u> All operational features and functions of this system were tested and found to be operating properly in accordance with the requirements of:
   - _____ NFPA 14
   - _____ IFC and IBC
   - _____ Manufacturer's Instructions
   - _____ Other (specify) _____

   Print Name: _____
   Signed: _____ Date: _____
   Organization: _____

Prefinal and Certification of Occupancy Inspection Requirements For Contractors
Contractors Worksheet

Standpipe System Test Requirements

1. All certification forms and documents are required to be on the site for review:
   - ___ Plans
   - ___ Permit
   - ___ A system hydrostatic test is required before calling for an inspection as well as the completion of with the items on this pretest form. Use the Acceptance Inspection worksheet for the pretest.
   - ___ Installation certification is completed, use the form contained in this book.
2. ___ A person familiar with installation must be present to perform the test.
3. ___ Owner's representative approval is needed for the time and date of testing.
4. ___ All areas are accessible.
5. ___ Hydrostatic testing and the flow test should be done during the same inspection.
6. ___ If Items 1-5 are incomplete, the inspection will be cancelled and another inspection request is required. A reinspection fee may be assessed.

Prior to the next approval test:

7. ___ When there are device additions, contractor must provide:
   - ___ As-builts and new calculations shall be submitted for review and approval.
   Note: New plan review will be submitted as "supplemental information" and proof of the additional review fee payment is required.
8. ___ A reinspection fee may be assessed if the system and paperwork are not ready.

Fire Plan Review and Inspection Guidelines

Report of Inspection, Testing & Maintenance
of Standpipe Systems

ALL QUESTIONS ARE TO BE ANSWERED AND ALL BLANKS TO BE FILLED
(Weekly inspection tasks are NOT included in this report)

Inspecting Firm: _____ Inspection Contract#: _____
Name of Inspected Property: _____
Inspector Name: _____ Date: _____
Inspection Frequency: ☐ Monthly   ☐ Quarterly   ☐ Annually   ☐ Other

Monthly Inspection for Standpipe Systems

		Y	N/A	N			Y	N/A	N
A.1.0	System in service on inspection				A.5.2	Alarm valve trim valves are in appropriate open or closed position			
A.2.0	Supply pressure gauge (where present) _____ psi				A.5.3	Alarm valve retarding chamber or alarm drain not leaking			
A.2.1	System water pressure gauge (where present) _____ psi				A.6.0	**ALARM PANEL CLEAR**			
A.2.2	System air pressure gauge (where present) _____ psi				A.7.0	**COMMENTS:**			
A.2.3	Top floor pressure gauge (where present) _____ psi								
A.2.4	Gauges appear to be in good condition (where present)								
A.3.0	Control valves in normal open or closed position								
A.3.1	Control valves properly locked or supervised								
A.3.2	Control valves accessible								
A.3.3	Control valves provided with appropriate wrenches								
A.3.4	Control valves free from external leaks								
A.3.5	Control valve identification signs in place								
A.3.6	Control valve signs indicate area served								
A.4.0	Backflow prevention assembly valves are locked or electrically supervised in open position								
A.4.1	Reduced pressure backflow prevention assembly not in continuous discharge								
A.5.0	Alarm valve gauges indicate normal supply water pressure								
A.5.1	Alarm valve free of physical damage								

Quarterly Testing for Standpipe Systems

		Y	N/A	N			Y	N/A	N
B.1.0	System in service before testing				B.5.0	Dry pipe valve priming water tested			
B.1.1	Pertinent parties notified before testing				B.5.1	Dry pipe valve priming water level normal			
B.1.2	Adequate drainage provided before flow testing				B.5.2	Dry pipe valve priming water level returned to normal			
B.2.0	Water flow alarm (other than vane type) tested and is operational				B.5.3	Low air pressure alarm tested in accordance with mfg. inst.			
B.2.1	Test conducted with inspectors test connection				B.6.0	Pertinent parties notified of test conclusion			
B.2.2	Test conducted with bypass connection (freezing weather)				B.7.0	**ALARM PANEL CLEAR**			
B.2.3	Test conducted per manufacturer's instructions				B.8.0	**SYSTEM RETURNED TO SERVICE**			
B.2.4	Alarm devices appear free of physical damage				B.9.0	**COMMENTS:**			
B.3.0	Supervisory switches initiated distinct signal during first two hand wheel revolutions or before valve stem moved one-fifth from normal position *(semi-annual testing requirement)*								
B.3.1	Signal restored only when valve returned to normal position								
B.4.0	Main drain test conducted downstream from backflow preventer								
B.4.1	Main drain test conducted downstream from pressure reducing valve								
B.4.2	Supply water gauge reading before flow (static) _____ psi								
B.4.3	Gauge reading during stable flow (residual) _____ psi								
B.4.4	Time for supply pressure to return to normal _____ sec								

(All "NO" answers to be explained.)
INSPECTOR'S INITIAL _____ OWNER/DESIGNATED REP. INITIAL _____ DATE _____

(AFSA Form 108A)

Report of Inspection, Testing & Maintenance of Standpipe Systems...*continued*

Inspecting Firm: _____ Inspection Contract#: _____
Name of Inspected Property: _____
Inspector Name: _____ Date: _____
Inspection Frequency: ☐ Monthly   ☐ Quarterly   ☐ Annually   ☐ Other

Quarterly Inspection for Standpipe Systems

	Y	N/A	N		Y	N/A	N
C.1.0 System in service on inspection				C.10.0 Hose storage devices operate with ease			
C.2.0 Alarm device free from physical damage				C.10.1 Hose storage devices undamaged			
C.3.0 FDC is visible				C.10.2 Hose storage devices unobstructed			
C.3.1 FDC is accessible				C.10.3 Hose racks in cabinets swing out at least 90°			
C.3.2 FDC swivels/couplings undamaged/rotate smoothly				C.11.0 Hose cabinets free of corrosion and parts undamaged			
C.3.3 FDC plugs/caps in place/undamaged				C.11.1 Hose cabinets open with ease			
C.3.4 FDC gaskets in place and in good condition				C.11.2 Hose cabinet doors open fully			
C.3.5 FDC identification sign in place				C.11.3 Hose cabinet door glazing free of cracks or breaks			
C.3.6 FDC check valve not leaking				C.11.4 Hose cabinet locks functioning properly (break-glass type)			
C.3.7 FDC automatic drain valve in place and operating properly				C.11.5 Hose cabinet glass break devices in place and attached			
C.3.8 FDC clapper is in place and operating properly				C.11.6 Hose cabinets identified as containing fire equipment			
C.3.9 FDC interior inspected where caps missing				C.11.7 Hose cabinets unobstructed			
C.3.10 FDC obstructions removed as necessary				C.11.8 Hose cabinet valves, hose, nozzles, fire extinguishers, etc., easily accessible			
C.4.0 Hose connection valve caps in place/undamaged				C.12.0 ALARM PANEL CLEAR			
C.4.1 Hose connection valve outlet threads undamaged				C.13.0 COMMENTS:			
C.4.2 Hose connection valve handles in place/undamaged							
C.4.3 Hose connection valve gaskets undamaged and free of deterioration							
C.4.4 Hose connection valve not leaking							
C.4.5 Hose connection valves unobstructed							
C.4.6 Hose connection valve restricting devices in place							
C.5.0 Hose connection pressure reducing hose valve (PRV) handwheels in place and undamaged							
C.5.1 Hose connection PRV outlet threads undamaged							
C.5.2 Hose connection PRVs not leaking							
C.5.3 Hose connection PRV reducers and caps in place and undamaged							
C.6.0 Hose rack assembly PRV handwheels in place undamaged							
C.6.1 Hose rack assembly PRVs not leaking							
C.7.0 Standpipe piping undamaged							
C.7.1 Standpipe piping support devices in place and undamaged							
C.7.2 Standpipe piping supervisory devices undamaged							
C.8.0 Hoses free of mildew, cuts, abrasions, and deterioration							
C.8.1 Hose couplings undamaged							
C.8.2 Hose gaskets in place and free of deterioration							
C.8.3 Hose coupling threads compatible							
C.8.4 Hoses connected to nipple or valve							
C.8.5 Hose tests not outdated							
C.8.6 Hose properly racked or rolled							
C.9.0 Hose nozzles in place							
C.9.1 Hose nozzle gaskets in place and free of deterioration							
C.9.2 Hose nozzles unobstructed							
C.9.3 Hose nozzles operate smoothly							
C.9.4 Hose nozzle clips in place and correctly contain nozzles							

INSPECTOR'S INITIAL _____   (All "NO" answers to be explained.)   OWNER/DESIGNATED REP. INITIAL _____   DATE _____

(AFSA Form 108A)

Report of Inspection, Testing & Maintenance of Standpipe Systems...*continued*

Inspecting Firm: _____ Inspection Contract# _____
Name of Inspected Property: _____
Inspector Name: _____ Date: _____
Inspection Frequency: ☐ Monthly ☐ Quarterly ☐ Annually ☐ Other

Annual Inspection and Testing for Standpipe Systems

		Y	N/A	N			Y	N/A	N
D.1.0	System in service before inspection and testing				D.6.3	Forward flow test conducted without measuring flow (device</=2" and outlet sized to flow system demand)			
D.1.1	Pertinent parties notified before inspection and testing								
D.1.2	Adequate drainage provided before flow testing				D.6.4	Backflow prevention assembly internal inspection conducted (where shortages last more than 1 year and rationing enforced by AHJ)			
D.1.3	Cabinet inspected in accordance with NFPA 1962								
D.1.4	Hose storage device inspected in accordance with NFPA 1962				D.6.5	Forward flow test satisfied by annual fire pump flow test			
D.1.5	Hose inspected in accordance with NFPA 1962				D.6.6	Backflow preventer performance test conducted as required by the AHJ			
D.1.6	Dry pipe valve internally inspected (during trip test)								
D.2.0	Main drain test conducted				D.7.0	Dry pipe valve trip tested *(at full flow every third year)*			
D.2.1	Main drain test conducted at low point drain or main drain test connection where supply main enters building (when provided)								
					D.7.1	Separate records of initial air and water pressure, tripping air pressure, and dry pipe valve operating conditions available on premises for comparison			
D.2.2	Supply water gauge reading before flow (static) ____ psi								
D.2.3	Gauge reading during stable flow (residual) ____ psi				D.7.2	Current trip test results compared to previous trip test results			
D.2.4	Time for supply pressure to return to normal ____ sec				D.7.3	Current results correlate with previous results			
D.3.0	Hose nozzle tested in accordance with NFPA 1962				D.7.4	Tag showing date of trip test and name of person and organization conducting test attached to valve			
D.3.1	Hose storage device tested in accordance with NFPA 1962								
D.3.2	Hose tested in accordance with NFPA 1962 *(3-year testing requirement)*				D.7.5	Low temperature alarms tested at beginning of heating season			
D.4.0	Control valves (including backflow and PIVs) operated through full range and returned to normal position				D.7.6	Automatic air pressure maintenance device tested in accordance with mfg. inst.			
					D.8.0	Hose connection PRVs flow tested at partial flow adequate to move valve from seat			
D.4.1	PIVs opened until spring or torsion felt in rod				D.8.1	Hose rack assembly PRVs flow tested at partial flow adequate to move valve from seat			
D.4.2	PIVs and OS&Ys backed 1/4 turn from full open								
D.5.0	Main drain test conducted				D.9.0	Pertinent parties notified of inspection and testing conclusion			
D.5.1	Supply water gauge reading before flow (static) ____ psi				D.10.0	**ALARM PANEL CLEAR**			
D.5.2	Gauge reading during stable flow (residual) ____ psi				D.11.0	**SYSTEM RETURNED TO SERVICE**			
D.5.3	Time for supply pressure to return to normal ____ sec				D.12.0	**COMMENTS:**			
D.6.0	Backflow prevention assembly forward flow test conducted					_____			
D.6.1	System demand flow was achieved through the device					_____			
D.6.2	Forward flow test conducted at maximum rate possible (only where connections do not permit full flow test)					_____			

Annual Maintenance for Standpipe Systems

		Y	N/A	N			Y	N/A	N
E.1.0	System in service before conducting maintenance				E.5.4	Time for supply pressure to return to normal ____ sec			
E.2.0	Pertinent parties notified before conducting maintenance				E.6.0	Low point drains drained prior to onset of freezing weather conditions			
E.3.0	Manual, semiautomatic, or dry standpipe – hose connection valve operates smoothly				E.7.0	Pertinent parties notified after conclusion of maintenance			
E.4.0	Operating stems of OS&Y (including backflow) valves lubricated				E.8.0	**ALARM PANEL CLEAR**			
					E.9.0	**SYSTEM RETURNED TO SERVICE**			
E.4.1	Valves completely closed and reopened				E.10.0	**COMMENTS:**			
E.5.0	Adequate drainage provided before flow testing					_____			
E.5.1	Main drain test conducted					_____			
E.5.2	Supply water gauge reading before flow (static) ____ psi					_____			
E.5.3	Gauge reading during stable flow (residual) ____ psi					_____			

INSPECTOR'S INITIAL _____   (All "NO" answers to be explained.) OWNER/DESIGNATED REP. INITIAL _____ DATE _____

(AFSA Form 108A)

Fire Plan Review and Inspection Guidelines

Report of Inspection, Testing & Maintenance of Standpipe Systems...*continued*

Inspecting Firm: _____ Inspection Contract# _____
Name of Inspected Property: _____
Inspector Name: _____ Date: _____
Inspection Frequency: ☐ Monthly   ☐ Quarterly   ☐ Annually   ☐ Other

Five-Year Inspection and Maintenance for Standpipe Systems

		Y	N/A	N			Y	N/A	N
F.1.0	System in service before conducting inspection and maintenance				F.5.1	Dry pipe valve strainers, filters, and restriction orifices in good condition			
F.2.0	Pertinent parties notified before inspection				F.6.0	Pertinent parties notified of inspection and maintenance conclusion			
F.3.0	Alarm valve internally inspected				F.7.0	**SYSTEM RETURNED TO SERVICE**			
F.3.1	Alarm valve strainers, filters, and restriction orifices inspected				F.8.0	**COMMENTS:**			
F.3.2	Alarm valve strainers, filters, and restriction orifices in good condition					_____			
F.3.3	Alarm valve internal components cleaned/replaced as necessary					_____			
F.4.0	Check valves internally inspected					_____			
F.4.1	Check valve internal components operate correctly					_____			
F.4.2	Check valve internal components move freely					_____			
F.4.3	Check valve internal components in good condition					_____			
F.4.4	Check valve internal components cleaned/repaired/replaced as necessary					_____			
F.5.0	Dry pipe valve strainers, filters, and restriction orifices inspected					_____			

Five-Year Testing for Standpipe Systems

		Y	N/A	N			Y	N/A	N
G.1.0	System in service before testing				G.5.4	Flow test results indicate the water supply supplies design pressure at required flow			
G.1.1	Pertinent parties notified before testing				G.6.0	Pressure control valves full flow tested			
G.1.2	Adequate drainage provided before flow testing				G.6.1	Test results compared to previous test results			
G.2.0	System gauges tested by comparison with calibrated gauge				G.6.2	Adjustments made in accordance with mfg. inst.			
G.2.1	System gauges accurate within 3% of full scale				G.7.0	Hose connection PRVs full flow tested			
G.2.2	System gauges recalibrated as necessary				G.7.1	Test results compared to previous test results			
G.2.3	System gauges replaced as necessary				G.7.2	Adjustments made in accordance with mfg. inst.			
G.2.4	System gauges test/replacement date:				G.8.0	Hose rack assembly PRVs full flow tested			
G.3.0	Standpipe hose tested in accordance with NFPA 1962				G.8.1	Test results compared to previous test results			
G.4.0	Hydrostatic test conducted on dry standpipe system				G.8.2	Adjustments made in accordance with mfg. inst.			
G.4.1	Hydrostatic test requirements and performance discussed with AHJ prior to testing				G.9.0	Pertinent parties notified of test conclusion			
G.4.2	Hydrostatic test conducted on dry portions of wet standpipe system				**G.10.0**	**ALARM PANEL CLEAR**			
G.4.3	Hydrostatic test conducted on manual standpipe system (except manual wet standpipes that are part of combined sprinkler/standpipe systems)				**G.11.0**	**SYSTEM RETURNED TO SERVICE**			
G.4.4	Hydrostatic test pressure measured at low elevation point of tested system				**G.12.0**	**COMMENTS:**			
G.4.5	Hydrostatic pressure tested at requirements in effect at time of installation					_____			
G.4.6	No leakage observed on inside standpipe systems during hydrostatic test					_____			
G.5.0	Flow test conducted through hydraulically most remote hose connection (where practical) of automatic standpipe system					_____			
G.5.1	Flow test requirements and performance discussed with AHJ prior to testing					_____			
G.5.2	AHJ consulted for appropriate test location (where hyd. most remote not practical)					_____			
G.5.3	Flow test conducted at requirements in effect at time of installation					_____			

(All "NO" answers to be explained.)
INSPECTOR'S INITIAL _____   OWNER/DESIGNATED REP. INITIAL _____   DATE _____

(AFSA Form 108A)

Reprinted with permission from the American Fire Sprinkler Association, Dallas, Texas.

Fire Plan Review and Inspection Guidelines

Fire Alarm Systems
Contents

1. Guideline for Proper Sound Pressure Levels.................231

2. High-rise Fire Alarm Requirements, 2002 NFPA 72. 233

3. Plan Review Worksheet, 2002 NFPA 72 235

4. Voltage Drop Calculation Form......................................241

5. Acceptance Test Checklist, 2002 NFPA 72. 242

6. Plan Review Worksheet, optional 2007 NFPA 72 244
   For jurisdictions that permit the use of the 2007 standard in lieu of the ICC referenced 2002

7. Acceptance Test Checklist, optional 2007 NFPA 72 251
   For jurisdictions that permit the use of the 2007 standard in lieu of the ICC referenced 2002

8. Installer Certification Form. 253

9. Prefinal and Certificate of Occupancy Inspection
   Requirements..254

10. Fire Alarm Sound Pressure (dBA) Level Log.................255

Guidelines for Proper Sound Pressure Levels

The following applies:
- fire alarm audible signaling devices must be heard in every occupied space, NFPA 72 and IFC 907.10.2.

- audible alarm signal is distinct and a on-off three-pulse temporal pattern, which repeats itself for 3 cycles, 2002 NFPA 72: 4.4.3.6 and A.6.8.6.4.1.

- audible alarm signals: shall be a distinct sound with the sound pressure level at least 15 dBA above the ambient sound levels or 5 dBA above any maximum sound level lasting at least 60 seconds, whichever is greater in every occupied space within a building, IFC 907.10.2.

- the maximum sound pressure level shall be 120 dBA, and where the average ambient noise exceeds 105 dBA visual signaling devices shall be provided in accordance with NFPA 72, IFC 907.10.2.

- the minimum sound pressure level for R-1 and I-1 shall be 70 dBA, mechanical rooms shall be 90 dBA, other occupancies shall be 60 dBA, exception: visual signaling devices are permitted in lieu of audible signaling devices in I-2 critical care areas, IFC 907.10.2.

- in public mode, establish a minimum sound pressure level of 5 dBA's above maximum noise level that occurs more than 60 seconds or 15 dBA's above the average ambient noise level, whichever is greater, 5 ft. above the floor level in an occupied area, 2002 NFPA 72: 7.4.2.1.

- in private mode, establish a minimum sound pressure level of 5 dBA's above maximum noise level that occurs more than 60 seconds or 10 dBA's above the average ambient noise level, whichever is greater, 5 ft. above the floor in an occupied area. This requirement can be reduced or eliminated by the AHJ if visual signaling is provided per 7.5, 2002 NFPA 72: 7.4.2.1.

- 2002 NFPA 72 Table A.7.4.2 provides average ambient sound pressure levels for different types of facilities. This list should serve only as a guideline. If ambient sound pressure levels within a facility are suspected to exceed the guideline, then design accordingly.

- existing facilities: if a facility is being remodeled or it is suspected the original sound pressure level design is not in compliance with the requirements in affect a the time of construction, an AHJ may consider recommending a design professional to perform a sound pressure level analysis and design that portion of the system using results of the analysis, this is a recommendation that has developed from field experience.

- a loss of sound pressure (attenuation) occurs when passing through construction elements: typically, 20-25 dBA's are lost through solid core doors, 10 to 15 dBA's are lost through hollow core doors, 35 to 41 dBA's are lost through stud walls, an expanded list can be obtained from NFPA's Fire Protection Engineering Handbook Table 4-1.24.

- for sleeping areas start with a 75 dBA minimum level for the occupiable area at the pillow level, (no requirement for bathrooms) 2002 NFPA 72 7.4.4.

- noted as a rule of thumb in the NFPA 72 Handbook, 6 dB's are subtracted from the signal rating each time the distance from the signal is doubled (start at 10 ft. from the device).

- decibel levels decrease as voltage decreases.

NOTE:
1. Because manufacturers of audible signaling devices with identical dB ratings vary as to their effectiveness to penetrate walls, doors, etc., it is difficult to predict if an area has sufficient coverage. A manufacturer may have a sound pressure level curve that will assist in design efforts.

2. Wall insulation affects audibility levels. In the past, fire alarm designs may have met sound pressure level requirements but a facility may over-insulate causing typical designs to be inadequate.

These two items create some unknowns for the designers and the reviewers. Plan examiners try to be conservative during reviews and assume the design will be sufficient when it is installed and tested.

High-rise Building Fire Alarm Requirements
2006 IBC, 2006 IFC, and 2002 NFPA 72

A high-rise building requires: 1) fire alarm system, 2) emergency voice/alarm signaling system, and 3) fire department communication system.

2006 IBC and IFC, 907.2.12.1 Automatic Fire Detection
Functions:
- shall be connected to the fire alarm system
- shall operate the emergency voice and alarm signaling system
- shall be located in every mechanical room, electrical, transformer, telephone equipment, or similar room which is not provided with automatic sprinkler protection
- shall be in elevator machine rooms and in elevator lobbies
- shall be located in the main return air and exhaust air plenum of each air-conditioning system with a capacity greater than 2,000 cubic feet per minute
- shall be located at each connection to a vertical duct or riser serving two or more stories from a return-air duct or plenum of each air-conditioning system, Group R-1 and R-2 occupancies can locate listed smoke detectors in each return air riser limited to 5,000 cubic feet per minute and not serving more than 10 air inlet openings

2006 IBC and IFC, 907.2.12.2 Emergency Voice/Alarm Communication System
Functions:
- any fire detector, sprinkler or flow device or manual pull box shall sound an alert tone followed by voice instructions on a general or selective basis to the following areas: elevator lobbies, corridors, rooms and tenant areas exceeding 1,000 sq. ft. and areas of rescue assistance; exception: in I-1 and I-2 occupancies the alarm can sound at a constantly attended location and occupant notification shall be occur over the overhead page
- paging zones shall be provided for elevator groups, exit stairways, each floor, and areas of refuge
- a manual override is provided for all paging zones
- the system is capable of broadcasting live voice messages to all paging zones on a selective or all-call basis
- standard for design and installation will be NFPA 72

2006 IBC and IFC, 907.2.12.3 Fire Department Communication System
Functions:
- the system will provide two-way communication
- standard for design and installation will be NFPA 72
- the system will communicate between the fire command center and elevators, elevator lobbies, emergency/standby power rooms, fire pump rooms, areas of refuge and enclosed exit stairways
- communication devices shall be provided in each stairway for each floor level

2006 IBC and IFC, 911.1 Fire Command Center
Functions:
- separate from rest of building by 1-hour fire barrier
- the room to be a minimum 96 square feet in area with a minimum dimension of 8 feet

- the command center shall contain: emergency voice and alarm communications unit, fire department communications unit, fire alarm annunciator unit, annunciator indicating elevator status, air-handling controls and status, firefighter's smoke control panel, controls to unlock stairway doors, sprinkler valve and water flow display panel, emergency power status indicators, telephone for fire department use, fire pump status indicator, building plans, worktable, generator controls, public address system where specified by other sections of the IBC

2002 NFPA 72
Consult NFPA 72 for emergency voice/alarm communication functions for partial or selective evacuation of a building.

Functions and design criteria:
- selective (automatic or manual) paging capability
- survivability of fire system circuits, 6.9.4.3
- speaker amplifiers, tone-generating equipment, and two-way telephone circuits integrity, 6.4.7.2 and 6.9.4.4
- fire command center that is remote from the central control equipment; the interconnecting wiring shall be protected, 6.9.4.6
- evacuation or relocation notification will be provided, 6.9.5.3
- prerecorded, synthesized, or live messages shall meet intelligibility requirements, 7.4.1.4 and consult A.7.4.1.4
- for when the voice and alarm are controlled separately and if the voice systems fails consult 6.9.5.4
- live voice instruction overrides, 6.9.5.5
- number of speakers provided in each notification zone, when required in each enclosed stairway and the speakers are on a separate notification zone for manual paging only, 6.9.7
- evacuation signal zoning, 6.9.8
- Two-way phone service (for fire department use), 6.9.9

ADA accessibility requirements; 2006 IBC and ICC/ANSI A117.1-2003
ICC/ANSI A117.1-2003
Accessible audible and visual alarms and notification appliances shall be installed in accordance with 2002 NFPA 72.
- Visual alarm signal appliances are required in restrooms, and any other general use areas (e.g., meeting rooms, classrooms, etc), hallways, lobbies, and other common use areas.

General Notes:

Speakers can replace horn strobe units; but strobes must still be provided in accordance with ADA accessibility requirements.

Speakers must meet audibility requirements based on audible tone not on fluctuating sound pressure of voice or the recorded message.

Very few smoke detectors are required, possibly only elevator machine rooms and in elevator lobbies. The other equipment rooms do not require smoke detection if the rooms when protected by automatic sprinklers.

Smoke detectors are not required in corridors or intervening spaces.

Fire Plan Review and Inspection Guidelines

Fire Alarm System
Plan Review Worksheet
2006 IFC and 2002 NFPA 72

Date of Review: _____ Permit Number: _____

Business/Building Name: _____ Address of Project: _____

Designer Name: _____ Designer's Phone: _____

Contractor: _____ Contractor's Phone: _____

FA Manufacturer: _____ FA Model: _____ Occupancy Classification: _____

Reference numbers following worksheet statements represent an NFPA code section unless otherwise specified.

Worksheet Legend: ✓ or **OK** = acceptable **N** = need to provide **NA** = not applicable

1. _____ Three sets of drawings are provided.
2. _____ Equipment is listed for intended use and compatible with the system, specification data sheets are required, 4.3.1.

Drawings Shall Detail the Following Items, IFC 907.1.1 and NFPA 72 4.5.1:

3. _____ A common scale is used and plan information is legible.
4. _____ Rooms are labeled and room dimensions are provided.
5. _____ Equipment symbol legend is provided.
6. _____ The type of fire alarm circuits (Class A or B) is indicated, IFC 907.9.
7. _____ When detectors are used, device locations, mounting heights, and building cross sectional details are shown on the plans.
8. _____ The type of devices installed are indicated.
9. _____ Wiring for alarm initiating and signaling devices is detailed.
10. _____ The location of the Fire Alarm Control Unit (FACU) and when required, the Remote Annunciator panel are located near the main entrance or as approved by the AHJ, 4.4.6.1.1. and 7.10.
11. _____ If more than one building is served by a system, each building is indicated separately on the FACU or annunciator and is identified on the shop drawings, 4.4.6.4.
12. _____ Sectional views of structure, roof, and ceiling, and rooms with beam or solid joists and drop ceilings, etc are illustrated unless the building or lease space has smooth ceilings.
13. _____ The riser diagram illustrates the number and type of devices installed on each circuit, identification of fire alarm zones (if the system is not addressable), and the primary and secondary power supplies. The primary power supply shall be a minimum 120 volt alternating current branch circuit labeled *Fire Alarm Circuit* whose access is limited to authorized personnel, 4.4.1.4.2.

Point to Point System Wiring Diagram:

14. _____ Interconnection and wire routing of identified devices and controls for each circuit.
15. _____ Indicate the number of conductors and wire gauge for each circuit run.
16. _____ Identify separate zones, circuits, and end of line locations.

Alarm Indicating Circuit Voltage Drop Calculations:

17. _____ Indicate the number of signaling devices, current consumption, the end of line voltage for each circuit, and the lowest nameplate operating voltage range for audible and visual notification devices.
18. _____ Approximate length of each circuit and resistance of wire, use National Electrical Code conductor resistance values or the manufacturer data sheet.
19. _____ Provide calculations for the acceptable circuit limits including:
   A. Standby power consumption of all current drawing devices times the hours required by NFPA 72 (24 hours) including power consumption of the control panel modules.
   B. Power consumption of all devices on standby power.
   C. The power consumption of all current consuming devices multiplied by the minutes required by NFPA (5 minutes for fire alarms or 15 minutes for emergency voice/alarm communication service).

Secondary Power Supply:

20. _____ The secondary power supply capacity for the type of system is rated for a minimum of 24 hours and will alarm for 5 minutes, 4.4.1.5.3.1.

Fire Plan Review and Inspection Guidelines

21. _____ The secondary power supply for voice/alarm system is rated for a minimum of 24 hours and will alarm for 15 minutes, 4.4.1.5.3.1(A)
22. _____ When used as a source of secondary power, batteries shall be sized for 100 percent of maximum normal load.

Performance Based Design

23. _____ Documents are provided outlining each performance objective, applicable scenarios, any calculations, modeling and other technical support in establishing the proposed fire design and life safety performance in accordance with 5.3. Readers should consult the ICC/Society of Fire Protection Engineers *Code Official's Guide to Performance-based Design Review*.

Initiating Devices:

24. _____ Smoke and heat detection device coverage is designed in accordance with total coverage (5.5.2.1), partial coverage (5.5.2.2), selective coverage (5.5.2.3), or nonrequired coverage (5.5.2.4).
25. _____ Detection devices: wiring details for devices are provided.
26. _____ Detection devices: type and location for the occupancy type is in accordance with IFC 907.
27. _____ Duct detector locations in HVAC ducts. The air flow rate per minute ratings are provided: including the manufacturer data sheet and a matrix or note detailing what size sampling tubes are to be used for each duct size, 5.14.5.
28. _____ Heat detectors: listing and spacing data sheets are provided.
29. _____ Heat detector spacing: identify which ceilings are smooth, sloped, have solid joist or beam construction, 5.6.5
30. _____ Heat detector heat classification color is written at detector location on the plan, 5.6.2.
31. _____ Heat detector spacing for rooms with smooth ceilings shall comply with spacing requirements in Section 5.6.5.1.
32. _____ Heat detector spacing for irregularly shaped areas shall comply with the spacing requirements in Section 5.6.5.2.
33. _____ Heat detector spacing for ceilings 10 ft. to 30 ft. shall be in compliance with Table 5.6.5.5.1, 5.6.5.5.1.
34. _____ Heat detector spacing at right angles to solid joist construction is not greater than 50 percent of the smooth ceiling spacing, 5.6.5.2.
35. _____ Heat detector spacing at right angles to beams projecting greater than 4 in. below the ceiling do not exceed 66 percent of the smooth ceiling spacing noted in 5.6.5.1.1 and .2 or if beams project greater than 18 in. below the ceiling and are spaced greater than 8 ft. on center then each bay is a separate area, 5.6.5.3.
36. _____ Heat detector spacing for sloped ceilings shall comply with Section 5.6.5.4.
37. _____ Unless listed for such use, smoke detectors shall not be installed in an environment where the temperature, relative humidity or air velocity exceeds the prescribed limits in Section 5.7.1.8
38. _____ Smoke detector spacing is in accordance with the listing data sheet.
39. _____ Smoke detector location and spacing shall be based on anticipated smoke flows due to the plume and ceiling jet produced by an anticipated fire, which should take into account: 1) ceiling shape and surface, 2) ceiling height, 3) configuration of contents, 4) combustion characteristics of fuel load, 5) compartment ventilation, 6) ambient pressure, pressure, altitude, and humidity. Provide document indicating these variables were considered, 5.7.3.1.2. The fire code official may require supporting documentation.
40. _____ Smoke detectors in high air movement areas are not located in the supply air stream and shall be spaced in accordance with Table 5.7.5.3.3 and Figure 5.7.5.3.3.
41. _____ If two smoke detectors are used to initiate an alarm, verify that at least two detectors are provided in each protected area that and alarm verification is not being used, 6.8.5.4.3.
42. _____ Room cross sectional details are provided for smoke detector designs listed in worksheet items 44 and 45.
43. _____ Smoke detector spacing for smooth ceiling will use 30 ft. spacing as a guide, manufacturer's data sheet listing criteria shall be followed. Other spacing is permitted depending on ceiling height, etc., for detecting flaming fires the guidelines of Annex B can be used, all points of a ceiling are within .7 (70%) of the selected spacing, 5.7.3.2.3.
44. _____ Smoke detector spacing for solid joist and beam construction: 1) solid joists are considered as beams, 2) level ceilings with a height of 12 ft. or less and beam depth of 1 ft. or less, spacing will run parallel to the beam run and will use smooth ceiling spacing and ½ spacing shall be applied in a direction perpendicular to the beam run and detectors can be mounted on ceiling or bottom of beam 3) beams greater than 1 ft. in depth then detectors will be mounted on the ceiling in each beam pocket, 4) for sloped ceilings, the detectors will be located on bottom of the joists, 5.7.3.2.4.

45. _____ Air sampling smoke detector design calculations are within the maximum air sample transport time of 120 seconds, system calculations and a manufacturer design manual is provided, 5.7.3.3.2.
46. _____ Air sampling smoke detector sampling pipe network is detailed on the plans with pipe size and lengths, with calculations showing flow characteristics of the piping network and each sampling port, 5.7.3.3.4.
47. _____ Air sampling smoke system: provided are details of pipe mounting system and signage for each pipe at changes of direction or pipe branches, each side of wall penetration, and at least every 20 ft., 5.7.3.3.8
48. _____ Projected beam smoke detector locations are detailed on the plans and the manufacturer's design data sheets are provided, 5.7.3.4.1
49. _____ Projected beam smoke detectors: stratification for a high ceiling was considered in the beam detector's use and documentation is provided indicating this evaluation was completed, 5.7.3.4.2.
50. _____ Projected beam smoke detectors shall be equivalent to a row of spot-type detectors on level or slope ceilings, 5.7.3.4.5.
51. _____ Smoke detector spacing located on peaked ceilings shall be spaced and located within 3 ft. of the peak, measured horizontally, and additional detectors, if any, shall be based on the horizontal projection of the ceiling, shed ceilings shall have detectors located on the ceiling within 3 ft. of the high side of the ceiling measured horizontally, and additional detectors, if any, shall be based on the horizontal projection of the ceiling, and room cross sectional are provided, 5.7.3.5 and .6.
52. _____ Smoke detector spacing under raised floors or above suspended ceilings shall be treated as separate rooms and spacing is in accordance with 5.7.3.7.
53. _____ Smoke detector spacing: when partitions are less then 18 in. from the ceiling a documented evaluation of its effects on smoke travel is provided, 5.7.3.8.
54. _____ Smoke detector used in plenums are listed for anticipated environment and shall not be used in lieu of open area detectors, 5.7.4.
55. _____ Smoke detectors are not installed in areas that exceed the prescribed limits of temperature, humidity or air movement in Section 5.7.1.8.
56. _____ Smoke detectors in high air movement areas are spaced in accordance with Table 5.7.5.3.3 and Figure 5.7.5.3.3
57. _____ Smoke detection is provided in areas not continuously occupied where the FACU and other control units are located, 4.4.5.
58. _____ Radiant energy-sensing fire detectors, detector device is detailed and the equipment data sheets are provided.
59. _____ Radiant energy-sensing fire detector data sheets show the detector matches the spectral emissions of the fire or fires to be detected and how false alarms will be minimized, 5.8.2.2.
60. _____ Radiant energy-sensing fire detector spacing will be in accordance with its listing or inverse square calculations and the number and location of detectors is based on complete unobstructed view coverage of the area, 5.8.3.
61. _____ Radiant energy-sensing flame and spark/ember detectors location and spacing are based on a documented and submitted engineering evaluation to include fire size, fuel involved, detector sensitivity, detector field of view, distance from fire to detector, radiant energy absorption, extraneous radiant emissions, purpose of coverage, and the response time required, 5.8.3.2 and 5.8.3.3.
62. _____ Other fire detectors not previously covered are installed in accordance with listing requirements, an engineering survey which includes structural features, occupancy and use, ceiling height, ceiling configuration, ventilation, ambient conditions, fuel load and content configuration, 5.9.
63. _____ Smoke detectors used for elevator recall: detectors in the elevator lobby, elevator machine room, hoist ways, and control room are connected to the facility fire alarm system, 6.15.3.1.
64. _____ Smoke detectors for elevators, in nonfire alarmed buildings, shall be connected to a dedicated fire control unit and labeled as such, all of which is detailed on the plans, 6.15.3.2.
65. _____ Smoke detectors for elevators shall initiate the fire alarm and have a distinct visual indicator at the control panel and annunciator, the activation of fire alarm indicating devices are not required if the signal transmits to a constantly attended location 6.15.3.8 and .9.
66. _____ For elevator recall the primary and alternate floors for recall are noted on the plans.
67. _____ Smoke detectors for elevators, a lobby detector is located within 21 ft. of the centerline of each elevator door within the elevator bank controlled by the detector, 6.15.3.5.
68. _____ Sprinkler waterflow alarm device is shown on the plan as part of an initiation circuit, 5.10 and IFC 907.7.
69. _____ Other automatic extinguishing systems are shown on the plan as part of an initiation circuit, 5.11 and IFC 907.14.
70. _____ Smoke detectors used in air duct systems are listed for such use and are appropriate for air velocities, temperatures, and humidity expected, 5.14.5.6.

71. _____ Smoke detectors used in smoke control systems: duct detectors for preventing recirculation of smoke beyond a room or space from which the smoke is generated have their location detailed and are in the return air duct or plenum upstream of any filters of the air-handling system when the air system exceeds 2,000 cfm, Exception: detectors are not required in the return air if all portions of the building that are served by the air system are protected by area smoke detection, NFPA 72 5.14.3 and IMC 606.2.
72. _____ Smoke detectors used for smoke control systems: multi-air systems that share common supply or return air ducts or plenums with a capacity exceeding 2,000 cfm the return air system shall be provided with smoke detectors in accordance with Item 74 above, consult the list of exceptions, IMC 606.2.2.
73. _____ Smoke detectors used for smoke control systems: return air risers serving 2 or more stories and serve any portion of a return air system exceeding 15,000 cfm has smoke detectors at each story, IMC 606.2.3.
74. _____ Smoke detectors used for smoke control systems, access to detector is detailed, IMC 606.3 and IFC 907.13.
75. _____ Smoke detectors used for smoke control systems, detectors are connected to fire alarm system and the visual/audible supervisory signals are shown located at a constantly attended location, Exceptions: 1) supervisory signal not required at constantly attended location if the duct smoke detectors activate the fire alarm system, 2) building without a fire alarm, the plans show the detector activates a visual/audible signal in an approved location (front entry) and the same for showing detector trouble conditions and it is shown to be signed/lettered as an air duct detector trouble, IMC 606.4.1 and IFC 907.12.
76. _____ Positive Alarm Sequence if used is approved by the fire code official, and must comply with 6.8.1.3.
77. _____ Fire safety control functions: door release smoke detector locations are detailed and in compliance with 5.14.6 and 6.15.6.
78. _____ Fire safety control functions: exit door unlocking devices are connected to the fire alarm system and release on alarm activation, 6.15.7 and IFC 907.2.15.
79. _____ Fire safety control functions: fan controls or door controls are interconnected with fire alarm system and detailed, any listed relays that initiate control are within 3 ft. of the control circuit or appliance and the relay data sheet is provided, and wiring is monitored for integrity, 6.15.
80. _____ Fire safety control functions: fire pump is supervised by fire alarm system, 6.8.5.8.
81. _____ Notification zones and circuits coincide with building outer walls, fire or smoke compartment boundaries, and floor separations, 6.8.6.2
82. _____ Zones: each floor will be zoned separately, not to exceed 22,500 sq. ft. nor exceed 300 ft. in length in any direction and each zone is clearly identified on the plans, IFC 907.9.
83. _____ Zones: each floor is considered a zone and if fire or smoke barriers are used for relocating occupants from one zone to another on the same floor, then each zone shall be annunciated separately and all zones are clearly identified on the plans, 4.4.6.
84. _____ Zones: a zone indication panel and controls are provided and the panel location is approved, IFC 907.9.1.
85. _____ Zones: visual annunciators display all of the zones in alarm, 4.4.6.1.3.
86. _____ Zones: each floor of a high-rise building is separately zoned and each zone includes smoke detectors, sprinkler water flow devices, manual pull boxes, and other approved automatic detection devices or suppression systems on that floor, IFC 907.9.2.
87. _____ Emergency voice/alarm communication system complies with 6.9 and IFC 907.2.12.2.
88. _____ Emergency voice/alarm communication system used for partial evacuation or relocation of occupants has its circuits provided with 2-hour protection by rated cable or a rated enclosure or an alternate approved by the AHJ, 6.9.4.3.
89. _____ Emergency voice/alarm communication system: Fire Command Center complies with 6.9.6 and IFC 505.1.
90. _____ Manual fire alarm boxes are located not less than 42 in. and not greater than 48 in. from the floor, IFC 907.42.
91. _____ Manual fire alarm boxes are red in color, IFC 907.4.3.
92. _____ Manual fire alarm boxes: shall be on each floor level, within 5 ft. of each exit door, at every floor exit, on both sides of grouped openings exceeding 40 ft. in width and within 5 ft. of the opening, and within 200 ft. of travel, 5.12 and IFC 907.4.1.
93. _____ Manual fire alarm boxes: if the system is not monitored by a supervising station, the plan notes signs are required at the pull station "Local Alarm Only-Call the Fire Department," unless it is manufactured in the device, IFC 907.4.4.
94. _____ 24 hour monitoring is required, the type of supervisory service is noted on the plans, IFC 907.15 and NFPA 72 Chapter 8.

95. _____ 24 hour monitoring: sprinkler alarm, supervisory, and trouble signals are distinctly different, 4.4.3.3, IFC 903.4.1.
96. _____ 24 hour monitoring: sprinkler alarm, supervisory, and trouble signals are transmitted to a central station, or remote supervisory station, or proprietary supervisory station, 4.4.3.6 and Chapter 8, IFC 903.4.1.
97. _____ 24 hour monitoring service: transmitting device is detailed and its listing data sheet is provided.
98. _____ 24 hour monitoring: for digital alarm communicator transmitters (DACT), dual monitoring control is required in case the primary transmission method fails, 8.5.3.2.1.5.

Alarm Signaling Devices:
99. _____ Fire alarm audible device design in public mode shall provide at least a minimum sound pressure level of 15 dBA above ambient sound pressure level or 5 dBA above maximum sound pressure level in every occupied space, and the sound pressure level rating of each audible device is noted adjacent each audible device on the plans, 7.4.2 and IFC 907.10.2.
100. _____ Fire alarm audible device design in private mode shall provide at least a sound pressure level of 45dBA and a minimum sound pressure level of 10 dBA above the ambient sound pressure level or 5 dBA above maximum sound pressure level, and the sound pressure level rating is noted on the plans, 7.4.3 and IFC 907.10.2.
101. _____ The device sound pressure level rating shall be not less than 70 dBA in R and I-1 occupancies, 90 dBA in mechanical rooms, 60 dBA in other occupancies and not more than 120 dBA, IFC 907.10.2.
102. _____ In areas where average ambient noise level is greater than 105 dBA, visible signals are provided, IFC 907.10.2.
103. _____ Sleeping areas shall provide at least a minimum sound pressure level of 15 dBA above ambient sound pressure level or 5 dBA above maximum sound pressure level or a sound pressure level of at least 75 dBA at the pillow, 7.4.4.1.
104. _____ Visual signaling device are permitted in I-2 critical care areas in lieu of audible devices, IFC 907.10.2.
105. _____ Notification devices are located not less than 90 in. above the floor and are greater than 6 in. from the ceiling unless listed for ceiling installations, 7.4.6.1.
106. _____ Mounting heights are permitted to be adjusted so as to ensure the sound pressure level requirements are met, 7.4.6.5.
107. _____ The alarm signal device is designed to transmit a three-pulse temporal pattern, 4.4.3.1, 6.8.6.4.1, and A.6.8.6.4.1.
108. _____ Speakers listed for notification use shall not be used for non-emergency use except as permitted by the, the two exceptions in Section 6.8.4.5.
109. _____ Visual alarm signaling appliances are provided in public and common areas, e.g. restrooms, meeting rooms and classrooms, hallways, and lobbies, IFC 907.10.1.
110. _____ When employee work areas are provided audible fire alarms then the fire alarm system is designed with at least an additional 20 percent capacity to permit the addition of future visual alarm signaling appliances, IFC 907.10.1.2.
111. _____ The equipment data sheet for visual alarm signaling appliances confirms the flash rate does nor exceed 2 flashes per second, 7.5.2.1.
112. _____ Visual alarm signaling appliances: Shop drawings illustrate the wall mounting is between 80 in. and 96 in. above the floor level, 7.5.4 and ceiling mounting is in accordance with Table 7.5.4.1.1(b).
113. _____ Visual alarm signaling appliances: The device spacing and effective intensity for an area comply with Figure 7.5.4.1.1, Tables 7.5.4.1.1 (a, b), 7.5.4.1.1 and 7.5.4.1.2.
114. _____ When 2 or more visual alarm signaling appliances installed in corridors are within the same field of view they shall be synchronized, 7.5.4.2.7.
115. _____ Visual alarm signaling appliances located in corridors greater than 20 ft. wide, the device spacing shall comply with Tables 7.5.4.1.1 (a, b) and Figure 7.5.4.1.1, or 7.5.4.2.3.
116. _____ Visual alarm signaling appliances located in corridors are within 15 ft. of the ends of corridor and do not exceed 100 ft. separation, 7.5.4.2.5.
117. _____ Visual alarm signaling appliances in sleeping areas shall be located in accordance with Table 7.5.4.4.2 and 7.5.4.4.
118. _____ Visual alarm signaling appliances located in rooms with ceilings exceeding 30 ft. in height, ceiling visual devices will be suspended below 30 ft. or wall mounting and spacing shall be per Table 7.5.4.1.1(a). Center of room ceiling mounted visual device comply with Table 7.5.4.1.1(b), and Sections 7.5.4.1.6 and 7.5.4.1.7.

Other Requirements:
119. _____ Speaker amplifier, tone generating equipment, and emergency phone circuit integrity are monitored, 4.4.7.2.1.

Fire Plan Review and Inspection Guidelines

120. ____ Class A circuit wiring is routed separately from the redundant circuit and redundant circuits are not in same cable assembly, conduit or raceway, 6.4.2.2.2.
121. ____ The sprinkler supervisory switch is connected to the fire alarm system and the audible signals shall be different between tamper switch and flow alarm, 4.4.3.3 and 4.4.3.6.
122. ____ For voice alarm systems a sample of the evacuation message is submitted for approval, 6.9.
123. ____ Voice alarm systems speakers shall be listed for fire alarm use, 4.3.1.
124. ____ A minimum of 2 loudspeakers are located in each paging zone of the building equipped with a voice alarm system, 6.9.7.2.
125. ____ Telephone communications: equipment is listed for two-way communication, 4.3.1, 6.9.9.1.
126. ____ Telephone communications: the number of handsets provided for telephone jack systems is provided.
127. ____ Telephone communications: fire emergency phone jack locations are shown on plans.
128. ____ Telephone communications: the system can permit up to 5 phones to operate simultaneously, 6.9.9.5.
129. ____ Telephone communications: 2 or more phone handsets are provided in the Fire Command Room, 6.9.9.14.
130. ____ Wireless systems are listed and meet the requirements of 6.16.

Additional Comments:

Review Date: _____   Approved or Disapproved   FD Reviewer: _____

Review Date: _____   Approved or Disapproved   FD Reviewer: _____

Review Date: _____   Approved or Disapproved   FD Reviewer: _____

Fire Plan Review and Inspection Guidelines

Voltage Drop Calculations for Notification Appliance Circuit (NAC): _____

Each NAC shall have its voltage drop determined. This sheet shall be used for one NAC but every NAC should have a sheet completed and submitted with each permit application.

STEP 1: complete the following to provide data for determining the resistance of the conductor in Step 2

Wire length is from fire alarm control panel to the end of the fire alarm circuit = _____ ft. X 2 = _____ ft.

Wire Size = #_____ AWG (American Wire Gauge)

Resistance (R) = _____ OHMS for a given 1,000 ft. of the conductor specified

Step 2: complete the following to determine the total resistance (OHMS) for a NAC

(R) = Total Wire Resistance

From Step 1 divide the OHMS by 1,000, which will convert the conductor resistance to OHMS in each linear foot of wire

Determine OHMS per foot = $\frac{_____ ft.}{1,000}$ = _____ OHMS/ft.

Take the total feet of wire from Step 1 and OHMS/ft. from the line above and put both in the equation below

Circuit resistance = _____ ft. X _____ OHMS/ per ft. = _____ (R) Total OHMS

Step 3: complete the following to determine the total alarm notification device amperage and devices may be rated in milliamps

(I) = Alarm Appliance Amperage

A. No. of Alarm Appliances = _____   B. Current amperage each _____ = A x B _____ (I)
A. No. of Alarm Appliances = _____   B. Current amperage each _____ = A x B _____ (I)
A. No. of Alarm Appliances = _____   B. Current amperage each _____ = A x B _____ (I)
A. No. of Alarm Appliances = _____   B. Current amperage each _____ = A x B _____ (I)

Total _____ (I)

Step 4: complete the following to determine the total voltage drop for the branch circuit

Voltage (E) = (I) X (R) from totals in Steps 2 and 3 above

(E) = _____ (I) X _____ (R)

= _____ (E) (shall not exceed 4.4)

Step 5: complete the following to determine if enough voltage is available to operate fire alarm notification devices

Maximum allowable voltage drop: notification devices cannot drop below their Nameplate Operating Voltage (NOV) range. As of 5/1/2004 UL 1971 and 464 required indicating devices to operate within their NOV. The UL NOV standard is 16VDC to 33VDC, consult the 2002 NFPA 72 Handbook 7.3 for more information. Fire Alarm Control Units (FACU) are tested to UL 864 and are required to operate at the end of useful battery life, 20.4 V.

Allowable voltage drop is 20.4 V (FACU) - 16 VDC (NOV) = 4.4 V

If (E) from Step 4 exceeds 4.4 V then the NAC is not compliant with NFPA 72

Take (E) from Step 4 and put in the equation below

Voltage Drop = 20.4 V - _____ (E) = _____ V (shall not be less than 16V)

Fire Plan Review and Inspection Guidelines

Fire Alarm System Acceptance Inspection
2006 IFC and 2002 NFPA 72

Date of Inspection: _____ Permit Number: _____

Business/Building Name: _____ Address of Project: _____

Contractor: _____ Contractor's Phone: _____

Reference numbers following worksheet statements represent an NFPA code section unless otherwise specified.

Pass | Fail | NA   **General**

1. ____|____|____ Obtained a copy of the fire alarm installation certification and a Record of Completion from installer, 4.5.2.
2. ____|____|____ Approved plans are on site.
3. ____|____|____ Fire alarm control unit (FACU) and remote annunciator (RA) are installed consistent with approved plans, 4.4.6.1.1. and 7.10.
4. ____|____|____ A zone and legend map is provided at the (RA) or an approved location.
5. ____|____|____ Fire alarm zones are properly identified on the FACU and RA panels.
6. ____|____|____ The Fire alarm system power supply is a dedicated 120 AC branch circuit, which is labeled.
7. ____|____|____ Type and gauge of wire or cable(s) for each circuit are consistent with the plans.
8. ____|____|____ Device location and installation is consistent with the plans.
9. ____|____|____ Pull stations are installed at the proper height and location, 42 in. to 48 in. and within the 200 ft. maximum travel distance.
10. ____|____|____ A Contractor Sound Pressure Level (dBA) Pretest Room Log is provided and verified with the use of a sound meter during a sound test.

Operational

11. ____|____|____ Fire alarm signaling devices sound throughout the occupancy providing a sound pressure level at least a minimum of 15 dBA above the average ambient noise level or 5 dBA above the maximum noise level. For bedrooms with closed door provide at least 75 dBA at the pillow, IFC 907.10.2.
12. ____|____|____ Fire alarm signaling devices are a three-pulse temporal pattern unless they were permitted to match existing audible devices, 6.8.6.4.1.
13. ____|____|____ Fire alarm visual notification device intensity (cd) ratings and settings, mounting height (80 in. to 96 in.), and location, are consistent with the plans.
14. ____|____|____ Fire alarm notification devices will activate by operation of an alarm initiating device.
15. ____|____|____ HVAC duct detectors are supervised by the fire alarm system, detectors are tested to verify if it can sample the air stream, and that fans shutdown and visual and audible status alarm functions, Table 10.4.2.2.13(g).
16. ____|____|____ 24 hour monitoring service agency received various signals during system tests.
17. ____|____|____ Verify that the correct and distinctive signals are received (alarm, trouble, and supervisory alarms), 4.4.3.3, 10.4.1.1.
18. ____|____|____ Two monitoring circuits are provided, both circuits send correct signals to the supervising station within 90 seconds, Table 10.4.2.2.16.
19. ____|____|____ Verify proper operation of magnetic door-releasing hardware and/or ventilation shutdown.
20. ____|____|____ Sprinkler tamper switch indicates a trouble signal at the annunciator panel.
21. ____|____|____ Fire alarm emergency phone jacks, if provided, are operational.
22. ____|____|____ For air sampling and flame detectors, test the device in accordance with the manufacturer instructions.
23. ____|____|____ Restorable detectors and pull stations are tested.
24. ____|____|____ Trouble signal is created for each circuit.
25. ____|____|____ Remote annunciator receives the correct information.
26. ____|____|____ Battery load test: the system is switched to battery operation 24 hours before the test and in the presence of the inspector the notification devices are activated and operate for 5 minutes and 15 minutes for emergency voice alarms.
27. ____|____|____ Check battery charger, measure load voltage and open circuit voltage.
28. ____|____|____ Test ground-fault monitoring circuit, if provided.
29. ____|____|____ Under primary and secondary power, perform these tests:

242

Fire Plan Review and Inspection Guidelines

    ___ |____| ___   A. power light on and in normal condition, trouble signal when on secondary power.
    ___ |____| ___   B. supervisory signals: fire pump power loss or phase reversal, water level/temp, pressure switches, control valves, etc.
    ___ |____| ___   C. silence switch functions.
    ___ |____| ___   D. a second alarm initiating zone overrides silence switch.
    ___ |____| ___   E. trouble signals and FACU panel lights operate for each circuit tested, disconnect wires from devices and primary power supply to simulate trouble conditions.
    ___ ___ ___   F. on secondary power, measure standby and alarm current demand.
    ___ |____| ___   G. trouble and alarm reset switches operate.
    ___ |____| ___   H. emergency voice alarms: the message is clear and distinct.
    ___ |____| ___   I. initiating devices tested, audible sound pressure levels, and visuals operate.
    ___ |____| ___   J. panel lamp test switch operates: if provided.
    ___ |____| ___   K. field zones and device address signals corresponded with panel zones and addresses.
    ___ |____| ___   L. elevator(s) recall to designated floor and alternate floor in accordance with the Elevator Code.

30. ___ |____| ___ Other systems activate fire alarm: kitchen hood suppression system, clean agent, HVAC duct detectors, etc.

31. ___ |____| ___ Corrected shop drawings shall be required when system installation is not consistent with the plans.

32. ___ |____| ___ Circuit loop resistance is within specifications and a test may be required if the system wiring has changed from the plans.

33. ___ |____| ___ Time from activating an initiating device to activation of a safety function or notification shall not exceed 10 seconds.

34. ___ |____| ___ Heat and spot smoke detectors are not within 4 in. of the sidewall, or if on the sidewall, the detector is 4 in. to 12 in. from the ceiling, 5.6.3.1, 5.7.3.2.1.

35. ___ |____| ___ Visual devices in a room or adjacent space with more than 2 devices within the field of view the flash are synchronized, 7.5.4.1.2(3). Devices In a corridor with more than 2 devices within the field of view and a maximum spacing of 100 ft., are synchronized, 7.5.4.2.5 and 7.5.4.2.7.

36. ___ |____| ___ Supplemental (extra) visual devices are permitted to be mounted less than 80 in. above the floor, 7.7.2.

37. ___ |____| ___ Ceiling mounted devices are listed for use and spaced in accordance with Table 7.5.4.1.1(b) and the approved plans.

*Note: additional testing criteria is found in NFPA 72: Chapter 10.

Additional Comments:

Inspection Date: _____   Approved or Disapproved   FD Inspector: _____

Inspection Date: _____   Approved or Disapproved   FD Inspector: _____

Inspection Date: _____   Approved or Disapproved   FD Inspector: _____

Fire Plan Review and Inspection Guidelines

Fire Alarm System
Plan Review Worksheet
2006 IFC and 2007 NFPA 72

This worksheet is for jurisdictions that permit the use of the 2007 edition of NFPA 72 in lieu of the 2002 edition of NFPA 72.

Date of Review: _____ Permit Number: _____

Business/Building Name: _____ Address of Project: _____

Designer Name: _____ Designer's Phone: _____

Contractor: _____ Contractor's Phone: _____

FA Manufacturer: _____ FA Model: _____ Occupancy Classification: _____

Reference numbers following worksheet statements represent an NFPA code section unless otherwise specified.

Worksheet Legend: ✓ or OK = acceptable   N = need to provide   NA = not applicable

1. ____ Three sets of drawings are provided.
2. ____ Equipment is listed for intended use and compatible with the system, specification data sheets are required, 4.3.1, 4.4.2.

Drawings Shall Detail the Following Items, IFC 907.1.1 and NFPA 72 4.5.1.1:

3. ____ Scale: a common scale is used and plan information is legible.
4. ____ Rooms are labeled and room dimensions are provided.
5. ____ Equipment symbol legend is provided.
6. ____ The type of fire alarm circuits (Class A or B) is indicated, IFC 907.9.
7. ____ When detectors are used, device locations, mounting heights, and building cross sectional details are shown on the plans.
8. ____ The type of devices installed is indicated on the plans.
9. ____ Wiring for alarm initiating and alarm signaling indicating devices are detailed.
10. ____ The location of the Fire Alarm Control Unit (FACU) and when required, the Remote Annunciator panel are located near the main entrance or as approved by the AHJ, 4.4.6.3.
11. ____ If more than one building is served by a system, each building is indicated separately on the FACU or annunciator and it is noted as such on the plans, 4.4.6.6.2.
12. ____ Type and gauge(s) of conductors _____.
13. ____ Sectional views of structure, roof, and ceiling, and rooms with beam or solid joists and drop ceilings, etc unless plans declare them smooth ceiling.
14. ____ The riser diagram illustrates the number and type of devices installed on each circuit, identification of fire alarm zones (if the system is not addressable), and the primary and secondary power supplies. The primary power supply shall be a minimum 120 volt alternating current branch circuit labeled *Fire Alarm Circuit* whose access is limited to authorized personnel, 4.4.1.4, 4.4.1.4.2.

Point to Point System Wiring Diagram:

15. ____ Interconnection and wire routing of identified devices and controls for each circuit.
16. ____ Indicate the number of conductors and wire gauge for each circuit run.
17. ____ Identify separate zones, circuits, and end of line locations.

Alarm Indicating Circuit Voltage Drop Calculations:

18. ____ Indicate the number of signaling devices, current consumption, the end of line voltage for each circuit, and the lowest nameplate operating voltage range for audible and visual notification devices.
19. ____ Indicate the approximate length of each circuit and resistance of wire using the National Electrical Code conductor ampacity values or provide manufacturer data sheet.
20. ____ Provide calculations for the acceptable circuit limits including:
    A. Standby power consumption of all current drawing devices times the hours required by NFPA 72 (24 hours) including power consumption of the control panel modules.
    B. Power consumption of all devices on standby power.
    C. The power consumption of all current consuming devices multiplied by the minutes required by NFPA (5 minutes for fire alarms or 15 minutes for emergency voice/alarm communication service).

Primary and Secondary power:
21. ____ The secondary power supply has a minimum capacity of 24 hours and will alarm for 5 minutes, 4.4.1.5.3.1.
22. ____ The secondary power supply for a emergency voice/communication alarm system has a minimum capacity of 24 hours capacity and will alarm for 15 minutes, 4.4.1.5.3.1(A).
23. ____ If batteries are used as a means of secondary power, they shall be sized to at least 100 percent of maximum normal load.

Performance Based Design
24. _____ Documents are provided outlining each performance objective, applicable scenarios, any calculations, modeling and other technical support in establishing the proposed fire design and life safety performance in accordance with 5.3. Readers should consult the Society of Fire Protection Engineers *Code Official's Guide to Performance-based Design Review*.

Initiating Devices; The Following Items are detailed on the Plans:
25. _____ Smoke and heat detection device coverage is designed in accordance with total coverage (5.5.2.1), partial coverage (5.5.2.2), selective coverage (5.5.2.3) or nonrequired coverage (5.5.2.4).
26. _____ Detection devices: wiring details for devices are provided.
27. _____ Detection devices: type and location for the occupancy type is in accordance with IFC 907.
28. _____ Duct detector locations in air/heat ducts. The air flow rate per minute ratings are provided: including the manufacturer data sheet and a matrix or note detailing what size sampling tubes are to be used for each duct size, 5.16.5.
29. _____ Heat detectors are listed and equipment data sheets are provided.
30. _____ Heat detector spacing: identify which ceilings are smooth, sloped, have solid joist or beam construction, 5.6.5.
31. _____ Heat detector heat classification color is indicated by detector location on the plan, 5.6.2.
32. _____ Heat detector spacing for rooms with smooth ceilings shall comply with the spacing requirements in Section 5.6.5.1.
33. _____ Heat detector spacing for irregularly shaped areas shall comply with the spacing requirements in Section 5.6.5.2.
34. _____ Heat detector spacing for ceilings 10 ft. to 30 ft. is in compliance with Table 5.6.5.5.1.
35. _____ Heat detector located at right angles to solid joist construction is not greater than 50 percent of the smooth ceiling spacing, 5.6.5.2.
36. _____ Heat detector spacing at right angles to beams projecting greater than 4 in. below the ceiling do not exceed 66 percent of the smooth ceiling spacing noted in 5.6.5.1.1 and .2 or if beams project greater than 18 in. below the ceiling and are spaced greater than 8 ft. on center then each bay is a separate area, 5.6.5.3.
37. _____ Heat detector spacing for sloped ceilings: for peaked ceilings a row of detectors are spaced and located at or within 3 ft. of the ceiling peak and additional detectors, if any, shall be spaced based on the horizontal projection of the ceiling; for shed ceilings the sloped ceiling will have detectors located within 3 ft. of the high side of the ceiling measured horizontally and additional detectors, if any, shall be spaced based on the horizontal projection of the ceiling; for roof slopes less than 30 degrees, detectors shall be spaced using the height at the peak and slopes greater than 30 degrees use the average slope height for detectors other than those at the peak, 5.6.5.4.
38. _____ Unless listed for such use, smoke detectors shall not be installed in an environment where the temperature, relative humidity, and the air velocity do not exceed the prescribed limits in Section 5.7.1.8.
39. _____ Smoke detector spacing is in accordance with the listing data sheet.
40. _____ Smoke detector location and spacing shall be based on anticipated smoke flows due to the plume and ceiling jet produced by an anticipated fire, which should take into account: 1) ceiling shape and surface, 2) ceiling height, 3) configuration of contents, 4) combustion characteristics of fuel load, 5) compartment ventilation, 6) ambient pressure, pressure, altitude, and humidity. Provide document confirming these variables were considered, 5.7.3.1.2. The fire code official may require supporting documentation.
41. _____ Smoke detectors in high air movement areas are not located in the supply vent air stream and shall be spaced in accordance with Table 5.7.5.3.3 and Figure 5.7.5.3.3.
42. _____ If two smoke detectors are used to initiate an alarm, verify that at least two detectors are provided in each protected area that and alarm verification is not being used, 6.8.5.4.3.
43. _____ Room cross sectional details are provided for smoke detector designs listed in worksheet Items 45 and 46.

Fire Plan Review and Inspection Guidelines

44. _____ Smoke detector spacing for smooth ceiling is based on a minimum 30 ft. spacing and the manufacturer's data sheet. 5.7.3.2.3.
45. _____ Smoke detector spacing for solid joist and beam construction: for level ceilings, 5.7.3.2.4.
    _____ A. If the beam depth is less than 10 percent of ceiling height then use smooth ceiling spacing criteria
    _____ B. If the beam depth is more than 10 percent of the ceiling height and beam spacing is more than 40 percent of the ceiling height, smoke detectors shall be located in each beam pocket
    _____ C. For waffle or pan-type ceiling with beams up to 24 in. and up to 12 ft. center-to-center spacing then use smooth ceiling spacing including spacing criteria for irregular areas. For these ceilings detectors can be placed on the ceiling or at the bottom of the beams
    _____ D. Corridors up to 15 ft. in width with beams perpendicular to the corridor length require smooth ceiling spacing including spacing criteria for irregular areas. The detectors can be placed on the ceiling or bottom of the beams
46. _____ Smoke detector spacing for solid joist and beam construction: for sloped ceilings, 5.7.3.2.4.3.
    _____ A. For beams running parallel to the slope use level beam ceiling spacing criteria
    _____ B. Ceiling height is determined as the average height over the length of the slope
    _____ C. Smoke detectors are not required at the low end of the slope using the 50 percent spacing requirement and where the ceiling slope is more than 10 degrees
    _____ D. Spacing is based on the horizontal projection of the ceiling
47. _____ For beams running perpendicular to the sloped ceilings, the detectors are spaced the same as level beamed ceilings, 5.7.3.2.4.4.
48. _____ Detectors are on the bottom of the solid joists of sloped ceilings, 5.7.3.2.4.5.
49. _____ Air sampling smoke detector design calculations are within the maximum air sample transport time of 120 seconds, 5.7.3.3.2.
50. _____ Air sampling smoke detector sampling pipe network is detailed on the plans with pipe size and lengths, with calculations showing flow characteristics of the piping network and each sampling port, 5.7.3.3.4.
51. _____ Air sampling smoke system: provided are details of pipe mounting system and signage for each pipe at changes of direction or pipe branches, each side of wall penetration, and at least every 20 ft., 5.7.3.3.8
52. _____ Projected beam smoke detector locations are detailed on the plans and the manufacturer's design data sheets are provided, 5.7.3.4.1.
53. _____ Smoke stratification was considered during the selection of the projected beam smoke beam detectors. The fire code official is authorized to require documentation of this evaluation, 5.7.3.4.2 and 4.5.1.1.
54. _____ Projected beam smoke detectors shall be equivalent to a row of spot-type detectors on level or slope ceilings, 5.7.3.4.5.
55. _____ Smoke detector spacing located on peaked ceilings shall be spaced and located within 3 ft. of the peak, measured horizontally, and additional detectors, if any, shall be based on the horizontal projection of the ceiling, shed ceilings shall have detectors located on the ceiling within 3 ft. of the high side of the ceiling measured horizontally, and additional detectors, if any, shall be based on the horizontal projection of the ceiling, and room cross sectional are provided, 5.7.3.5 and .6.
56. _____ Smoke detector spacing under raised floors or above suspended ceilings shall be treated as separate rooms and spacing is in accordance with 5.7.3.7.
57. _____ Smoke detector spacing: when partition distance to the ceiling is within 15 percent of the ceiling height, treat each partitioned area as a separate room, 5.7.3.8.
58. _____ Smoke detectors used in plenums are listed for anticipated environment and shall not be used in lieu of open area detectors, 5.7.4.
59. _____ Smoke detectors are not installed in areas that exceed the prescribed limits of temperature, humidity or air movement in Section 5.7.1.8.
60. _____ Smoke detectors in high air movement areas are spaced in accordance with Table 5.7.5.3.3 and Figure 5.7.5.3.3
61. _____ Smoke detection is provided in areas not continuously occupied where the FACU and other control units are located, 4.4.5.
62. _____ Radiant energy-sensing fire detectors, detector device is detailed and the manufacturer's data sheets are provided.
63. _____ Radiant energy-sensing fire detector data sheets show the detector matches the spectral emissions of the fire or fires to be detected and how false alarms will be minimized, 5.8.2.2.
64. _____ Radiant energy-sensing fire detector spacing will be in accordance with its listing or inverse square law and the number of detectors is based on complete unobstructed view coverage of the area, 5.8.3.

65. _____ Radiant energy-sensing flame detectors and spark/ember, location and spacing are based on a engineering evaluation to include fire size, fuel involved, detector sensitivity, detector field of view, distance from fire to detector, radiant energy absorption, extraneous radiant emissions, purpose of coverage, and the response time required, 5.8.3.2 and 5.8.3.3.
66. _____ Video image flame, combination, multi-criteria, and multi-sensor detectors are in compliance with 5.8.5 and 5.9.
67. _____ Other fire detectors not previously covered are installed in accordance with listing requirements, an engineering evaluation which includes structural features, occupancy and use, ceiling height, ceiling configuration, ventilation, ambient conditions, fuel load and content configuration, 5.10.
68. _____ For elevator recall service, smoke detectors shall be provided in the elevator lobby, elevator machine room, hoist ways, and control room. The detectors shall be connected to the fire alarm system, 6.16.3.1.
69. _____ In a building not equipped with an automatic fire alarm system, smoke detectors installed for elevator recall service shall be connected to a dedicated fire alarm control unit and labeled as such, all of which is detailed on the plans, 6.16.3.2.
70. _____ Smoke detectors for elevator recall service shall initiate the fire alarm and have a distinct visual indicator at the FACU and annunciator. Activation of fire alarm indicating devices are not required if the smoke detector signal is transmitted to a constantly attended location 6.15.3.8 and 6.15.3.9.
71. _____ For elevator recall the primary and alternate floors for recall are noted on the plans.
72. _____ Smoke detectors for elevators shall be located in accordance with the requirements in 6.16.3.5.
73. _____ Sprinkler waterflow alarm device is shown on the plan as part of an initiation circuit, 5.11 and IFC 907.7.
74. _____ Other automatic extinguishing systems are shown on the plan as part of an initiation circuit, 5.11 and IFC 907.14.
75. _____ Smoke detectors used in air duct systems are listed for such use and are appropriate for air velocities, temperatures, and humidity expected, 5.16.5.6.
76. _____ Smoke detectors used in smoke control systems: duct detectors for preventing recirculation of smoke beyond a room or space from which the smoke is generated have their location detailed and are in the return air duct or plenum upstream of any filters of the air-handling system when the air system exceeds 2,000 cfm, Exception: detectors are not required in the return air if all portions of the building that are served by the air system are protected by area smoke detection, 5.16.4.2 and IMC 606.2.
77. _____ Smoke detectors used for smoke control systems: multi-air systems that share common supply or return air ducts or plenums with a capacity exceeding 2,000 cfm the return air system shall be provided with smoke detectors in accordance with item 78. above, consult the list of exceptions, IMC 606.2.2.
78. _____ Smoke detectors used for smoke control systems: return air risers serving 2 or more stories and serve any portion of a return air system exceeding 15,000 cfm has smoke detectors at each story, IMC 606.2.3.
79. _____ Smoke detectors used for smoke control systems, access to detectors are detailed, IMC 606.3 and IFC 907.13.
80. _____ Smoke detectors used for smoke control systems, detectors are connected to fire alarm system and the visual/audible supervisory signals are shown located at a constantly attended location, Exceptions: 1) supervisory signal not required at constantly attended location if the duct smoke detectors activate the fire alarm system, 2) building without a fire alarm, the plans show the detector activates a visual/audible signal in an approved location (front entry) and the same for showing detector trouble conditions and it is shown to be signed/lettered as an air duct detector trouble, IMC 606.4.1 and IFC 907.12.
81. _____ Positive Alarm Sequence if used is approved by fire code official, and must comply with 6.8.1.3.
82. _____ Fire safety control functions: door release smoke detector locations are detailed and in compliance with 5.16.6 and 6.16.6.
83. _____ Fire safety control functions: exit door unlocking devices are connected to the fire alarm system and release on alarm activation, 6.16.7 and IFC 907.2.15.
84. _____ Fire safety control functions: fan controls or door controls are interconnected with fire alarm system and detailed; any listed relays that initiate control are within 3 ft. of the control circuit or appliance and the relay data sheet is provided, and wiring is monitored for integrity, 6.16.2.
85. _____ Fire safety control functions: fire pump is supervised by fire alarm system, 6.8.5.9.
86. _____ Combination system design (fire alarm with non-fire alarm systems) complies with 6.8.4.
87. _____ Notification zones and circuits coincide with building outer walls, fire or smoke compartment boundaries, and floor separations, 6.8.6.3.
88. _____ Zones: each floor will be zoned separately, not to exceed 22,500 sq. ft. nor exceed 300 ft. in length in any direction and each zone is clearly identified on the plans, IFC 907.9.

89. _____ Zones: each floor is considered a zone and if fire or smoke barriers are used for relocating occupants from one zone to another on the same floor, then each zone shall be annunciated separately and all zones are clearly identified on the plans, 4.4.6.
90. _____ Zones: a zone indication panel and controls are provided and the panel location is approved, IFC 907.9.1.
91. _____ Zones: each floor of a high-rise building is separately zoned and each zone includes smoke detectors, sprinkler water flow devices, manual pull boxes, and other approved automatic detection devices or suppression systems on that floor, IFC 907.9.2.
92. _____ Emergency voice/alarm communication system complies with 6.9.
93. _____ Circuits required for the emergency voice/alarm communication system shall be protected in accordance with Sections 6.9.10.4.2 and 6.9.10.4.3.
94. _____ If a Fire Command Center is required to contain an emergency voice/alarm communication system it shall comply with 6.9.6 and IFC Section 509.
95. _____ Manual fire alarm boxes: cross sectional detail shows mounting is not less than 42 in. and not greater than 48 in. from the floor, IFC 907.42.
96. _____ Manual fire alarm boxes are noted on the plans as being red in color, IFC 907.4.3.
97. _____ Manual fire alarm boxes: shall be on each floor level, within 5 ft. of each exit door, at every floor exit, on both sides of grouped openings exceeding 40 ft. in width and within 5 ft. of the opening, and within 200 ft. of travel, 5.13 and IFC 907.4.1. Manual boxes are not required for E occupancies that are sprinklered throughout and the fire alarm is initiated by the sprinkler water flow an by a manual means located in a normally occupied location, IFC 907.4.1
98. _____ Manual fire alarm boxes: if the system is not monitored by a supervising station, the plan notes signs are required at the pull station "Local Alarm Only-Call the Fire Department," unless it is manufactured in the device, IFC 907.4.4.
99. _____ 24 hour monitoring is required, the type of supervisory service and the service company name is noted on the plans, IFC 907.15 and NFPA 72 Chapter 8.
100. _____ 24 hour monitoring: sprinkler alarm, supervisory, and trouble signals are distinctly different, 4.4.3.3, IFC 903.4.1.
101. _____ 24 hour monitoring: sprinkler alarm, supervisory, and trouble signals are transmitted to a Central, Remote or Proprietary supervisory station, 4.4.3.2, 4.4.3.5, and Chapter 8, IFC 903.4.1.
102. _____ 24 hour monitoring service: transmitting device is detailed and it is listed.
103. _____ 24 hour monitoring: When a digital alarm communicator transmission (DACT) is used dual monitoring of the device is required, 8.6.3.2.1.4 (B).

Notification Appliances:
104. _____ Fire alarm audible devices in public mode shall provide at least a minimum sound pressure level of 15 dBA above the average ambient sound pressure level or 5 dBA above maximum sound pressure level that lasts for 60 seconds, whichever is greater, in every occupied space 7.4.2 and IFC 907.10.2.
105. _____ Fire alarm audible devices in private mode shall provide at least a minimum sound pressure level of 10 dBA above the average ambient sound pressure level or 5 dBA above maximum sound pressure level that lasts for 60 seconds, whichever is greater, 7.4.3 and IFC 907.10.2.
106. _____ Fire alarm audible device sound pressure level rating shall be not less than 70 dBA in R and I-1 occupancies, 90 dBA in mechanical rooms, 60 dBA in other occupancies and not more than 120 dBA, IFC 907.10.2
107. _____ Visible signaling devices are provided in areas where average ambient noise level is greater than 105 dBA, IFC 907.10.2.
108. _____ Sleeping areas shall have a minimum sound pressure level of 15 dBA above the average ambient pressure level or 5 dBA above maximum sound pressure level that lasts for 60 seconds or a sound pressure level of at least 75 dBA, whichever is greater, measured at the pillow, 7.4.4.1.
109. _____ For narrow band tone signaling the calculations, noise data, documentation and sound pressure design is in compliance with 7.4.5.
110. _____ The design for exit marking audible notification appliances complies with 7.4.6.
111. _____ Exit marking audible notification appliances are located at each area of refuge, exit and exit discharge, 7.4.6.4.
112. _____ Visual signaling devices are permitted in I-2 critical care areas in lieu of audible devices, IFC 907.10.2.
113. _____ Audible design: devices are not less than 90 in. above the floor and are greater than 6 in. from the ceiling unless listed for ceiling mount, 7.4.6.1. Use mounting height criteria from 7.5.4 for audible/visible appliances, 7.4.7.3.
114. _____ Audible design: mounting heights different than noted in worksheet item 113 is permitted if the sound pressure level requirements are met, 7.4.6.5.

115. ____ Audible notification devices shall be programmed for a three-pulse temporal pattern, 4.4.3.1, 6.8.6.5.1, and A.6.8.6.5.1.
116. ____ Audible design: speakers listed for notification use shall not be used for non-emergency use, consult the two exceptions, 6.8.4.5.
117. ____ Visual alarm notification appliances are provided in public and common areas, e.g. restrooms, meeting rooms and classrooms, hallways, and lobbies, IFC 907.10.1.
118. ____ In employee work areas equipped with audible fire alarms the fire alarm system shall be designed with an additional 20 percent capacity to permit the addition of future visual alarm notification appliances, IFC 907.10.1.2.
119. ____ Visual alarm notification appliances shall be limited to a flash rate of 1 to 2 flashes per second based on the listed device's voltage range, 7.5.2.1.
120. ____ Wall-mounted visual alarm notification appliances are located between 80 in. and 96 in. above the floor level, 7.5.4. Ceiling-mounted visual alarm notification appliances are located in accordance with Table 7.5.4.3.1(b).
121. ____ Visual alarm notification appliances: device spacing and effective intensity (cd) for an area are in compliance with Fig. 7.5.4.3.1, Tables 7.5.4.3.1 (a) and 7.5.4.3.1 (b), and Sections 7.5.4.3.1 and 7.5.4.3.2.
122. ____ In corridors with a width of 20 feet or less, visual alarm notification appliances within corridors with 2 or more devices that are in the field of view shall be synchronized, 7.5.4.4.7.
123. ____ Visual alarm notification appliances: for corridors greater than 20 ft. wide, device spacing is in accordance with Tables 7.5.4.3.1 (a, b) and Figure 7.5.4.3.1, 7.5.4.4.4.
124. ____ In corridors with a width of 20 feet or less, visual alarm notification devices located within 15 ft. of the ends of corridor and the spacing between each device is 100 ft. or, 7.5.4.4.5.
125. ____ Visual alarm notification appliances located in sleeping areas shall comply with the requirements of Section 7.5.4.6.
126. ____ Visual alarm notification appliances installed in rooms with ceilings exceeding 30 ft. in height, the visual devices shalll be suspended below 30 ft. or wall mounting and spaced in accordance with Table 7.5.4.3.1(a). Center of room ceiling mounted visual device complies with Table 7.5.4.3.1(b), 7.5.4.3.6 and .7.
127. ____ A performance based design that provides a minimum illumination of 0.0375 lumens/Ft.^{2} is permitted provided the design satisfies the requirements in Section 7.5.4.5.
128. ____ Textual audible appliances meet the sound pressure level as required in 7.4.2, 7.4.3, and IFC 907.10.2., NFPA 7.8

Other Requirements:
129. ____ Miscellaneous: speaker amplifier, tone generating equipment, and emergency phone circuit integrity are monitored, 4.4.7.2.
130. ____ Miscellaneous: class A circuit wiring, each circuit out and back is routed separately from the redundant circuit, the redundant circuits are not in same cable assembly, conduit or raceway, 6.4.2.2.2.
131. ____ Miscellaneous: the sprinkler supervisory switch is connected to the fire alarm system; the audible signals shall be different between tamper switch and flow alarm, show how that is accomplished, 4.4.3.3 and 4.4.3.6.
132. ____ Emergency voice/ alarm communications systems: a sample of the evacuation message is submitted for approval, A.6.9.
133. ____ Emergency voice/ alarm communications systems: speakers are located in compliance with IFC 907.2.12.2 and NFPA 72 Chapter 7, 6.9.7.
134. ____ Fire department communications system: equipment is listed for two-way communication, 4.3.1, 6.10.1.1.
135. ____ Fire department communications system: the design is in compliance with 6.10.1.1 through 6.10.1.16. The manufacturers' data sheets are provided to verify compliance.
137. ____ Fire department communications system: the number of handsets provided is indicated.
138. ____ Fire department communications system: fire alarm fire emergency phone jack locations are illustrated on the shop drawings and complies with IFC 907.2.12.3.
139. ____ Fire department communications system: the system is designed to allow 5 phones to operate simultaneously, 6.10.1.6.
140. ____ Telephone communications: 2 or more phone handsets are provided in the Fire Command Room, 6.10.1.15.
141. ____ Circuits service the fire department communications system shall be protected in accordance with the one of the five approved methods in Section 6.10.1.6.

Fire Plan Review and Inspection Guidelines

142. ____ Wireless systems (low power radio) are listed for use and meet the requirements of 6.16.
143. ____ Relays or appliances used to initiate other fire safety functions are listed and within 3 ft. of the controlled circuit or appliance, 6.16.2.2.
144. ____ The wiring between the relay or appliance and FACU is supervised for integrity, 6.16.2.4.

Additional Comments:

Review Date: _____   Approved or Disapproved   FD Reviewer: _____

Review Date: _____   Approved or Disapproved   FD Reviewer: _____

Review Date: _____   Approved or Disapproved   FD Reviewer: _____

Fire Plan Review and Inspection Guidelines

Fire Alarm System Acceptance Inspection
2006 IFC and 2007 NFPA 72

This worksheet is for jurisdictions that permit the use of the 2007 NFPA 72 in lieu of IFC's referenced 2002 NFPA 72.

Date of Inspection: _____ Permit Number: _____

Business/Building Name: _____ Address of Project: _____

Contractor: _____ Contractor's Phone: _____

Reference numbers following worksheet statements represent an NFPA code section unless otherwise specified.

Pass | Fail | NA **General**

1. ___ | ___ | ___ Obtained a copy of the fire alarm installation certification and a Record of Completion from installer, 4.5.2.1.
2. ___ | ___ | ___ Approved plans are on site.
3. ___ | ___ | ___ Fire alarm control unit (FACU) and remote annunciator (RA) are installed consistent with approved plans, 4.4.6.1.1. and 7.10.
4. ___ | ___ | ___ A zone and legend map is provided at the RA or an approved location.
5. ___ | ___ | ___ Fire alarm zones are properly identified on the FACU and RA panels.
6. ___ | ___ | ___ The fire alarm system power supply is a dedicated 120 AC branch circuit, which is labeled, 4.4.1.4.2.2.
7. ___ | ___ | ___ Type and gauge of wire or cable(s) for each circuit are consistent with the plans.
8. ___ | ___ | ___ Device location and installation are consistent with the plans.
9. ___ | ___ | ___ Pull stations are installed at the proper height and location, 42 in. to 48 in. and within the 200 ft. maximum travel distance, 5.13 and IFC 907.4.1, .2.
10. ___ | ___ | ___ A Contractor Sound Pressure Level (dBA) Pretest Room Log is provided and verified with the use of a sound meter during a sound pressure test.

Operational

11. ___ | ___ | ___ Fire alarm audible notification devices sound throughout the occupancy providing a sound pressure level at least a minimum of 15 dBA above the average ambient noise level or 5 dBA above the maximum noise level. For bedrooms with closed door provide at least 75 dBA at the pillow, 7.4.4.1, IFC 907.10.2.
12. ___ | ___ | ___ Fire alarm audible devices are a three-pulse temporal pattern unless they were permitted to match existing audible devices, 6.8.6.5.1.
13. ___ | ___ | ___ Fire alarm visual notification device intensity (cd) ratings and settings, mounting height (80 in. to 96 in.), and location, are consistent with the plans, 7.5.4.1.
14. ___ | ___ | ___ Emergency voice/ alarm communications systems is tested and documentation is provided documenting the verbal statement(s) is distinguishable and understandable, Table 10.4.2.2.15(b).
15. ___ | ___ | ___ In sprinklered buildings, the fire alarm notification devices will activate by operation of the sprinkler flow alarm.
16. ___ | ___ | ___ HVAC duct detectors are supervised by the fire alarm system, detectors are all tested to verify if it can sample the air stream, fans shutdown upon activation and visual and audible status alarm functions, Table 10.4.2.2.14(g).
17. ___ | ___ | ___ A central, remote or proprietary monitoring service received various signals during system tests.
18. ___ | ___ | ___ Verify that the correct and distinctive signals are received (alarm, trouble, and supervisory alarms), 4.4.3.3, 10.4.1.1
19. ___ | ___ | ___ Two monitoring circuits are provided, both circuits send correct signals to monitoring company within 90 seconds, Table 10.4.2.2.16.
20. ___ | ___ | ___ Verify proper operation of magnetic door-releasing hardware and/or ventilation shutdown.
21. ___ | ___ | ___ Sprinkler tamper switch activation transmits a trouble signal at the annunciator panel.
22. ___ | ___ | ___ Fire department communications system, if provided, is operational.
23. ___ | ___ | ___ For air sampling and flame detectors, test the device in accordance with the manufacturer instructions.
24. ___ | ___ | ___ Restoreable heat and smoke detectors, and pull stations are tested.
25. ___ | ___ | ___ Trouble condition is created for each circuit and the FACU responses appropriately.
26. ___ | ___ | ___ Remote annunciator displays the correct zone and device information.

Fire Plan Review and Inspection Guidelines

27. ____|____|____ Battery load test: the system is switched to battery operation 24 hours before the test and in the presence of the inspector the notification devices are activated and operate for 5 minutes or 15 minutes for emergency voice alarms.
28. ____|____|____ Check battery charger, measure load voltage, and open circuit voltage.
29. ____|____|____ Test ground-fault monitoring circuit, if provided.
30. ____|____|____ Under primary and secondary power, perform these tests:
　　____|____|____ A. power light on and in normal condition, trouble signal when on secondary power.
　　____|____|____ B. supervisory signals: fire pump power loss or phase reversal, water level/temp, pressure switches, control valves, etc..
　　____|____|____ C. silence switch functions.
　　____|____|____ D. a 2nd alarm initiating zone overrides silence switch.
　　____|____|____ E. trouble signals and FACU panel lights operate for each circuit tested; disconnect .wires from devices and primary power supply to simulate trouble conditions.
　　____|____|____ F. on secondary power, measure standby and alarm current demand.
　　____|____|____ G. trouble and alarm reset switches operate.
　　____|____|____ H. emergency voice alarms: the message is clear and distinct.
　　____|____|____ I. initiating devices tested, audible sound pressure levels, and visuals operate.
　　____|____|____ J. panel lamp test switch operates: if provided.
　　____|____|____ K. field zones and device address signals corresponded with panel zones and addresses.
　　____|____|____ L. elevator(s) recall to designated floor and alternate floor in accordance with the 6.16.3.
31. ____|____|____ Other systems activate fire alarm: kitchen hood suppression system, clean agent, etc.
32. ____|____|____ As-builts are required when system installation is not consistent with the plans.
33. ____|____|____ Circuit loop resistance is within specifications and a test may be required if the system wiring has changed from the plans.
34. ____|____|____ Heat and spot smoke detectors are not within 4 in. of the sidewall, or if on the sidewall, the detector is 4 in. to 12 in. from the ceiling, 5.6.3.1, 5.7.3.2.1.
35. ____|____|____ Visual devices in a room or adjacent space with more than 2 devices within the field of view the flash are synchronized, 7.5.4.1.2(3). Devices in a corridor with more than 2 devices within the field of view and a maximum spacing of 100 ft., are synchronized, 7.5.4.2.5 and 7.5.4.2.7.
36. ____|____|____ Visual devices are wall mounted 80 in. to 96 in. above the floor level unless otherwise permitted by the approved plans and the fire code official, 7.5.4.1.
37. ____|____|____ Supplemental (extra) visual devices are permitted to be mounted less than 80 in. above the floor, 7.7.2.
38. ____|____|____ Ceiling mounted devices are listed for use and spaced in accordance with Table 7.5.4.1.1(b) and the approved plans.

*Note: additional testing criteria is found in NFPA 72: Chapter 10.

Additional Comments:

Inspection Date: _____ Approved or Disapproved FD Inspector: _____

Inspection Date: _____ Approved or Disapproved FD Inspector: _____

Fire Plan Review and Inspection Guidelines

Fire Alarm Installation Certification

Permit #: _____   Date: _____

	Property Protected	System Installer	System Supplier
Business Name:	_____	_____	_____
Address:	_____	_____	_____
	_____	_____	_____
Representative:	_____	_____	_____
Telephone:	_____	_____	_____

Location of Plans: _____

Location of Owner's Manual: _____

1. <u>Certification of System Installation:</u> Complete this section after system is installed, but prior to conducting operational acceptance tests. Check wiring for opens, ground faults, and improper branching.

    This system installation was inspected and found to comply with the installation requirements of:
    _____ NFPA 72
    _____ Article 760 of NEC
    _____ Manufacturer's Instructions
    _____ Other (specify; FM, UL, etc.) _____

    Print Name: _____
    Signed: _____ Date: _____
    Organization: _____

2. <u>Certification of System Operation:</u> All operational features and functions of this system were tested and found to be operating properly in accordance with the requirements of:
    _____ NFPA 72
    _____ Design Specifications
    _____ Manufacturer's Instructions
    _____ Other (specify) _____

    Print Name: _____
    Signed: _____ Date: _____
    Organization: _____

Prefinal and Certification of Occupancy Inspection Requirements For Contractors
Contractors Worksheet

Fire Alarm Test Requirements

1. All certification forms and documents are required to be on the site for review:
   - ___ Plans
   - ___ Permit
   - ___ Prefinal detector test and a device location inspection are required as well as the completion of the items on this pretest form. Use the Acceptance Inspection worksheet for the pretest.
   - ___ Installation certification is completed, use the form contained in this book or NFPA's "Record of Completion" form, in NFPA 72.
2. ___ Person familiar with installation must be present to perform the test.
3. ___ Owner's representative approval needed for time and date of testing.
4. ___ All rooms or areas are unlocked and accessible.
5. ___ Device placement is in accordance with the plan, sound pressure level verification to have been done and recorded by the contractor.
6. ___ If Items 1-5 are incomplete, the inspection will be cancelled and another inspection request will be required. A reinspection fee may be assessed.

Prior to the next approval test:

7. ___ When additional devices, e.g., initiating or indicating devices, etc., are installed, the contractor must provide:
   - ___ As-builts and new calculations for review and approval.
   - Note: A new plan review will be submitted as "supplemental information" and proof of the additional review fee payment is required.
8. ___ A reinspection fee may be assessed if the system and required documentation are not ready.

Fire Alarm System Sound Pressure (dBA) Level Log

Date: _____ Permit Number: _____

Address: _____ Type of Sound meter: _____

Room: number or location	Installer: dBA level	Fire Inspector: dBA level

Fire Plan Review and Inspection Guidelines

Fire Pump
Contents

1. Plan Review Worksheet, 2003 NFPA 20. 259

2. Acceptance Inspection Worksheet 264

3. Plan Review Worksheet, 2007 NFPA 20. 267
   For jurisdictions that permit the use of the 2007 standard in lieu of the ICC referenced 2003

4. Pump Field Acceptance Form 272

5. Installer Certification Form 274

6. Contractor Prefinal and Certificate of Occupancy Inspection Requirements 275

7. Testing Guideline. 276

Fire Pump
Plan Review Worksheet
2006 IFC 913 and 2003 NFPA 20

Date of Review: _____ Permit Number: _____

Business/Building Name: _____ Address of Project:: _____

Designer Name: _____ Designer's Phone: _____

Contractor: _____ Contractor's Phone: _____

Pump Manufacturer: _____ Pump Model: _____

Controller Manufacturer: _____ Controller Model: _____

Driver Type: _____ Driver Manufacturer: _____

Driver Model: _____ Occupancy Classification: _____

Reference numbers following worksheet statements represent an NFPA code section unless otherwise specified.

Worksheet Legend: ✓ or OK = acceptable   N = need to provide   NA = not applicable

1. _____ Three sets of drawings are provided.
2. _____ Equipment is listed for intended use, product listing data sheets are provided, 5.1.2.1, 5.7.
3. _____ A copy of a fire hydrant flow test summary sheet is provided, which includes static and residual pressures, flow rate, and the location of the flow and test hydrant(s).

Drawings Shall Show the Following:

4. _____ Scale:_____, a common architectural scale is used and plan information is legible.
5. _____ Plan view and cross sectional views of installed equipment are provided, 5.2.
6. _____ Room dimensions are provided.
7. _____ Equipment symbol legend is provided.
8. _____ Suction pipe flushing requirements from Table 14.1.1.1 are on the plans.
9. _____ Plot plan illustrating connection to the water supply pipe and pipe diameter, and the pipe routing from the source to the fire pump.
10. _____ Driver, pump, and controller manufacturer, respective models, and driver type are specified.
11. _____ Copy of the factory pump test curve is provided, 5.5.
12. _____ Pump GPM rating: _____ Head rating: _____ RPM: _____ are provided.
13. _____ A pressure gauge complying with 5.10.1 is detailed as installed near the discharge casting.
14. _____ A compound and vacuum pressure gauge complying with 5.10.2.1 are detailed as installed to the suction pipe, (does not apply to vertical shaft turbine-type pumps taking suction from a well or open pit).
15. _____ A automatic relief valve complying with 5.11.1 is detailed as installed on the discharge side of the pump before the discharge check valve and it discharges to the drain. This requirement does not apply to engine-driven pumps provide cooling water from its discharge to the engine.
16. _____ A nonsprinklered fire pump room is separated by a 2 hour fire barrier. A sprinklered fire pump room in a nonsprinklered building is separated by a 2 hour fire barrier. A sprinklered fire pump room in a sprinklered building is separated by a 1 hour fire barrier. An exterior fire pump building is separated from adjacent buildings by at least 50 ft., Table 5.12.1.1.
17. _____ The fire pump room containing a diesel pump driver and fuel storage tanks is protected by an automatic sprinkler system in compliance with NFPA 13, 5.12.1.3.
18. _____ Outdoor fire pump unit is placed at least 50 ft. from any building that would be an exposure, 5.12.1.2.
19. _____ When required by the environment or engine manufacturer, the fire pump room has a heat source in accordance with Section 5.12.2.
20. _____ Emergency lighting for the fire pump room is provided in accordance with Section 5.12.4.
21. _____ Ventilation is provided in the pump room, 5.12.5.
22. _____ The pump room floor is adequately pitched to provide drainage and it drains to a frost free location, 5.12.6.
23. _____ The coupling guards for the flexible couplings or shaft connections between the pump driver and pump are noted or illustrated, 5.12.7.

24. _____ If used, the operating angle of a flexible connecting shaft is detailed and does not exceed the manufacturer listing requirements, 7.5.1.7.2.
25. _____ Size and type of pump suction and discharge pipe used are specified and detailed.
26. _____ Steel pipe size is specified for aboveground pipe, 5.13.1.
27. _____ The method of joining the steel pipe is specified, 5.13.2.

Suction Pipe and Fittings:
28. _____ Size and the arrangement of the suction pipe complies with Table 5.25 (a or b), 5.14.3.4.
29. _____ The suction piping arrangement and OS & Y gate valve comply with Section 5.14.5.
30. _____ The installation of elbows and tees shall be in accordance with Section 5.14.6.3.
31. _____ If provided, eccentric taper reducer or increaser (for suction pipe and pump flange size differentials) is detailed and complies with Section 5.14.6.4.
32. _____ Open source water supplies shall be equipped with a suction screen in accordance with Section 5.14.8 and 5.14.9.
33. _____ Screens for open water source are removable and the screen material is specified, 5.14.8.6.
34. _____ Screens have at least a 62.5 percent open area, 5.14.8.11.
35. _____ When devices are installed in the suction piping they shall comply with Section 5.14.9.
36. _____ A vortex plate is provided on the suction fitting that obtains water from a stored water supply, 5.14.10.

Discharge Pipe and Fittings:
37. _____ The size of the pump discharge pipe and fittings are in accordance with Table 5.25 (a or b), 5.15.5.
38. _____ A listed check valve or backflow prevention device is installed within the pump discharge assembly, 5.15.6.
39. _____ Indicating gate or butterfly valve is on system side of the check valve, 5.15.7.

General:
40. _____ When required, a pressure relief valve for centrifugal pump is provided in accordance with Section 5.18.1.
41. _____ The pressure relief valve's listing data sheet is provided and the valve is either spring loaded or pilot-diaphragm type, 5.18.4.
42. _____ Pressure relief valve discharge is designed in accordance with Section 5.18.5.
43. _____ The size of the discharge pipe is in accordance with Table 5.25 (a or b), 5.18.5. If the pipe has more than 1 elbow, enlarge the pipe one size.
44. _____ The test header pipe diameter, number of hose valves, or the flow meter size and piping are detailed in compliance with Table 5.25, 5.19.
45. _____ When provided, the location of the backflow prevention device is detailed, and the equipment data sheet and friction loss information are provided, 5.26.
46. _____ The pressure maintenance pump location and piping are detailed and equipment data sheets are provided, 5.24.
47. _____ A check valve in the pressure maintenance pump discharge pipe is detailed and the location of gate or butterfly valves for allowing component repair are detailed, 5.24.
48. _____ Where located, check valves and backflow prevention devices or assemblies are located a minimum 10 pipe diameters from the pump suction flange, 5.26.3.
49. _____ For seismic design areas, the fire pump, driver and associated equipment and piping is provided seismic bracing in accordance with Section 5.27 and seismic calculations for each method of protecting equipment are provided.

Centrifugal Pumps:
50. _____ The selected centrifugal pump is specified and meets the design requirements of 6.1.1.
51. _____ The application of a centrifugal pump complies with the requirements of Section 6.1.2.
52. _____ When required, the automatically controlled centrifugal pump has a float operated air release valve at least ½ in. diameter, 6.3.3.
53. _____ When a pipe strainer is required, the distance from the suction flange, construction materials, and free area are in compliance with Section 6.3.4.
54. _____ The foundation and setting for the pump are detailed and in compliance with Section 6.4.
55. _____ The method of securing the pump base plate to the foundation is detailed, 6.4.3.

Vertical Shaft Turbine-Type Pumps:
56. _____ Detailed for well installations is the submergence level of the second pump impeller level being at least 10 ft. below the water level and 1 ft. submergence is added for each 1,000 ft. of elevation, 7.2.2.1.1, 7.2.2.1.2.

57._____ Detailed for wet pit installations is the submergence level of the second pump impeller level being below the lowest pumping level of the open body of supply water. A greater submergence is required for pumps rated 2,000 GPM or greater. Obtain submergence depth requirement data sheet from manufacturer 7.2.2.2.1 to 7.2.2.2.3.

58._____ The well casing, screen, and suction strainer are detailed, 7.2.3, 7.3.4.

59._____ A report verifying the well can produce the appropriate quantity of water supply for the specified pump is provided, 7.2.3.1.

60._____ The dimensions of the well, its casing and casing materials, well screen, fill gravel around the well screen, method of sealing the well bottom are detailed, 7.2.4.

61._____ Specified is whether the well is in consolidated or unconsolidated formations, 7.2.4.

62._____ A certified performance test report of the well is provided, 7.2.7.

63._____ The tubular well for fire pumps 450 GPM or less is designed in compliance with 7.2.4 except 7.2.4.11 through 7.2.4.15, 7.2.4.16.2.

64._____ The suction strainer has a free area at least 4 times the area of the suction connection and the screen can restrict passage of a .5 in. sphere, 7.3.4.

65._____ The air relief valve and size, water level detector, pressure gauges, relief valves, hose valve header, valves or metering device locations are detailed and in conformance with Section 7.3.5.

66._____ The well is equipped with a water level detector, 7.3.5.3.

67._____ The pump foundation, support, anchoring, etc. design is detailed on the plans and in compliance with Section 7.4.3.

Positive Displacement Pumps:

68._____ The pump is listed for its intended use and the listing verifies the pump's performance curves, 8.1.2.

69._____ When installed on a closed head fire system a listed dump valve type is specified, and detailed in accordance with Section 8.1.6.

70._____ When provided, foam concentrate and additive pumps installations are detailed in conformance with 8.2. Pump data sheets are provided.

71._____ When provided, water mist pump installations are detailed in conformance with 8.3. Pump data sheets are provided.

72._____ Detailed are compound suction and discharge pressure gauges, and a listed safety relief valve locations, 8.4.

73._____ A pump suction strainer is provided and is in compliance with the requirements of 8.4.5.

74._____ The pump foundation, support, anchoring, etc. design is detailed on the plans and in compliance with Section 8.7.

75._____ A means for flow testing is provided and the piping schematic is provided, 8.9.

Driver Information:

76._____ Type: _____ Manufacturer: _____ Model: _____ Rated H.P.: ____ RPM: _____ are provided.

77._____ If the pump uses a diesel driver, calculations indicating the number of hours of fuel supply are provided.

Controller Information:

78._____ Manufacturer: _____ Model: _____ are provided.

Electric Drive and Electrical, complies with National Electrical Code Article 695:

79._____ An electrical circuit schematic is provided.

80._____ When provided, the electrical schematic shall detail the design for the secondary power circuit and transfer equipment is provided, 9.2.4.

81._____ A second power source is provided in accordance with 9.2.4 when electric motors are used and the building height exceeds fire apparatus pumping capability, 9.2.1.2.

82._____ Supply conductors shall directly connect to a listed combination fire pump controller and transfer switch or to a disconnecting means and one or more overcurrent protective devices, 9.2.5.4.

83._____ Circuit conductors feeding fire pump(s) shall be dedicated and protected from fire, structural failure, or operational accident, 9.3.1.

84._____ Circuits that supply the electric motor and that directly connect the power source to a listed pump controller are designed with a means of power continuity in accordance with 9.3.2.2.

85._____ The electric motor is listed for fire pump service and meets the construction, horsepower and locked rotor current requirement of Section 9.5.1.1.

86._____ When an on-site generator is required to meet the power reliability requirements of NFPA 20, it has the capacity to run under the loads identified in Section 9.6.1. The loads are specified and provided.

87. _____ Required generator(s) shall comply with the foundation requirements in Section 6.4 and be of the level, type, and class specified in Section 9.6.2. The system shall be designed in accordance with NFPA 110, and have a minimum fuel supply to operate the fire pump at its 100 percent rated capacity, 9.6.2.
88. _____ Transfer of power shall occur in the pump room, 9.6.4.
89. _____ The controller installation is detailed. It is located near and in sight of the motors it controls and energized controller components are provided working clearances in accordance with the National Electrical Code Article 110. NFPA 20 10.2.
90. _____ The fire pump controller is listed for use with a electric motor-driven fire pump and labeled in accordance with Section 10.1.2.1.
91. _____ The controller and accessories are mounted on a single noncombustible support foundation, 10.3.2.
92. _____ Enclosures for the controller and accessories are in compliance with Section 10.3.3.
93. _____ Controllers shall be provided with voltage surge arrestor, isolating switch, circuit breaker, locked rotor protection, and motor contacts in accordance with Sections 10.4.1 through 10.4.5.
94. _____ Provided and detailed is an alarm circuit and a signal device at a constantly attended location when the pump room is not constantly attended. The alarm signal transmission occurs in accordance with Sections 10.4.7.2(A) through 10.4.7.2(D), 10.4.7.
95. _____ When required, the dedicated fire pump transfer switch location is detailed, the listing data sheets are provided, and the design complies with Section 10.8.3.
96. _____ When required, one dedicated transfer switch is assigned to a fire pump, 10.8.2.3.

Diesel Driver:

97. _____ The engine is a compression ignition type and is listed for fire pump service, 11.1.2.1 and 11.2.1.
98. _____ The engine meets the rating requirements of Section 11.2.2.
99. _____ The engine connection to the fire pump is noted and designed in compliance with Section 11.2.3.
100. _____ The engine is equipped with a governor complying with the requirements of Section 11.2.4.1.
101. _____ When the engine uses a variable speed pressure limiting control system it is noted on the plans and complies with Section 11.2.4.2.
102. _____ The engine is equipped with overspeed shutdown device that complies with Section 11.2.4.3.
103. _____ The engine is equipped with an instrument panel containing: tachometer, oil pressure gauge, and temperature gauge, 11.2.4.4- 11.2.4.7.
104. _____ Batteries have two means of recharging detailed and the chargers are listed for fire pump service and comply with the requirements listed in Sections 11.2.5.2.3 and 11.2.5.2.4.
105. _____ Detailed or noted on the plans is that each engine has two batteries that are rack supported, and current carrying-parts (cables) are not less than 12 in. above the floor, 11.2.5.2.1.1. and 11.2.5.2.5 -.6
106. _____ The engine cooling system is closed-circuit liquid type and is specified as radiator or heat exchange type, 11.2.6.
107. _____ Adequate ventilation is provided for the pump room and the engine, 11.3.2.
108. _____ Fuel supply tank capacity calculations are provided and are at least 1 gallon per horsepower plus 5 percent volume for expansion and 5 percent volume for sump, 11.4.3 and the fuel supply tank design complies with IFC 34.
109. _____ Fuel piping is designed in compliance with Section 11.4.6.
110. _____ The controller is listed for use with diesel engine-driven fire pumps and labeled in accordance with Section 12.1.3.
111. _____ The controller installation is detailed. It is located near and in sight of the engine it controls and energized controller components are provided working clearances in accordance with the National Electrical Code Article 110, 12.2.2 -.4.
112. _____ The controller and accessories are mounted on a single noncombustible support foundation, 12.3.2.
113. _____ Enclosures for the controller and accessories are in compliance with Section 12.3.3.
114. _____ Provided and detailed is an alarm circuit and a signal device(s) in the engine room. The visible indicators and a common alarm signal occurs in accordance with events listed in Sections 12.4.1.3(1) through 12.4.1.3(11), 12.4.
115. _____ When the pump room is not constantly attended, the alarm and signal devices are remote from the controller, in a constantly attended location, and are detailed and designed in accordance with Section 12.4.2.
116. _____ Engine exhaust is vented to the exterior and where the exhaust will not harm persons or endanger buildings, 11.5.
117. _____ Engine exhaust piping connections, diameter, clearances to combustible materials, and termination points are detailed and design in accordance with Section 11.5.3.

Additional Comments:

Review Date: _____   Approved or Disapproved   FD Reviewer: _____

Fire Pump Acceptance Inspection
2006 IFC and 2003 NFPA 20

Date of Inspection: _____ Permit Number: _____

Business/Building Name: _____ Address of Project:: _____

Contractor: _____ Contractor's Phone: _____

Reference numbers following worksheet statements represent an NFPA code section unless otherwise specified.

Pass | Fail | NA   **General**
1. ____|____|____ Received fire pump system installation and test certification from installer.
2. ____|____|____ Received fire pump manufacturer pump curve certification test form.
3. ____|____|____ The pump suction pipe is flushed in accordance with NFPA Table 14.1.1(a) or (b).
4. ____|____|____ After the suction pipe is flushed, a 2 hour hydrostatic test is performed on suction and discharge piping, 200 PSI or 50 PSI above maximum static pressure, whichever is greater, 14.1.2.
5. ____|____|____ Pressure and flush test certification is provided before performing the field acceptance test, 14.1.3.
6. ____|____|____ The approved plan is on site.
7. ____|____|____ Fire pump and controller, piping, gauges, jockey pump, and other component locations and design are the same as shown on the approved set of plans.
8. ____|____|____ Fire pump has name plate.
9. ____|____|____ Wire installation to motor, control inner wiring, and jockey pump wiring is correct.
10. ____|____|____ A pressure gauge not less than 3 ½ in. diameter is near the pump discharge casting, and the pressure range is at least twice the rated working pressure of the pump but not less than 200 PSI, 5.10.1.
11. ____|____|____ A compound pressure/vacuum gauge not less than 3 ½ in. diameter is connected to the suction pipe and the pressure range is twice the rated maximum suction pressure of the pump but not less than 100 PSI, 5.10.2. This does not apply to vertical shaft-turbine pumps taking a water supply from an open pit or well.
12. ____|____|____ When provided, all valves (suction valve, discharge valve, bypass valves, backflow prevention device or assembly isolation valves) shall be supervised open by an off-site monitoring company, a local signal, locked open, or by seals, 5.16.
13. ____|____|____ Pump room has lighting, emergency lighting, heat, ventilation, and floor drain, 5.12.2 through 5.12.6.
14. ____|____|____ A circulation relief valve is provided on the pump of at least, ¾ in. for less than 2,500 GPM and 1 in. for 3,000 to 5,000 GPM, and it discharges to a drain, 5.11.1. This does not apply to pumps providing cooling water from its discharge to the engine driver.
15. ____|____|____ Coupling guards are provided for driver to pump connecting flexible couplings or flexible connecting shafts, 5.12.7.
16. ____|____|____ The operating angle of a flexible connecting shaft does not exceed the manufacturer listing requirements, 7.5.1.7.2.
17. ____|____|____ When installed, the eccentric taper reducer for suction has the taper on the bottom, 5.14.6.4.
18. ____|____|____ Suction screening is provided for open source water supplies, verify that its size matches what is detailed on the approved set of plans, 5.14.8.
19. ____|____|____ When a vortex plate is provided for taking suction from stored water supply, verify that it size and location matches what is detailed on the approved set of plans, 5.14.10.
20. ____|____|____ A check valve is installed in pump discharge assembly, 5.15.6.
21. ____|____|____ An indicating gate or butterfly valve is installed on fire protection system side of the check valve, 5.15.7.
22. ____|____|____ For a centrifugal pump and when provided a pressure relief valve is located between the pump and pump discharge check valve, verify that its location matches what is on the approved set of plans, 5.18.3.
23. ____|____|____ The test header and the number of hose valves are provided, in accordance with Table 5.25 and their location matches the approved set of plans, 5.19.3.

Fire Plan Review and Inspection Guidelines

24. ____|_____|____ The construction of the fire pump room (1 or 2 hour fire-resistive) matches the approved set of plans, 5.12.1, 5.12.2.
25. ____|_____|____ The pressure maintenance (jockey) pump has a check vale in its discharge piping and isolation valves (indicating butterfly or gate) location match the approved set of plans, 5.24.4.

Operational Tests are Performed by the Contractor or Manufacturer

26. ____|_____|____ Flow tests for positive displacement pumps is performed and recorded in accordance with Sections 14.2.7.3.3 and A.14.2.7.3, using a flow meter in a test loop that discharges the flow back to the supply.
27. ____|_____|____ For the load start test, the fire pump, without interruption, will be brought to rated speed providing a discharge equal to peak load, 14.2.7.4.

Controller

28. ____|_____|____ The fire pump controller is tested in accordance with the manufacturer's requirements and Section 14.2.8.
29. ____|_____|____ A minimum 6 manual starts and 6 automatic starts are performed, split the tests between each set of engine batteries and emergency power (only if emergency power is required for operating the pump) and simulate loss of the primary power source to verify the transfer to secondary power source, 14.2.8.2 and 14.2.9.
30. ____|_____|____ Each start is no less than a 5 minute run time, and total pump operation shall not be less than 1 hour, 14.2.8.4 and 14.2.12.
31. ____|_____|____ Simulate primary power loss and allow automatic transfer to secondary power supply (only if emergency power is required for operating the pump) while pump is operating at peak load, 14.2.9.1.
32. ____|_____|____ Engines with electronic fuel management control systems will test both primary and alternate control systems, 14.2.13.
33. ____|_____|____ Pump packing drips.
34. ____|_____|____ No overheating.
35. ____|_____|____ No excessive vibration.
36. ____|_____|____ Pump starts on water flow.
37. ____|_____|____ Pump starts on pressure drop.
38. ____|_____|____ Casing relief valve operates.
39. ____|_____|____ Pressure relief valve operates.
40. ____|_____|____ Jockey pump stop point pressure is recorded.
41. ____|_____|____ Jockey pump start point pressure is recorded.
42. ____|_____|____ Fire pump start point pressure is recorded (usually 5 PSI above jockey stop PSI).
43. ____|_____|____ Pump flow tests are conducted at churn (no flow), rated (100 percent of rated capacity), and peak (150 percent of rated capacity) loads, 14.2.7.2.1. Additional test points can be taken.

Electric Driven Pump

44. ____|_____|____ Supervised alarms operate when motor stops running, loss of phase, electric phase reversal and controller trouble.
45. ____|_____|____ Simulated test for phase reversal is conducted.
46. ____|_____|____ Switching from normal power to emergency and back to normal at peak load does not trip the breaker.
47. ____|_____|____ Pump started once from manual emergency handle operation.
48. ____|_____|____ Pump start up on emergency power occurs automatically.

Diesel Driven Pump

49. ____|_____|____ Audible alarms operate when overspeed (120 percent) causes shutdown, low oil PSI, high temp, battery failure, charger failure, low air or hydraulic PSI, and failure to automatically start.
50. ____|_____|____ Audible or visual alarms provided at constantly attended location when engine stops running, controller main switch is turned off, or there is trouble on the controller or engine.
51. ____|_____|____ Instrumentation panel includes tachometer, oil PSI gauge, and temperature gauge.
52. ____|_____|____ Battery chargers and ampmeters function.
53. ____|_____|____ Charger listed for fire pump service use, batteries rack is supported and secured at least 12 in. above the floor level.
54. ____|_____|____ Timer set for 30 minute each week run time cycle.

55. ____|____|____ For automatic shut down after an automatic start, the shutdown occurs in accordance with Section 12.5.5.2.

Well Test and Inspection for Vertical Turbine Pumps

56. ____|____|____ The well's production capability is verified by a continuous 8 hour test at 150 percent of the pump rated capacity. Test readings are taken every 15 minutes and the test data provides the static and pumping water levels at 100 and 150 percent of the pump's rated capacity, 7.2.7.

Additional Comments:

Inspection Date: _____ Approved or Disapproved FD Inspector: _____

Inspection Date: _____ Approved or Disapproved FD Inspector: _____

Inspection Date: _____ Approved or Disapproved FD Inspector: _____

Fire Plan Review and Inspection Guidelines

Fire Pump
Plan Review Worksheet
2006 IFC 913 and 2007 NFPA 20
This worksheet is for jurisdictions that permit the use of the 2007 NFPA 72 in lieu of IFC's referenced 2003 NFPA 20.

Date of Review: _____ Permit Number: _____

Business/Building Name: _____ Address of Project:: _____

Designer Name: _____ Designer's Phone: _____

Contractor: _____ Contractor's Phone: _____

Occupancy Classification: _____ _____

Reference numbers following worksheet statements represent an NFPA code section unless otherwise specified.

Worksheet Legend: ✓ or **OK** = acceptable   **N** = need to provide   **NA** = not applicable

1. _____ Three sets of drawings are provided.
2. _____ Equipment is listed for intended use, product listing data sheets are provided, 5.1.2.1, 5.7.
3. _____ A copy of a fire hydrant flow test summary sheet is provided, which includes static and residual pressures, flow rate, and location of test hydrant(s).

Drawings Shall Show the Following:

4. _____ Scale:_____ , a common architectural scale is used and the plan information is legible.
5. _____ Plan view and cross sectional views of installed equipment are provided, 5.2.
6. _____ Room dimensions are provided.
7. _____ Equipment symbol legend is provided.
8. _____ Suction pipe flushing requirements from Table 14.1.1.1 are on the plans.
9. _____ Plot plan illustrating connection to the water supply including the pipe diameter and the pipe routing (including any elevation changes) from the source to the fire pump.
10. _____ Driver, pump, and controller manufacturer, respective models, and driver type are specified.
11. _____ Copy of the factory pump test curve is provided, 5.5.
12. _____ Pump GPM rating: _____ Head rating: _____ RPM: _____ are provided.
13. _____ A pressure gauge complying with 5.10.1 is detailed as installed near the discharge casting.
14. _____ A compound and vacuum pressure gauge complying with 5.10.2.1 are detailed as installed to the suction pipe, (does not apply to vertical shaft turbine-type pumps taking suction from a well or open pit).
15. _____ A automatic relief valve complying with 5.11.1 is detailed as installed on the discharge side of the pump before the discharge check valve and it discharges to the drain. This requirement does not apply to engine-driven pumps provide cooling water from its discharge to the engine.
16. _____ A nonsprinklered fire pump room is separated by a 2 hour fire barrier. A sprinklered fire pump room in a nonsprinklered building is separated by a 2 hour fire barrier. A sprinklered fire pump room in a sprinklered building is separated by a 1 hour fire barrier. An exterior fire pump building is separated from adjacent buildings by at least 50 ft., Table 5.12.1.1.2.
17. _____ Fire pumps in high-rise buildings are separated by 1-hour fire-rated construction, 5.12.1.1.1.
18. _____ The fire pump room containing a diesel pump driver and fuel storage tanks is protected by an automatic sprinkler system in compliance with NFPA 13. 5.12.1.3.
19. _____ A fire pump installed outdoor is located at least 50 ft. from any building that would be an exposure, 5.12.1.2.
20. _____ When required by the environment or engine manufacturer, the fire pump room has a heat source in accordance with Section 5.12.2.
21. _____ Emergency lighting for the fire pump room is provided in accordance with Section 5.12.4.
22. _____ Ventilation is provided in the pump room, 5.12.5.
23. _____ The pump room floor is adequately pitched to provide drainage and it drains to a frost free location, 5.12.6.
24. _____ The coupling guards for the flexible couplings or shaft connections between the pump driver and pump are noted or illustrated and in compliance with Section 8 of ANSI B15.1, 5.12.7.
25. _____ If used, the operating angle of a flexible connecting shaft is detailed and does not exceed the manufacturer listing requirements, 7.5.1.8.2.
26. _____ Size and type of pump suction and discharge pipe used are specified and detailed.
27. _____ Steel pipe size is specified for aboveground pipe, 5.13.1.

28. _____ The method of joining the steel pipe is specified, 5.13.2.

Suction Pipe and Fittings:
29. _____ The size and arrangement of the suction pipe complies with Table 5.25 (a or b), 5.14.3.4.
30. _____ The suction piping arrangement and OS & Y gate valve complies with Section 5.14.5.
31. _____ The installation of elbows and tees shall be in accordance with Section 5.14.6.3.
32. _____ If provided, eccentric taper reducer or increaser (for suction pipe and pump flange size differentials) is detailed and complies with Section 5.14.6.4.
33. _____ Open source water supplies shall be equipped with a suction screen in accordance with Section 5.14.8 and 5.14.9.
34. _____ Screens for open water source are removable and the screen material is specified, 5.14.8.6.
35. _____ Screens have at least a 62.5 percent open area, 5.14.8.11.
36. _____ When devices are installed in the suction piping they shall comply with Section 5.14.9.
37. _____ A vortex plate is provided and detailed on the suction fitting that obtains water from a stored water supply, 5.14.10.

Discharge Pipe and Fittings:
38. _____ The size of the pump discharge pipe and fittings are in accordance with Table 5.25 (a or b), 5.15.5.
39. _____ A listed check valve or backflow prevention device is in the pump discharge assembly, 5.15.6.
40. _____ Indicating gate or butterfly valve is on system side of the check valve, 5.15.7.

General:
41. _____ When required, a pressure relief valve for centrifugal pump is provided in accordance with Section 5.18.1.
42. _____ The pressure relief valve's listing data sheet is provided and the valve is either spring loaded or pilot-diaphragm type, 5.18.4.
43. _____ Pressure relief valve discharge is designed in accordance with Section 5.18.5.
44. _____ The size of the discharge pipe is in accordance with Table 5.25 (a or b), 5.18.5. If the pipe has more than 1 elbow, enlarge the pipe one size.
45. _____ The test header pipe diameter, number of hose valves, or the flow meter size and piping are detailed in compliance with Table 5.25, 5.19.
46. _____ When provided, the location of the backflow prevention device is detailed, and the listing data sheet and friction loss information are provided, 5.26.1.
47. _____ The pressure maintenance (jockey or make-up) pump location and piping are detailed and specification data sheets are provided, 5.24.
48. _____ A check valve in the pressure maintenance pump discharge pipe is detailed and the location of gate or butterfly valves for allowing component repair are detailed, 5.24.
49. _____ Where located, check valves and backflow prevention devices or assemblies are located a minimum 10 pipe diameters from the pump suction flange, 5.26.3.
50. _____ For seismic design areas, the fire pump, driver and associated equipment and piping is provided earthquake protection in accordance with Section 5.27 and seismic calculations for each method of protecting equipment are provided.
51. _____ Packaged fire pump, house, and skid/unit are in compliance with 5.28 and design details are provided.
52. _____ Pressure sensing line details are provided and the lines are located between the pump discharge check valve and discharge control valve, 5.29.
53. _____ Break tank use is as a backflow prevention device or to eliminate city water pressure fluxuations or to augment the city water supply, 5.30.1.
54. _____ Break tank capacity provides at least 15 minutes of water at 150 percent of the pump's rated capacity, 5.30.2.
55. _____ The design of the break tank refill equipment is in accordance with 5.30.3.
56. _____ The break tank installation complies with NFPA 22, 5.30.4.

Centrifugal Pumps:
57. _____ The selected centrifugal pump is specified and meets the design requirements of 6.1.1.
58. _____ The application of the centrifugal pump complies with the requirements of Section 6.1.2.
59. _____ When required, the automatically controlled centrifugal pump has a float operated air release valve at least ½ in. diameter, 6.3.3.
60. _____ The foundation and setting for the pump are detailed and in compliance with Section 6.4.
61. _____ The method of securing the pump base plate to the foundation is detailed, 6.4.3.

Fire Plan Review and Inspection Guidelines

Vertical Shaft Turbine-Type Pumps:

62._____ Detailed for well installations is the submergence level of the second pump impeller level being at least 10 ft. below the water level and 1 ft. submergence is added for each 1,000 ft. of elevation, 7.2.2.1.1, 7.2.2.1.2.

63._____ Detailed for wet pit installations is the submergence level of the second pump impeller level being below the lowest pumping level of the open body of supply water. A greater submergence is required for pumps rated 2,000 GPM or greater. Obtain submergence depth requirement data sheet from manufacturer 7.2.2.2.1 to 7.2.2.2.4.

64._____ The well casing, screen, and suction strainer are detailed, 7.2.3, 7.3.4.

65._____ A report verifying the well can produce the appropriate quantity of water supply for the specified pump is provided, 7.2.3.1.

66._____ The dimensions of the well, its casing and casing materials, well screen, fill gravel around the well screen, method of sealing the well bottom are detailed, 7.2.4.

67._____ Specified is whether the well is in consolidated or unconsolidated formations, 7.2.4.

68._____ A certified performance test report of the well is provided, 7.2.7.

69._____ The tubular well for fire pumps 450 GPM or less is designed in compliance with 7.2.3 and .4 except 7.2.4.11 through 7.2.4.15, 7.2.4.16.2.

70._____ The suction strainer has a free area at least 4 times the area of the suction connection and the screen can restrict passage of a .5 in. sphere, 7.3.4.2.

71._____ The air relief valve and size, water level detector, pressure gauges, relief valves, hose valve header, valves or metering device locations are detailed and in conformance with Section 7.3.5.

72._____ The well is equipped with a water level detector, 7.3.5.3.

73._____ The pump foundation, support, anchoring, etc. design is detailed on the plans and in compliance with Section 7.4.3.

Positive Displacement Pumps:

74._____ The pump is listed for its intended use and the listing verifies the pump's performance curves, 8.1.2.

75._____ When installed on a closed head fire system a listed dump valve type is specified, and detailed in accordance with Section 8.1.6.

76._____ When provided, foam concentrate and additive pumps installations are detailed in conformance with 8.2. Pump data sheets are provided.

77._____ When provided, water mist pump installations are detailed in conformance with 8.3. Pump data sheets are provided.

78._____ Detailed are compound suction and discharge pressure gauges, and a listed safety relief valve locations, 8.4.

79._____ A pump suction strainer is provided and is in compliance with the requirements of 8.4.5.

80._____ The pump foundation, support, anchoring, etc. design is detailed on the plans and in compliance with Section 8.7.

81._____ A means for flow testing is provided and the piping schematic is provided, 8.9.

Driver Information:

82._____ Type: _____ Manufacturer: _____ Model: _____ Rated H.P.: ____ RPM: _____ are provided.

83._____ If the pump uses a diesel driver, calculations indicating the number of hours of fuel supply are provided.

Controller Information:

84._____ Manufacturer: _____ Model: _____ are provided.

Electric Drive and Electrical, complies with National Electrical Code Article 695:

85._____ Normal power is arranged in compliance with one of the 5 choices offered in Section 9.2.2.

86._____ Detailed is only one disconnection means when using a power arrangement of 9.2.2(1) or (2) or (3) or (5), 9.2.3. The disconnecting means meets the 5 criteria in Section 9.2.3.1.

87._____ Detailed or noted on the plans is the size, verbiage, and location of the disconnect placard i.e., adjacent the fire pump controller, stating the location of the disconnect means and unlocking key if required, 9.2.3.2.

88._____ The disconnect means permitted in 9.2.3 has the supervision method, of the closed position, described on the plans, 9.2.3.3.

89._____ If a secondary power supply is provided, an electrical schematic for the circuit and transfer equipment is provided, 9.3.

90._____ A second power source is provided in accordance with 9.3 when the building height exceeds fire apparatus pumping capability and an electric driver is used, 9.3.1, 9.3.3.

91. _____ The electric motor is listed for fire pump service and meets the construction, horsepower and locked rotor current requirement of Section 9.5.1.1.
92. _____ When an on-site generator is required to meet the power reliability requirements of NFPA 20, it has the capacity to run under the loads identified in Section 9.6.1. The loads are specified and provided.
93. _____ Required generator(s) shall comply with the foundation requirements of Section 6.4, be of a level, type, class specified in Section 9.6.2. The system shall be designed in accordance with NFPA 110, and have a minimum fuel supply to operate the fire pump at its 100 percent rated capacity, 9.6.2.
94. _____ Transfer of power shall occur in the pump room, 9.6.4.
95. _____ The controller installation is detailed. It is located near and in sight of the motors it controls and energized controller components are provided working clearances in accordance with the National Electrical Code Article 110, NFPA 20 10.2.
96. _____ The fire pump controller is listed for use with an electric motor-driven fire pump and labeled in accordance with Section 10.1.2.1.
97. _____ The controller and accessories are mounted on a single noncombustible support foundation, 10.3.2.
98. _____ Enclosures for the controller and accessories are in compliance with Section 10.3.3.
99. _____ Controllers shall be provided with voltage surge arrestor, isolating switch, circuit breaker, locked rotor protection, and motor contacts in accordance with Sections 10.4.1 through 10.4.5.
100. ____ Provided and detailed is an alarm circuit and a signal device at a constantly attended location when the pump room is not constantly attended. The alarm signal transmission occurs in accordance with Sections 10.4.7.2.1– 10.4.7.2.4, 10.4.7.
101. ____ When required, the dedicated fire pump transfer switch location is detailed, the listing data sheets are provided, and the design complies with Section 10.8.3.
102. ____ When required, one dedicated transfer switch is assigned to a fire pump, 10.8.2.3.
103. _____ If used, the design details of a controller with variable speed pressure limiting control are in compliance with Sections 10.10.1– 10.10.11.

Diesel Drive:
104. ____ The engine is a compression ignition type and is listed for fire pump service, 11.1.3.1 and 11.2.1.
105. ____ The engine meets the rating requirements of Section 11.2.2.
106. ____ The engine connection to the fire pump is noted and designed in compliance with Section 11.2.3.
107. ____ The engine is equipped with a governor complying with the requirements of Section 11.2.4.1.
108. ____ When the engine uses a variable speed pressure limiting control system, it is noted on the plans and complies with Section 11.2.4.2.
109. ____ The engine is equipped with overspeed shutdown device that complies with Section 11.2.4.3.
110. ____ The engine is equipped with an instrument panel containing: tachometer, oil pressure gauge, and temperature gauge, 11.2.4.4 –11.2.4.7.
111. ____ Batteries have two means of recharging detailed and the chargers are listed for fire pump service and comply with the requirements listed in Sections 11.2.5.2.4 and 11.2.5.2.5.
112. ____ Detailed or noted on the plans is that each engine has two batteries that are rack supported, and current carrying-parts (cables) are not less than 12 in. above the floor, 11.2.5.2.2.1. and 11.2.5.2.6
113. ____ The engine cooling system is closed-circuit liquid type and is specified as radiator or heat exchange type, 11.2.6.
114. ____ Adequate ventilation is provided for the pump room and the engine, 11.3.2.
115. ____ Fuel supply tank capacity calculations are provided and are at least 1 gallon per horsepower plus 5 percent volume for expansion and 5 percent volume for sump, 11.4.3 and the fuel supply tank design complies with IFC 34.
116. ____ Fuel piping is designed in compliance with Section 11.4.6.
117. ____ The controller is listed for use with diesel engine-driven fire pumps and labeled in accordance with Section 12.1.3.
118. ____ The controller installation is detailed. It is located near and in sight of the engine it controls and energized controller components are provided working clearances in accordance with the National Electrical Code Article 110, 12.2.2 –12.2.2.4.
119. ____ The controller and accessories are mounted on a single noncombustible support foundation, 12.3.2.
120. ____ Enclosures for the controller and accessories are in compliance with Section 12.3.3.
121. ____ Provided and detailed is an alarm circuit and a signal device(s) in the engine room. The visible indicators and a common alarm signal occurs in accordance with events listed in Sections 12.4.1.3 and 12.4.1.4, 12.4.

122. ____ When the pump room is not constantly attended, the alarm and signal devices are remote from the controller, in a constantly attended location, and are detailed and designed in accordance with Section 12.4.2.
123. ____ Engine exhaust is vented to the exterior and where the exhaust will not harm persons or endanger buildings, 11.5.
124. ____ Engine exhaust piping connections, diameter, clearances to combustible materials, and termination points are detailed and design in accordance with Section 11.5.3.

Additional Comments:

Review Date: _____ Approved or Disapproved FD Reviewer: _____

Review Date: _____ Approved or Disapproved FD Reviewer: _____

Review Date: _____ Approved or Disapproved FD Reviewer: _____

Stationary Fire Pump - Field Acceptance Form

Business Name:	
Business Address:	
Testing/Inspection Co:	
Tested/Inspected By:	Date:
Fire Department Rep:	
Type of Test: ( ) Quarterly ( ) Annually ( ) Acceptance	

General Info

Pump

Manufacturer	Model	S/N
Rated GPM	Rated PSI/Head	Rated RPM

Driver

Manufacturer		Model	S/N	
Rated HP	Rated RPM	Actual RPM		Driver Type

Motor

If Electric: Rated voltage	Rated Amperage
If Diesel: Fuel Supply for 8 hrs. (1gal/HP)	Gallons

Controller

Manufacturer	Model	S/N
Fire Pump: Start_____PSI Stop_____PSI	Jockey Pump: Start_____PSI Stop_____PSI	City Static: Pressure_____PSI

Water Flow

Capacity	Streams			GPM	Suction	Discharge	Net	Amp/Leg			RPM	Run Time
	No.	Size	Pitot PSI		PSI	PSI	PSI	1	2	3		
No Flow												
100%												
150%												

Flow Meter

Capacity				GPM	Suction	Discharge	Net	Amp/Leg			RPM	Run Time
					PSI	PSI	PSI	1	2	3		
No Flow												
100%												
150%												

Pump is Connected to a Legally Required Standby Power

Manufacturer	Model

Capacity	Streams			GPM	Suction	Discharge	Net	Amp/Leg			RPM	Run Time
	No.	Size	Pitot PSI		PSI	PSI	PSI	1	2	3		
No Flow												
100%												
150%												

Additional Comments:

Fire Plan Review and Inspection Guidelines

Fire Pump Installation Certification

Permit #: _____ Date: _____

	Property Protected	System Installer	System Supplier
Business Name:	_____	_____	_____
Address:	_____	_____	_____
	_____	_____	_____
Representative:	_____	_____	_____
Telephone:	_____	_____	_____

Location of Plans: _____

Location of Owner's Manual: _____

1. <u>Certification of System Installation:</u> Complete this section after system is installed, but prior to conducting operational acceptance tests. Check wiring for opens, ground faults, and improper branching.

   This system installation was inspected and found to comply with the installation requirements of:
   _____ NFPA 20 and National Electric Code
   _____ Manufacturer's Instructions
   _____ Other (specify; FM, UL, etc.) _____

   Print Name: _____

   Signed: _____ Date: _____

   Organization: _____

2. <u>Certification of System Operation:</u> All operational features and functions of this system were tested and found to be operating properly in accordance with the requirements of:
   _____ NFPA 20 and National Electric Code
   _____ Design Specifications
   _____ Manufacturer's Instructions
   _____ Other (specify) _____

   Print Name: _____

   Signed: _____ Date: _____

   Organization: _____

Prefinal and Certificate of Occupancy Inspection Requirements For Contractors
Contractors Worksheet

Fire Pump Test Requirements

1. All certification forms and documents are required to be on the site for review:
   - ___ Plans
   - ___ Permit
   - ___ Prefinal test: churn, 100%, 150% flow test, alarms, and trouble signals. Use the Acceptance Inspection worksheet for the pretest.
   - ___ Installation certification form is completed, use the form contained in this book or NFPA's "Record of Completion" form, in NFPA 20.
2. ___ A person familiar with the installation must be present to conduct test.
3. ___ Owner's representative approval needed for time and date of testing.
4. ___ Room or area is unlocked and accessible.
5. ___ Equipment placement is in accordance with the plan.
6. ___ If Items 1-5 are incomplete, inspection is cancelled and another inspection request is required. A reinspection fee may be assessed.

Prior to the next approval test:

7. ___ When there are equipment additions or relocations, the contractor must provide:
   - ___ As-builts are provided for review and approval.

   Note: New plan review will be submitted as "supplemental information" and proof of the additional review fee payment is required.
8. ___ A reinspection fee may be assessed if the system and paperwork are not ready.

Fire Pump Testing Guideline

General:

This guideline was developed to assist both service contractors and fire personnel with the procedures and methods of witnessing, performing, and documenting the testing of fire pumps.

Fire pumps are required to be tested upon installation and annually. The performance evaluation test is in accordance with the requirements of NFPA 20, NFPA 25, and the International Fire Code.

Definitions:

<u>Field performance test</u>: a fire pump test made after a new installation and where major pump repairs have been performed. Tests should be witnessed by Fire Department personnel. Installing contractors or manufacturer representatives of the system should conduct the test.

Test Criteria:

NFPA Standard 20 has three specific flow check points that must be performed before a fire pump system can be approved. These are:

- At churn, shutoff, or zero flow, the net pressure should not exceed 140 percent of the rated pressure of the horizontal split-case pumps. The net pressure should not exceed 140 percent of the rated pressure of end-suction and vertical shaft turbine-type fire pumps.
- At 100 percent of the rated pump capacity, the net pump pressure should be 100 percent of the rated pressure.
- At 150 percent of the rated pump capacity, the net pump pressure should be at least 65 percent of the rated pump pressure.

Test Equipment:

- All test instruments shall have been calibrated within the last twelve months.
- Sufficient 2 ½-inch or greater diameter fire hose to reach from the pump test header to either the playpipe nozzles or portable monitor.
- One underwriter-style playpipe for each 250 GPM of pump capacity.
- One pitot tube with a calibrated 150 PSI pressure gauge.
- Two calibrated pressure gauges incremented for a 0-300 PSI range.
- One hand-held tachometer.
- Voltage amperage meter.
- Fire line barrier tape.
- Two portable radios.

Fire Pump Performance Test:
1. Inspect the room or building housing the fire pump to ensure compliance with the adopted codes and plan design.
2. Inspect the fire pump, driver, piping, jockey pump and the controller to ensure compliance with the adopted codes and plan design.
3. Survey the site of the pump test to determine how the pump discharge will drain or be disposed of. The discharge from the smallest fire pump can approach 750 GPM and contain rocks and debris. If necessary, secure the discharge area from vehicular or pedestrian traffic.
4. If the pump is monitored, notify the monitoring company that a fire pump test is being conducted and that an alarm condition will be transmitted. During the pump test, verify that an alarm condition is being received.
5. Isolate the fire pump from the fire protection system by closing the discharge OS&Y valve, and opening either the indicating control valve to either the pump test header or pump flow loop, depending on the type of installation.
6. If a test header is used, connect the playpipes with nozzles and hose lines to the pump test header. The playpipes should be secured to a test bracket near the test header. If a permanent playpipe bracket is not provided, a portable playpipe bracket can be used.
7. Replace the pump's suction and discharge gauges with calibrated test gauges. Teflon tape should be wrapped around the gauge threads to ensure an adequate seal.
8. Inspect the suction line to ensure that the control valves are in their proper positions and that the pump is primed with water. If the pump needs priming, follow the manufacturer's directions for priming the pump.
9. If the pump utilizes an electric driver, a volt-amperage meter will be used to measure the amperage for each of the three conductors connected to the driver. These conductors are located in the pump controller.
10. Depending on the type of equipment used, access to the driver end plate may be required to measure the motor's RPM.
11. Allow approximately 30 seconds between each successive manual start. During this portion of the test, verify that an alarm condition is being transmitted.
12. To begin the automatic start portion of the test, open the fire protection system control valve slowly allowing a drop in the system pressure. The pump should start automatically.
13. Upon completion of the automatic start portion of the test, slowly open the 2 inch main drive valve of the automatic sprinkler system to cause a water-flow condition. This test should activate the pump.
14. During the time period that Steps 11, 12, and 13 are being performed, check the pump for excessive noise, vibration, and leaks. Other areas to check include the following:
    - Operation of the circulating relief valve.
    - Pump packing is emitting one drop of water per second.
15. While operating, record the pressure registered at the discharge and intake gauges, the driver RPM, and if applicable, the amperage measurements.

16. Fire Pump Standpipe Test: The roof personnel will remove the hose valve caps and notify the test coordinator to start the fire pump and allow water to flow in the standpipe. This is to allow for flushing of the system. Personnel should be located behind the hose valve to avoid being struck by debris. Continue flushing until the water is clear. Upon completion, notify the Control Officer to shutdown the pump.

17. Connect the hoselines, playpipes, and gate valves to the hose valve connection. When discharging water, attempt to direct the water discharge toward the roof drains, or toward an open perimeter area. If discharge is above the building, secure the discharge area from vehicles and pedestrians.

18. If applicable, open the indicating control valve to the floor test loop. This is located between the suction and discharge piping below the pump bypass. Activate the flow meter; do not attempt to calibrate or adjust the device.

19. Open the fire protection system indicating control valve to charge the system.

    Fully open a 2-1/2 inch gate valve slowly and pitot the flow. Notify the Control Officer of the pitot measurement. During this period, the drive RPM, discharge pressure, suction pressure, flow meter measurement, and amperage (if applicable) should be recorded.

    Open each successive hoseline as instructed by the Control Officer and measure the flow using the pitot tube. After opening each hoseline, measure the flow from each hoseline previously opened. EXAMPLE: If opening hose valve number 3; pitot hose valves one and two after measuring the flow from hose valve three. Record the pitot measurements and calculate the total flow.

20. Review the discharge information of hoseline #1 in comparison to the flow meter reading. Evaluate the two measurements and, if necessary, have the test site representative calibrate the flow meter.

21. Determine the net pump pressure from the flow measurements taken in Step 19.

22. Adjust the gate valve on each hose valve so that the total flow equals 100 Percent of the pumps' rated capacity. Adjust the individual gate valves so that the pitot measurements equals the data calculated by the Control Officer.

    Once the specified pitot pressure is achieved, notify the Control Officer. Measure the driver RPM, discharge and suction pressures, flow meter reading, and amperage, if applicable.

23. Once the necessary measurements are recorded and the field calculated results satisfy the Control Officer, shut down the roof hoselines and remove all of the equipment.

24. Remove the calibrated test gauges and the volt-ampmeter after closing the indicating control valve to the flow test loop. Ensure that the fire protection system control valve is open and electrically supervised. Place the fire pump controller in the automatic mode and re-inspect the entire installation to verify that the system is operative and will function in case of fire.

25. Notify the property representative that the test is complete and that the test results will be forwarded. Notify the alarm monitoring agency and Fire Dispatch that the test is complete.

Miscellaneous Plan Review Subjects
Contents

1. Spray Finishing, Powder Coating and
   Electrostatic Worksheet 281

2. Motor Vehicle Fuel-dispensing Station Worksheet 284

3. Above-ground Fuel Tank, Plan Review Worksheet 287

4. Underground Fuel Tank Plan Review Worksheet..............289

5. Medical Gases, Plan Review and Inspection Worksheet...291

Fire Plan Review and Inspection Guidelines

**Spray Finishing, Powder Coating, and Electrostatic Apparatus
Plan Review or Acceptance Inspection Worksheet**
2006 IFC, Chapter 15 and 2003 NFPA 33

Date of Review: _____ Permit Number: _____

Business/Building Name: _____ Address of Project: _____

Designer Name: _____ Designer's Phone: _____

The numbers that follow worksheet statements represent a IFC code section unless otherwise stated.

Worksheet Legend: ✓ or OK = acceptable N = need to provide, NA = not applicable

1. _____ Three sets of drawings are provided for the plan review process.

The Following Information Shall be Submitted for Review:
2. _____ Manufacturer's drawings and data sheets are provided for premanufactured booths.
3. _____ "No Smoking" signs are provided, 1503.2.6.
4. _____ "No Welding" signs are provided, 1503.2.7.

Spray Booth or Room Construction:
5. _____ With the exception of automobile undercoating operations, limited spraying areas or resin application areas used for the manufacturing of reinforced plastics, spray room operations located in A, E, I, or R occupancies are located in a spray room protected with an approved sprinkler system and separated from other areas by construction specified in the IBC, 1504.2.
6. _____ Floors for spray room or booth are noncombustible or covered with a nonsparking material, 1504.3.1.1, 1504.3.2.3.
7. _____ The booth is constructed of steel not less than .0478-inch (18 gauge) and for two-layer metal assemblies each sheet is not less than .0359-inch (20 gage), 1504.3.2.1.
8. _____ Interior surfaces are smooth, 1504.3.2.2.
9. _____ Aluminum shall not be used for interior surfaces of spray booths or rooms, 1504.3.2.2
10. _____ Premanufactured spray booth exit doors shall have a minimum width of 30 in. and minimum height of 80 in., 1504.3.2.4. Spray room exits comply with IBC Chapter 10.
11. _____ The booth shall be separated at least 3 ft. from other operations and construction unless the booth is adjacent to a 1-hour fire-resistive wall or a noncombustible exterior wall, 1504.3.2.5.
12. _____ The aggregate square footage for multiple booths does not exceed 10 percent of the floor area or the basic area allowed for a Group H-2 occupancy. The area of a single spray booth shall not exceed 1,500 sq. ft., 1504.3.2.6.

Electrical Equipment and Wiring:
13. _____ Spray spaces and vapor areas have wiring and equipment designed for hazardous (classified) locations. Such locations are Class I, Division 1 or Class II, Division I locations, 1503.2.1.1.
14. _____ Electrical wiring and equipment outside of but within 5 ft. horizontally and 3 ft. vertically of openings in a spray booth or a spray room shall be approved for Class I, Division 1 or Class II, Division I locations, 1503.2.1.3.
15. _____ The lighting through glass panels or other transparent materials is fixed and protected by heat treated or wired glass, and any integral luminaires are listed for Class I, Division 2 or Class II, Division 2 locations, 1504.6.2.
16. _____ Luminaries located outside of the spray booth shall be equipped with vapor tight seals. Exterior luminaries shall be listed for ordinary hazard locations, 1504.6.2.2.

Mechanical Ventilation:
17. _____ Spray area ventilation is designed to be on at all times during spraying and for a period of time after spraying, 1504.7 and IMC 510.
18. _____ Spray equipment is interlocked with ventilation such that spraying can not occur unless ventilation is operating, 1504.7.1 and IMC 510.
19. _____ Air exhausted from a spray area shall be not recirculated unless the spraying operation occurs in a unmanned spray area, solid particulates are removed, the atmosphere in the spray area is maintained at less than 25 percent of the lower flammable limit, approved equipment is provided to monitor the vapor concentration and in the event the LFL is > 25%, and alarm is transmitted an alarm and shuts down the

Fire Plan Review and Inspection Guidelines

automatic spraying operation, 1504.7.2 and IMC 502.7. In occupied booths recirculation is permitted when all of the requirements of 1504.7.2 are satisfied and documentation is provided demonstrating the atmosphere does not pose a life safety hazard to personnel inside of the spray booth, room, or area.

20. ____ The ventilation design provides a minimum airflow velocity of 100 linear feet/minute (FPM) measured using the cross-sectional area of the spray booth, 1504.7.3 and IMC 502.7.
21. ____ Each booth and spray room have an independent exhaust duct system discharging outside unless multiple booths with a combined frontal area does not exceed 18 sq. ft. and if more than 1 fan serves one booth, fans are interconnected to operate simultaneously, 1504.7.5 and IMC 502.7.
22. ____ Ducts conveying flammable vapors are terminated 30 ft. from the property line, 10 ft. from openings, 6 ft. from walls and roofs, 30 ft. from combustible walls or openings into buildings which are in the direction of the exhaust discharge, and 10 ft. above grade, 1504.7.6 and IMC 502.7.
23. ____ Details of exhaust duct doors, panels, or other means that permit inspection, maintenance, cleaning or access to fire protection devices are provided, NFPA 33: 7.9.
24. ____ Other product conveying outlets terminate 10 ft. from the property line, 10 ft. from openings, 3 ft. from walls and roofs, and 10 ft. above grade, 1504.7.6 and IMC 502.7.
25. ____ Fan motor locations are detailed and verify the motors are not inside the booth or duct and spec sheets are provided verifying fans are nonferrous or nonsparking, 1504.7.7 and IMC 502.7.
26. ____ Air intake filters that are part of a wall or ceiling assembly are listed as Class I or II in accordance with UL 900. Equipment data sheets are provided, 1504.7.8.
27. ____ Filter supports are of noncombustible materials, 1504.7.8.1.
28. ____ Gauges and alarm locations for ensuring air velocity are maintained, and detailed, 1504.7.8.3.
29. ____ Booths using automatic dry filter rolls shall advance the filter when the air velocity is less than 100 lineal feet/minute. If the automatic filter roll fails to advance the spray operation shall shutdown, 1504.7.8.4.

Fire Protection:
30. ____ Booths, exhaust ducts, and both sides of dry filters shall be protected by a fire-extinguishing system, 1504.4.
31. ____ If automatic sprinkler protection is used, the sprinklers shall be protected from residue, 1504.5.2.
32. ____ Fire protection systems protecting automated spray operations shall be interlocked to stop the spray operations and workpiece conveyors serving the flammable vapor areas. If provided, activation of the fire protection system shall activate the fire alarm system, 1504.8.1.
33. ____ Each automated spray operation shall be equipped with a manual fire alarm and emergency shutdown station, 1504.8.1.1.
34. ____ Air makeup and spraying area exhaust systems shall not be interlocked with the fire alarm system and remain operational during a fire alarm condition, 1504.8.2.
35. ____ The size and placement of portable fire extinguishers shall be based on the requirements for an extra (high) hazard occupancy, IFC 1504.4.1, 906.3.

Spray Booths or Rooms with Drying Operations:
36. ____ Spraying equipment, drying apparatus, and ventilating system are equipped with interlocks to: (1504.6.1.2.1)
    ____ A. prevent spraying while drying.
    ____ B. purge spray vapors 3 minutes before drying.
    ____ C. cause drying shutdown on ventilation failure.
    ____ D. cause drying shutdown when booth exceeds 200^0F.
37. ____ In spray booths or rooms equipped with drying apparatus, the spray applicable shall be limited to low volume spraying equipment, 1504.6.1.1.

Storage, Handling and Use of Flammable and Combustible Liquids:
38. ____ The container sizes (closed type or provided with covers) that supply spray nozzles shall be limited to individual volume of 10 gallons, 1503.3.1.
39. ____ Bonding shall be provided when transferring flammable liquids from one container to another, 1503.3.3
40. ____ Piping systems for Class I and II liquids shall be permanently grounded, 1503.3.3.
41. ____ Class I liquids used for cleaning shall be used in equipment listed and approved for such purposes in accordance with the requirements in Section 1503.3.5.1. When Class I liquids are used for cleaning spray nozzles and equipment, the spray booth or room ventilation system shall be operated, 1503.3.5.1.

Powder Coating:
42. ____ Unless powder coating is performed in a listed spray booth assembly, powder coating shall be performed in a ventilated enclosed room or booth constructed of noncombustible materials or in a room designed in accordance with Section 1504.3 (IFC Section 1506.3).

43._____ An approved automatic fire-extinguisher systems shall be provided in powder coating areas in accordance with Chapter 9, 1506.4.

44._____ Automated powder application equipment shall be equipped with a supervised flame detection device that responds to an open flame within 0.5 second. Activation of the flame detection shall: 1) shut down electrical power and compressed air to the conveyor, ventilation system, and the powder coating equipment, 2) close segregation dampers in ductwork, 3) activate an audible alarm in the powder coating room or booth, 1506.4.1. Plans and equipment data sheets for the flame detection system shall be submitted in accordance with IFC Section 907.1.1.

45._____ Ventilation is designed to maintain the atmosphere less than 50 percent of the minimum explosive concentration of the dry powder. Plans should include the material data sheets for the powder(s) that will be applied and the ventilation design data, 1506.7.

46._____ The size and placement of portable fire extinguishers shall be based on the requirements for an extra (high) hazard occupancy,, IFC 1506.4.2, 906.3.

Electrostatic Apparatus:

47._____ With the exception of high-voltage grids, electrostatic atomizing heads and connections to the atomizing heads, transformers, power packs, and control apparatus shall be located outside of the spraying or vapor areas or the installed equipment shall comply with the hazardous (classified) location requirements in 1503.2.1, 1507.6.

48._____ Electrodes and electrostatic atomizing heads shall be insulated from ground, 1507.3.

49._____ Sufficient detail shall be provided that demonstrates a minimum clearance of 2 times the sparking distance is provided between materials being coated and electrodes, electrostatic spraying heard or their conductors. A sign indicating the minimum required separation distance shall be conspicuously posted 1507.2.

50._____ Electrostatic equipment shall be equipped with automatic shutdown without time delay designed to disconnect the power supply to the high-voltage transformer and signal the operator when the ventilation system fails or stops, the conveyor carrying articles stop past the high voltage grid, when an occurrence of a ground or imminent ground at any point can occur in the high-voltage system, or when the required clearance specified in Section 1507.2 is reduced, 1507.8.

51._____ Hand electrostatic equipment shall be interlocked such that the equipment will not operate unless ventilation is in operation, 1507.9.

52._____ For automated liquid electrostatic spraying a supervised flame detection system. Within 0.5 seconds of activation the flame detection system shall: 1) activate a local alarm. If so equipped, activation of the flame detection system in a building equipped with a fire alarm system shall activate alarm signals throughout the building 2) stop the liquid coating material delivery system, 3) terminate spray operations, 4) stop conveyors into and out of the flammable vapor area, 5) disconnect power to high-voltage elements in the spray area and the system, 1507.4.1. Plans shall be submitted in accordance with Section 907.1.1.

53._____ Detailed are the locations of the signs to: 1) designate the process zone as dangerous, 2) identify grounding requirements for all electrically conductive objects in the spray area, including persons, 3) restrict access to qualified persons only, 1507.5.2.

Additional Comments:

Review/Inspection Date: _____   Approved or Disapproved   FD Reviewer: _____

Review/Inspection Date: _____   Approved or Disapproved   FD Reviewer: _____

Review/Inspection Date: _____   Approved or Disapproved   FD Reviewer: _____

Fire Plan Review and Inspection Guidelines

Motor Vehicle Fuel-Dispensing Station
Plan Review and Acceptance Inspection Worksheet
2006 IFC Chapters 22 and 34, 2003 edition of NFPA 30 and 2003 edition of NFPA 30A

Date of Review: _____ Permit Number: _____

Business/Building Name: _____ Address of Project: _____

Designer Name: _____ Designer's Phone: _____

Contractor: _____ Contractor's Phone: _____

The numbers that follow worksheet statements represent a IFC code section unless otherwise stated.

Worksheet Legend: ✓ **or OK** = acceptable   **N** = need to provide,   **NA** = not applicable

Dispensing Area:
1. _____ Three sets of drawings are provided for the plan review process. A common scale is used and plan information is legible.
2. _____ Fuel tank installation shall comply with IFC Chapters 22 and 34. Plans shall be submitted for review, 2201.3.
3. _____ Dispensing devices are 10 ft. from the property line or the exterior of the building if it is combustible construction, 20 ft. from fixed sources of ignition, and extended hose and nozzle are at least 5 ft. from any building opening, 2203.1.
4. _____ Dispensers are installed on a minimum 6 in. high concrete island or provided with vehicle protection in accordance with IFC 312, 2206.7.3.
5. _____ A dispenser emergency valve is provided at the base of the dispenser, 2206.7.4.
6. _____ Dispenser hoses have a maximum length of 18 ft. and are equipped with breakaway devices, 2206.7.5.
7. _____ Listed automatic-closing-type hose nozzles are provided, 2206.7.6.1.
8. _____ Emergency shutdown devices are provided 20 ft. to 100 ft. from dispensers and the signage requirements are noted on the plan, 2203.2.
9. _____ The location and signage verbiage for operating instructions, emergency procedures and warning signs are provided on the plans, 2204.3.4 and 2204.3.5, 2205.6.
10. _____ Extinguisher classification and the mounting locations are detailed on the plans, a 2-A, 20-B:C is within 75 ft. of a pump or dispenser or fill pipe opening, 2205.5.

Tank Construction:
11. _____ Above-ground storage tanks (ASTs) for Class I liquids are listed as 2 hour fire resistive "Protected Tanks" or are in a special enclosure or vault in accordance with 2206.2.3, 2206.2.6.
12. _____ AST storing Class II or III liquids is listed per UL 142, Standard for Construction for Steel Flammable and Combustible Liquid Storage Tanks or other approved tank standards permitted by NFPA 30, 3404.2.7 and 2003 NFPA 30: 4.2.3.1.
13. _____ Tanks constructed with integral secondary containment are listed as meeting UL 142, 2080 or 2085, 3404.2.7, 2003 NFPA 30A: 4.2.3.1.
14. _____ Each tank has a permanent nameplate identifying its design standard, 3404.2.7.
15. _____ Special enclosures are detailed and comply with construction and size requirements, 2206.2.6.
16. _____ Protected AST volume does not exceed 12,000 gallons or 48,000 gallons aggregate, 2206.2.3.
17. _____ AST is located within secondary containment unless it is a listed secondary containment tank and a visual or automatic means of monitoring the secondary containment is identified, 2206.5.

Tank Venting:
18. _____ Normal vents for ASTs storing Class I liquids shall be equipped with either a pressure/vacuum vent or a listed flame arrestor, 3404.2.7.3.6.
19. _____ For protected ASTs, the normal vents shall be equipped with an approved flame arrester, 3404.2.9.6.3.
20. _____ Emergency relief venting is provided for above-ground storage tanks, tank compartments, and enclosed secondary containment spaces in accordance with Sections 3404.2.7.4, 2206.6.2.5.
21. _____ The tank's normal vent is terminated at least 12 ft. above adjacent grade and the vent is at least 5 ft. from building openings, or property lines, 3404.2.7.3.3.
22. _____ The tank's normal vent is not manifold with other vents, 3404.2.7.3.5.

Fire Plan Review and Inspection Guidelines

Above-ground Storage Tanks and Piping
Plan Review Worksheet
2006 IFC Chapters 27 and 34, and 2003 NFPA 30

Date of Review: _____ Permit Number: _____

Business/Building Name: _____ Address of Project: _____

Designer Name: _____ Designer's Phone: _____

Contractor: _____ Contractor's Phone: _____

Note: An aboveground fuel system also includes integral and base tanks for generators.

The numbers that follow worksheet statements represent a IFC code section unless otherwise stated.

Worksheet Legend: ✓ or OK = acceptable   N = need to provide   NA = not applicable

1. _____ Three sets of drawings.
2. _____ Equipment is listed for intended use and specification listing data sheets are provided.

Floor Plan Providing:
3. _____ A common scale is used and plan information is legible.
4. _____ Equipment symbol legend.
5. _____ Scaled floor or site plan showing room, room dimensions, equipment placement or location in relation to buildings, streets, property lines, or other public ways.

Other Information to Be Provided in a Specification Sheet or on the Plan:

Tank Construction:
6. _____ Tank is listed or designed in accordance with recognized engineering practices, 3404.2.7 and 2003 NFPA 30: 4.2.3.1 and 4.2.3.2.
7. _____ Tanks constructed with integral secondary containment are UL 142 listed or designed in accordance with recognized engineering practices, NFPA 4.2.3.1.1.
8. _____ Exterior protected above-ground storage tank has secondary containment drainage control or diking, 3404.2.9.6.4.
9. _____ Each tank has a permanent nameplate identifying its design standard, 3404.2.7.

Tank Venting:
10. _____ The use of a flame arrester or venting device in a vent line is in compliance with their listing also compliant with API 2028 for a flame arrestor, 3404.2.7.3.2.
11. _____ A tank's normal vent is not less than 12 ft. above adjacent grade nor located to trapped vapors under eaves, and at least 5 ft. from building openings, or property lines, 3404.2.7.3.3.
12. _____ A tank's normal vent is not manifolded, 3404.2.7.3.5.
13. _____ For shop-fabricated tanks, the emergency vent that is commercial has a stamp indicating opening pressure and flow rate, 3404.2.7.4, and 2003 NFPA 30: 4.2.5.2.9.
14. _____ Tank emergency vent does not vent inside a building, 3404.2.7.4.

Openings Other Than Vents:
15. _____ Filling, emptying, and vapor recovery openings are located outside the building, not less than 5 ft. from building openings or lot lines, 3404.2.7.5.2.
16. _____ For top load tanks, a metallic fill pipe is installed to minimize static electricity by terminating within 6 inches of the tank bottom, 3404.2.7.5.5. Tank openings are on the top only, 3404.2.9.6.9.
17. _____ A spill container with a capacity of not less than 5 gallons, is provided for each fill connection. Top fill containers are noncombustible, fixed to the tank and equipped with a manual drain valve that drains into the main tank, 3404.2.9.6.8.

Overfill Requirements:
18. _____ A tank storing Class I, II, IIIA liquids outside a structure is equipped with a device or means to prevent overflow in accordance with 3404.2.7.5.8.
19. _____ Outside tanks with a volume of more than 1,320 gallons that contain Class I, II, or IIIA liquids have an approved overfill prevention system, 3404.2.7.5.8.
20. _____ Tanks storing Class I, II, and IIIA liquids inside a building are equipped with a device to prevent overflow into the building and are not limited to a float valve, a preset meter in fill line, or a valve actuated by the weight of the tank's content, 3404.2.9.4.

Piping:
21. _____ Connections to tank that are below the liquid level are provided with an internal or external control valve near the tank shell, 3403.6.7.
22. _____ Tank piping is supported and protected from mechanical damage or fire exposure, 3403.6.4. and 3403.6.8.
23. _____ Pipe joints are liquid tight, welded, threaded or flanged, Class 1 liquid joints are welded if the joints are located inside the building, 3403.6.10.
24. _____ Pipe testing criteria is detailed on the plans, hydrostatic tested to 150 percent of the system design pressure or pneumatically tested to 110 percent of the system design pressure for a minimum of 10 minutes with no leakage, 3403.6.3.
25. _____ Piping is labeled in accordance with ANSI A13-1, 2703.2.2.1.(2).
26. _____ Fill pipe connection is designed to provide a direct connection to the vehicle's fuel delivery hose so fuel is not exposed to the air during filling, 3404.2.9.6.7.

Valves:
27. _____ Piping has sufficient number of control valves and check valves to control the flow of liquids, 3403.6.6.
28. _____ Any portion of the fill pipe is below the top of the tank, a check valve is installed at the fill pipe not more than 12 in. from the fill hose connection, 3404.2.9.6.7.

Tank Support:
29. _____ Tank foundation, support, and anchorages are designed in accordance with NFPA 30:4.2.4 and the IBC, IFC 3404.2.9.2.
30. _____ Tanks containing Class I, II, IIIA liquids that are elevated more than 12 inches above grade shall have a fire-resistance rating of not less than 2-hours in accordance with ASTM E 1529 unless one of the three exceptions to Section 3404.2.9.1.3 is applicable.

Miscellaneous:
31. _____ Plans show location and verbiage for signs prohibiting open flames and no smoking, 3403.5, 3404.2.3 2703.5, and 2703.6.
32. _____ Tanks exceeding 100 gallons have NFPA 704 placard location and content detailed on the plans, 3404.2.3.2.
33. _____ Tank subject to vehicular damage is protected by guard posts designed in accordance with Section 312
34. _____ Drainage control and diking are provided along with containment capacity calculations unless technical report is provided stating no hazard exists, or the tank is a listed tank with secondary containment, 3404.2.10.
35. _____ When provided the design of the dike systems complies with 3404.2.10.1 -3404.2.10.1.5.
36. _____ When required, ASTs are seismically anchored in accordance with the IBC, 3404.2.9.2, and NFPA 30 4.2.4.3.

Additional Comments:

Review Date: _____ Approved or Disapproved   FD Reviewer: _____
Review Date: _____ Approved or Disapproved   FD Reviewer: _____

Underground Storage Tank
Plan Review or Acceptance Inspection Worksheet
2006 IFC Chapter 34 and 2003 NFPA 30

Date of Review: _____ Permit Number: _____

Business/Building Name: _____ Address of Project: _____

Designer Name: _____ Designer's Phone: _____

The numbers that follow worksheet statements represent a IFC code section unless otherwise stated.

Worksheet Legend: ✓ or **OK** = acceptable   **N** = need to provide   **NA** = not applicable

1. _____ Three sets of drawings are provided for the plan review process.
2. _____ Equipment, pipe, tank, and valve listing data sheets are provided.

The Following Shall be on the Plans or Accompany the Plans:

3. _____ Piping, valves, and fittings are listed or approved for use with flammable or combustible liquids, 3403.6.2, and 2003 NFPA 30.
4. _____ Low melt point piping, valves, and fittings shall be protected against fire or located so leaks do not unduly expose people or buildings, or are located were the leak can be controlled by a valve, 3403.6.2.1 and nonmetallic piping meets the requirements for NFPA 30 5.3.6.
5. _____ Above ground piping, valves, and fittings are protected against vehicular damage, 3403.6.4 and 312.
6. _____ Pipe support methods are detailed if applicable, 3403.6.8.
7. _____ Pipe joints for are welded flanged or threaded unless nonmetallic pipe joints are used. Nonmetallic pipe and joints shall be approved and installed in accordance with the manufacturer's instructions, 3403.6.10.
8. _____ Listed flexible joints are provided where pipe connects to underground tank, where pipe ends at pump island and vent risers, and at points where differential movement in the piping can occur, 3403.6.9.
9. _____ Pipe test requirements are noted on plans, 3403.6.3.
10. _____ Before covering, pipe is hydrostatic tested to 150 percent of the maximum pressure of the system or pneumatically tested to 100 percent of the maximum pressure of the system for a minimum of 10 minutes, 3403.6.3.
11. _____ Normal vent pipe outlet is 12 ft. above grade and 5 ft. from property lines or building openings, designed not to collect or trap liquid and shall not be manifold unless for special purposes like vapor recovery or air pollution control, 3404.2.7.3.3. and 3404.2.7.3.5.
12. _____ Normal vent pipe is at least the size specified on the listing sheet.
13. _____ A metallic fill pipe extends within 6 in. of the bottom of the tank to minimize the generation of static electricity, 3404.2.7.5.5.
14. _____ Tanks that are only filled from the top and exceeding 1,000 gallons for Class 1 liquids are equipped with a tight fill device for connecting the fill hose to the tank, 3404.2.7.5.5.2.
15. _____ Tanks are at least 3 ft. from foundations, walls or property lines and at least 1 ft. from the nearest tank, 3404.2.11.2.
16. _____ The tank detail has it resting on a firm foundation and surrounded with 6 in. or more of sand, 3404.2.11.3.
17. _____ The tank is covered with earth, concrete or asphalt in accordance with NFPA 30: 4.3.3.2.2.
18. _____ A means of overfill prevention is provided in accordance with 3404.2.11.4 and NFPA 30: 4.6.1.
19. _____ Corrosion protection method such as corrosion resistant materials or cathodic protection is detailed, 3404.2.7.9 and NFPA 30: 4.2.6.1.
20. _____ Before covering, the storage tank is to be subjected to a tightness test in accordance with 3404.2.12.2 and NFPA 4.4.2.2 and 4.4.2.2.3.

Additional Comments:

Review Date: _____ Approved or Disapproved FD Reviewer: _____

Review Date: _____ Approved or Disapproved FD Reviewer: _____

Review Date: _____ Approved or Disapproved FD Reviewer: _____

Medical Gases
Plan Review and Inspection Worksheet
2006 IFC Chapter 27, Section 3006, and 2002 Standard NFPA 99

Date of Review: _____ Permit Number: _____

Business/Building Name: _____ Address of Project: _____

Designer Name: _____ Designer's Phone: _____

The numbers that follow worksheet statements represent a IFC code section unless otherwise stated.

Worksheet Legend: ✓ **or OK** = acceptable   **N** = need to provide   **NA** = not applicable

IFC Section 3006, nonflammable gas; Occupancy Uses: hospitals, dentistry, podiatry, veterinary, etc.
Scope: for oxidizer compressed gases (nitrous oxide and oxygen) exceeding 504 cu. ft., Table 105.6.8.

1. _____ Gases shall be stored in a separate room or enclosure of at least 1-hour fire resistive construction or a gas cabinet in accordance with 3006.2.1, 3006.2.2, 3006.2.3.
2. _____ Exterior rooms have at least one exterior wall, an interior door that is at least a 1-hour fire rated, self-closing smoke and draft control assembly for openings between the room and building interior, at least 2 vents on the exterior wall not less than 36 sq. in. each, 1 vent is within 6 in. of the floor and the other vent within 6 in. of the ceiling, and at least 1 sprinkler for container cooling, 3006.2.1.
3. _____ Interior rooms shall have automatic sprinklers, supply and exhaust ducts are within a 1-hour rated shaft enclosure from the room to the exterior, ventilation a minimum of 1 cfm/sq. ft., and be in compliance with the IMC, 3006.2.2.
4. _____ Gas cabinets operate at a negative pressure in relation to surrounding area, have an average ventilation velocity of 200 fpm across the face of access port/window and a minimum of 150 fpm at any point of access port/window, cabinet is connected to an exhaust system and internally sprinklered, 3006.2.3.
5. _____ Gas cabinets shall have a self-closing door, access port and noncombustible window, be constructed of at least 12 gauge steel, 3006.2.3 and 2703.8.6.1.
6. _____ Gas cabinets are limited to storing 3 cylinders, 2703.8.6.3.
7. _____ Gases are stored in dedicated areas with no other storage or uses, 3006.2.
8. _____ Piping, tubing, valves, and fittings are compatible with gas, of adequate strength, labeled per ANSI A13.1, have manual shutoff valve or auto emergency shutoff valves at point of use and at the source, and emergency valve is identified, 2703.2.2.1.
9. _____ Medical gas system is detailed and designed in accordance with NFPA 99 Chapter 5, IFC 3006.4.
10. _____ Provisions are made to securely fasten or rack gas cylinders, NFPA 99: 5.1.3.3.2(7).
11. _____ Rooms or enclosures shall have lockable doors, NFPA 99: 5.1.3.3.2(2).
12. _____ All gas piping shall be labeled at the source, the outlet, and not more than every 20 ft., and at least once in or above every room, and both sides of wall penetrations, NFPA 99: 5.1.11.1.
13. _____ Shut-off valve(s) are identified with name of gas or specific vacuum system, the room or areas served, and a warning not to close or open valves except in an emergency, NFPA 99: 5.1.11.2.
14. _____ Inspection, testing, and certification documents shall be provided as detailed in NFPA 99 Chapter 4 and some of the information to be included is: 3006.4.
    _____ Tests to be performed, documented, and certified by the installer: NFPA 99: 5.1.12.2
    _____   A. Blow down
    _____   B. Initial pressure test for pressure gases, 1.5 times the working pressure, minimum 150 PSIG
    _____   C. Pipe purge test
    _____   D. Cross-connection test, no cross connection exists between systems
    _____   E. Standing pressure test, 24 hrs at 20 percent above normal operating line pressure

Fire Plan Review and Inspection Guidelines

15. ____ System verification tests to be performed, documented, and certified by a qualified third party: NFPA 99: 5.1.12.3.
    - ____ A. Cross-connection test
    - ____ B. Valve test
    - ____ C. Outlet flow test
    - ____ D. Alarm testing
    - ____ E. Standing pressure
    - ____ F. Piping particulate test
    - ____ G. Pipe purge test
    - ____ H. Piping purity test
    - ____ I. Operational pressure test, pressure differential
    - ____ J. Medical gases concentration test
    - ____ K. Medical air purity test, compressor system
    - ____ L. Final tie-in and labeling

16. ____ Source equipment verification shall be performed, documented, and certified by a qualified party: NFPA 5.1.12.3.14.
    - ____ A. Gas supply sources
    - ____ B. Medical air compressor

Additional Comments:

Review/Inspection Date: _____   Approved or Disapproved   FD Reviewer: _____

Review/Inspection Date: _____   Approved or Disapproved   FD Reviewer: _____

Review/Inspection Date: _____   Approved or Disapproved   FD Reviewer: _____

Index

Aboveground Fuel Tank
    Plan Review Checklist, 287

Acceptance Inspection Checklist
    Aboveground fuel tank, 287
    Clean agent, 200
    Carbon dioxide (CO2), 178
    Electrostatic Apparatus, 281
    Emergency power, 85
    Fire alarm, 242 and 251
    Fire pump, 264
    Fuel dispensing station, 284
    Halon, 188
    Kitchen hood suppression system, 210
    Medical gases, 291
    Motor Vehicle Fuel Dispensing Station, 284
    Powder coating, 281
    Private fire service mains, 55
    Private water tank, 78
    Spray finish, 281
    Sprinkler 13, 101 and 112
    Sprinkler 13D, 130 and 134
    Sprinkler 13R, 118 and 124
    Standpipe, 220
    Underground storage tank, 289
    Water mist, 168

Acceptance Inspection Process, sample, 10

Access Roads
    Engineered turnarounds, 44
    Road turnarounds diagrams, 43
    Site plan review checklist, 39
    Turnaround Purpose, 41

Acronym List, 299

Carbon Dioxide (CO2)
    Acceptance inspection checklist, 178
    Contractor pretest checklist, 181
    Installer certification form, 180
    Plan review checklist, 175
    Voltage drop calc sheet, 177
    When is a permit required?, 7

Class K Fire Extinguisher, 209 and 211

Clean Agent
    Acceptance inspection checklist, 200
    Contractor pretest checklist, 203
    Installer certification form, 202

    Plan review checklist, 197
    Voltage drop calc sheet, 199
    When is a permit or plan review required?, 7

Code Study Information for Plan Review, 11

Contractor Pretest Checklists
    Clean agent, 203
    Carbon Dioxide (CO2), 181
    Emergency power, 92
    Fire alarm, 254
    Fire pump, 275
    Halon, 190
    Kitchen hood suppression system, 213
    Private fire service mains, 57
    Private water tank, 81
    Sprinkler 13, 13D, 13R, 160
    Standpipe, 223
    Water mist, 171

Electrostatic Apparatus, 281

Emergency Access Lock Box, 49

Emergency Access Roads, 41

Emergency Plans
    Summary guide, 23
    Detailed pamphlet, 24

Emergency and Standby Power, ESSP I and II
    Aboveground fuel tank plan review checklist, 287
    Acceptance inspection checklists, 85
    Contractor pretest checklist, 92
    Guidelines for annual service, 90
    Installer certification form, 89
    Plan review checklist, 85 and 87

Fire Access Road Turnarounds (see Access Roads)

Fire Alarm
    Acceptance inspection checklist, 242 and 251
    Contractor pretest checklist, 254
    Guide for Sound Pressure (dBA) level, 231
    High-rise requirements, 233
    Installer certification form, 253
    Plan review checklist, 235 and 244
    Sound Pressure Level (dBA) Log, 255
    Voltage drop calc sheet, 241
    When it a permit or plan review required?, 7

Fire Plan Review and Inspection Guidelines

Fire Extinguishers (see Portable Fire Extinguishers)

Fire Inspection Checklist for Building Permit, 17

Fire Plan Review Checklist, general, 13

Fire Pump
- Acceptance inspection checklist, 264
- Acceptance test form, 272
- Contractor pretest checklist, 275
- Installer certification form, 274
- Plan review checklist, 259 and 267
- Test procedure, 276

Fire Watch Policy, 22

Fuel Dispensing Station Plan Review/Inspection Checklist, 284

Generator (see Emergency Power)

Halon
- Acceptance inspection checklist, 188
- Contractor pretest checklist, 190
- Installer certification form, 189
- Plan review checklist, 185
- Semi-annual service report, 191
- Voltage drop calc sheet, 187
- When is a permit or plan review required?, 7

High-piled storage
- Storage design forms, 137

Hydrants (see Private Fire Service Mains)

Inspections
- Final fire inspection checklist for building permit, 17
- Intervals for protection systems, 20

Installer Certification Forms
- Clean agent, 202
- Carbon dioxide (CO2), 180
- Emergency power, 89
- Fire alarm, 253
- Fire pump, 274
- Halon, 189
- Kitchen hood suppression system, 212
- Private fire service mains, 56
- Private water tank, 80
- Sprinkler 13, 104
- Sprinkler 13D, 135
- Sprinkler 13R, 126
- Standpipe, 222
- Water mist, 170

Kitchen Hood Suppression
- Acceptance inspection checklist, 210
- Contractor pretest checklist, 213
- Installer certification form, 212
- Plan review checklist, 208
- Plan review drawing, 207
- Semi-annual service guide, 214
- When is a permit or plan review required?, 8

Lock Box Policy, 49

Medical Gases Plan Review/Inspection Checklist, 291

Motor Vehicle Fuel Dispensing Plan Review or Inspection Checklist, 284

Permits or Plan Review Required
- Clean agent, 7
- General purpose, 7
- Fire alarms, 7
- Kitchen hood suppression system, 8
- Sprinklers, 7
- Water mist, 7

Plan Review Checklists
- Aboveground fuel tank, 287
- Clean agent, 197
- Carbon dioxide (CO2), 175
- Electrostatic apparatus, 281
- Emergency and standby power, 85 and 87
- Fire alarm, 235 and 244
- Fire plan review, general, 13
- Fire pump, 259 and 267
- Fuel dispensing station, 284
- Halon, 185
- Kitchen hood suppression system, 208
- Medical gases, 291
- Motor vehicle fuel dispensing station, 284
- Powder coating, 281
- Private fire service mains, 53
- Private water tank, 75
- Site plans, 39
- Spray finish, 281
- Sprinkler 13, 95 and 105
- Sprinkler 13D, 127 and 131
- Sprinkler 13R, 115 and 120
- Sprinkler storage misc. up to 12 ft., 99, 110
- Sprinkler storage up to 25 ft., 138 and 143
- Sprinkler storage over 25 ft., 149 and 154
- Standpipe, 217
- Underground storage tank, 289
- Water mist, 163

Plan Review Deficiencies, common, 12

Plan Review Building Checklist Guide, 13

Policies (samples)
- Emergency access roads, 41
- Fire alarms, 5
- Fire pumps, 6
- Fire watch, 22

General purpose, 5
Lock box, 49
Plan review sample process, 9
Portable extinguishers, 6
Pretest systems, 9
Service reports for protection systems, 20
Sprinklers, 6
Turnaround and secondary access, 41
When are permits or plan review for systems required?, 7

Powder Coating Plan Review/Inspection Checklist, 281

Private Fire Service Mains
  Acceptance inspection checklist, 55
  Annual private hydrant service checklists, wet 63 and dry 64
  Contractor pretest checklist, 57
  Flow test info sheet, AFSA, 71
  Hydrant flow test procedure, 65
  Hydrant flush procedure, 58
  Hydrant orientation graphic, 62
  Hydrant placement, 59
  Hydrant site plan review checklist, 53
  Hydrant specifications (sample), 60
  IFC Number and Distribution App. Table, 61
  Installer certification form, 56
  Plan review checklist, 53
  Water flow summary sheet, AFSA, 72

Private Water Tank
  Acceptance inspection checklist, 78
  Contractor pretest checklist, 81
  Installer certification form, 80
  Plan review checklist, 75

Secondary Access, 41

Service Intervals for Fire Protection Systems, 20

Site Plan Review Checklist, 39

Spray Finish Plan Review/Inspection Checklist, 281

Sprinkler
  Acceptance inspection checklist 13, 101 and 112
  Acceptance inspection checklist 13D, 130 and 134
  Acceptance inspection checklist 13R, 118 and 124
  Contractor pretest checklist 13, 13D, 13R, 160
  Installer certification form 13, 104
  Installer certification form 13D, 135
  Installer certification form 13R, 126
  Main flushing procedures, 58
  Plan review checklist 13, 95 and 105
  Plan review checklist 13D, 127 and 131
  Plan review checklist 13R, 115 and 120

Plan review checklist misc. storage up to 12 ft., 99, 110
Plan review checklist storage up to 25 ft., 138 and 143
Plan review checklist storage > 25 ft., 149 and 154
Storage design forms, 137
When it a permit or plan review required?, 7

Sound Pressure (dBA) Level, Fire Alarm, 231 and 255

Standby Power, (see Emergency Power)

Standpipe
  Acceptance inspection checklist, 220
  Contractor pretest checklist, 223
  Inspection, test, maintenance forms, AFSA, 224-227
  Installer certification form, 222
  Plan review checklist, 217

Turnaround and Secondary Access, 41

Underground Storage Tank Plan Review/Inspection Checklist, 289

Voltage Drop Calc Sheets
  Carbon dioxide, 177
  Clean agent, 199
  Fire alarm, 241
  Halon, 187
  Water mist, 167

Water Mist
  Acceptance inspection checklist, 168
  Contractor pretest checklist, 171
  Installer certification form, 170
  Plan review checklist, 163
  Voltage drop calc sheet, 167
  When is a permit required?, 8

Water Tank, Private (see Private Water Tank)

Fire Plan Review and Inspection Guidelines

Acronym Table

1. AFFF — Aqueous Film Forming Faom
2. AHJ — Authority Having Jurisdiction
3. ANSI/ASME — American National Standards Institute/American Society of Mechanical Engineers
4. ASTM — American Society for Testing Materials
5. AWG — American Wire Gauge
6. AWWA — American Water Works Association
7. cd — Candella
8. dBA — Decible/ weighting filter "A"
9. EC — Extended Coverage
10. EH — Extra Hazard
11. EOL — End-of-line
12. ESFR — Early Suppression Fast Response
13. FACP — Fire Alarm Control Panel
14. FDC — Fire Department Connection
15. HC — Hydraulic Calculated
16. IBC — International Building Code
17. IFC — International Fire Code
18. IMC — International Mechanical Code
19. LEL — Lower Explosive Limit
20. LH — Light Hazard
21. LOAEL — Lowest Observed Adverse Effect Level
22. NAC — Notification Appliance Circuit
23. NEC — National Electrical Code
24. NFPA — National Fire Protection Association
25. NICET — National Institute for Certification in Engineering Technologies
26. NOAEL — No Observed Adverse Effect Level
27. NOV — Nameplate Operating Voltage
28. NST — National Standard Thread
29. OH — Ordinary Hazard
30. OSU — Oklahoma State University
31. PIV — Post Indicating Valve
32. PRV — Pressure Reducing Valve
33. PS — Pipe Schedule
34. PSIG — Pounds per Square Inch Gauge
35. QOD — Quick Opening Device
36. QRES — Quick Response Early Suppression
37. UL — Underwriters Laboratory
38. UPS — Uninterrupted Power Supply

 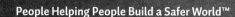

More Tools for Fire Inspectors

SIGNIFICANT CHANGES TO THE INTERNATIONAL FIRE CODE, 2006 EDITION
Identifies key changes made to the 2003 IFC and provides analyses of their effects on various applications. Includes hundreds of color photos and illustrations with descriptive narrative.
#7404S06

A GUIDE TO SMOKE CONTROL IN THE 2006 IBC
Authors Dr. John H. Klote and Douglas H. Evans provide a greater understanding of the intent and application of Section 909 of the 2006 International Building Code®.
#4006S06

2006 INTERNATIONAL FIRE CODE SPRINKLER PLAN REVIEW RECORDS
A thorough checklist of applicable code sections from the 2006 International Fire Code. It is an essential resource to conduct detailed, consistent plan reviews. Sold in sets of 25.
#0402PR06

STUDY COMPANIONS
2006 IFC STUDY COMPANION #4407S06
2006 IBC STUDY COMPANION #4017S06
Both comprehensive study guides provide practical learning assignments for independent study, key points for review, code text and commentary, and a helpful quiz at the end of each section.

ORDER NOW! 1-800-786-4452 | www.iccsafe.org/fprad

8-61804-07

Don't Miss Out On Valuable ICC Membership Benefits. Join ICC Today!

Join the largest and most respected building code and safety organization. As an official member of the International Code Council®, these great ICC® benefits are at your fingertips.

EXCLUSIVE MEMBER DISCOUNTS

ICC members enjoy exclusive discounts on codes, technical publications, seminars, plan reviews, educational materials, videos, and other products and services.

TECHNICAL SUPPORT

ICC members get expert code support services, opinions, and technical assistance from experienced engineers and architects, backed by the world's leading repository of code publications.

FREE CODE—LATEST EDITION

Most new individual members receive a free code from the latest edition of the International Codes®. New corporate and governmental members receive one set of major International Codes (Building, Residential, Fire, Fuel Gas, Mechanical, Plumbing, Private Sewage Disposal).

FREE CODE MONOGRAPHS

Code monographs and other materials on proposed International Code revisions are provided free to ICC members upon request.

ICC *BUILDING SAFETY JOURNAL*®

A subscription to our official magazine is included with each membership. The bi-monthly magazine offers insightful articles authored by world-renowned code experts, plus code interpretations, job listings, event calendars, and other useful information. ICC members may also enjoy subscriptions to a bi-monthly newsletter and an electronic newsletter.

PROFESSIONAL DEVELOPMENT

Receive "Member Only Discounts" for on-site training, institutes, symposiums, audio virtual seminars, and on-line training! ICC delivers educational programs that enable members to transition to the I-Codes®, interpret and enforce codes, perform plan reviews, design and build safe structures, and perform administrative functions more effectively and with greater efficiency. Members also enjoy special educational offerings that provide a forum to learn about and discuss current and emerging issues that affect the building industry.

ENHANCE YOUR CAREER

ICC keeps you current on the latest building codes, methods, and materials. Our conferences, job postings, and educational programs can also help you advance your career.

CODE NEWS

ICC members have the inside track for code news and industry updates via e-mails, newsletters, conferences, chapter meetings, networking, and the ICC Web site (www.iccsafe.org). Obtain code opinions, reports, adoption updates, and more. Without exception, ICC is your number one source for the very latest code and safety standards information.

MEMBER RECOGNITION

Improve your standing and prestige among your peers. ICC member cards, wall certificates, and logo decals identify your commitment to the community and to the safety of people worldwide.

ICC NETWORKING

Take advantage of exciting new opportunities to network with colleagues, future employers, potential business partners, industry experts, and more than 40,000 ICC members. ICC also has over 300 chapters across North America and around the globe to help you stay informed on local events, to consult with other professionals, and to enhance your reputation in the local community.

For more information about membership
or to join ICC, visit www.iccsafe.org/members
or call toll-free 1-888-ICC-SAFE (422-7233), x33804

People Helping People Build a Safer World™